California Natural History Guides: 40

WATER BIRDS OF CALIFORNIA

BY
HOWARD L. COGSWELL

Illustrated by Gene Christman

UNIVERSITY OF CALIFORNIA PRESS
BERKELEY · LOS ANGELES · LONDON

California Natural History Guides
Arthur C. Smith, General Editor

Advisory Editorial Committee:
Mary Lee Jefferds
A. Starker Leopold
Robert Ornduff
Robert C. Stebbins

ı

University of California Press
Berkeley and Los Angeles, California
University of California Press, Ltd.
London, England

© 1977, by The Regents of the University of California
ISBN 0-520-02994-1
Library of Congress Catalog Card Number: 73-93049
Printed in the United States of America

CONTENTS

CONTENTS

APPENDIX, 287

INDEX 393

INTRODUCTION

Birds fascinate people. Many are bright and colorful; they sing; their behavior is primarily based on senses of sight and hearing (as is man's); their family life is appealing. And many birds are active in the daytime and can easily be observed almost anywhere. Furthermore, birds fly — an ability man has achieved, in comparatively cumbersome fashion, only after centuries of effort. Beyond this, there are the seasonal comings and goings of many birds, often on long migrations about which there is much folklore and, in recent years, much new information.

The long history of nonprofessional ornithology has favored a great variety of popular books on birds in general, their behavior, their nests; guides for identification of the birds of whole continents, and distributional lists both for larger regions and for local areas. However, the interested amateur who does not amass a library and keep up with recent journal literature often has difficulty applying statements in the guides and distributional lists to local conditions. This is particularly true with descriptions of habitat and seasonal occurrence and abundance, which vary so much from region to region. This book, together with its companion volume on land birds to follow, is intended to serve this need as well as to provide sufficient notes on recognition and habits of the species occurring in California to enable a careful observer of birds to identify them and learn their salient behavioral features.

The present volume covers the water birds, i.e., those families of birds of which most or all members are associated with fresh- or salt-water habitats. There is one exception: the Osprey, a fish-catching type of hawk in a family of its own, is so obviously a hawk that it is included with the other hawks in the land bird volume. All other water bird species occurring regularly or frequently anywhere in California are treated herein, but greater detail is given for those which are numerous or widespread — especially those found readily along the central and southern coast and in the Central Valley. Brief notes are inserted on a number of other species that occur only occasionally, but repeatedly, in the state; while those for which there are less than five records are mentioned only in the Appendix with the detailed Graphic Calendars of all birds.

The 1957 American Ornithologists' Union *Check List of North American Birds,* as modified by subsequent supplements, has been used for this book. A number of changes, especially among shorebirds, were announced in 1973, and these have been incorporated here, with a few additional groupings of species recommended by Jehl (1968: *Relationships in the Charadrii (Shorebirds),* San Diego Soc. of Natural History, Memoir 3). The sequence of families and species followed in this guide is in accordance with their evolutionary relationships, beginning with the most primitive types and ending with those considered most recently evolved.

Included among the water birds are the *waterfowl,* a term properly applied only to one big family — the ducks, geese, swans, and mergansers. Along with a few rails, coots, and snipe, these birds make up the bulk of the game birds that thousands of hunters seek in our wetlands for a limited season each year. There is, however, no closed season on the use of binoculars, telescope, and camera, and the rewards are limited only by one's knowledge and diligence.

TECHNIQUES IN OBSERVING BIRDS

Primitive man's skill at close approach to wild creatures remains undeveloped in most people today. It is true that with powerful binoculars and telescopes many close views can be obtained without special precautions. However, to be really successful under varied circumstances one must frequently become a quiet stalker behind a screen of vegetation, keeping talk and other noise to a minimum. If lack of cover requires an approach in the open, as is so often true with water birds, it is usually better to walk slowly toward them in a direct line — since birds with eyes on the side of the head see such movement poorly in three dimensions. Stop frequently for observation, and make no quick movements, even of arms. Some-

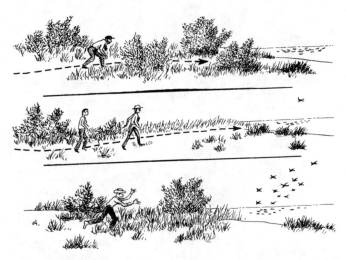

times, of course, it is advantageous to make a sudden movement so birds will fly up — either for identification on the wing or to get a better estimate of numbers. Since birds do have good color vision, it is best to wear neutral or dull-colored clothing.

Because our own perception of colors, and indeed of gradations of light and dark, is greatly hampered by backlighting, field trips should when possible be planned so as to have the sun more or less behind the observer. We just cannot see the colors or fine markings on a bird when it is silhouetted against a bright sky or glinting water. Many silhouettes are distinctive, however.

When time allows, the closest observations can often be made by sitting quietly in a good habitat, preferably among or behind screening vegetation, and allowing birds to come into the vicinity on their own after they have forgotten your noisy arrival. For many of the more wary birds treated in this volume, however, this requires a great deal of patience unless it is augmented by a well-constructed "blind" in or behind which one can move about a bit when muscles get cramped. Any such blind, of course, must be set in place long enough for the birds to have become used to its normally "harmless" nature.

It also helps to have another person leave the area conspicuously after you enter such a hideout, whereupon the alarmed
birds usually resume their undisturbed feeding or resting activities much sooner.

To find the most water birds, whether you are seated
quietly or on the move, glance about repeatedly at all likely
spots. In your active scanning, include points sufficiently far
away to avoid alarming most of the birds before you know
they are there. Many water birds are gregarious most of the
year, and alarming the near fringe of a large flock will commonly trigger a mass retreat by all of them.

Shorelines are especially good watching spots, and many
offshore snags or rocks may harbor birds. Sandy and other low
peninsulas, dikes, and islets are often favored as resting areas,
primarily because they have few intruders. For species that
swim or dive on open water, you should first scan the surface
of ponds, lakes, bays, and ocean from good vantage points,
either on shore or from a boat — but be cautious in coming up
on small coves suddenly. In marshy areas and some other
special situations, deliberate crashing through vegetation where
secretive birds (such as rails) hide may sometimes cause them
to fly up into view; a rock tossed into the marsh, or a sudden
loud noise of any kind — e.g., clapping hands several times —
may stimulate them to give identifying call-notes.

When birds are at some distance from the observer, identification will of course require magnification. Even then, there are often others still visible but so far away as to be indistinguishable as to species. The real bird student, however, readily picks out shearwaters a mile or more offshore from coastal headlands where the casual visitor sees only waves and close-in gulls, cormorants, and pelicans.

ON KEEPING RECORDS

The beginning bird student is necessarily greatly concerned with identification of the species he sees. A list is the obvious first step at record-keeping. The usual birdwatcher's "life list" includes the date and place where identification was first certain; most students soon take additional notes of some sort, including such items as field marks recognized, how many birds of each species were seen and in what habitat, etc. All this requires some kind of record book or card system.

Suggestions on note-taking in nature study in general, with appropriate emphasis on objectivity, can be found in Smith (1959) and Jaeger and Smith (1966), the introductory volumes to this Natural History Guide series. Stebbins (1959) suggests a notebook format of paragraph-style notes, with separate numbered pages for each species and other pages as a journal of the observer's route and time afield, habitats visited, etc. As your knowledge of birds grows, you will want to keep records on different aspects, so a flexible system such as this is highly recommended. Many bird students put important items such as locality, date, situation or habitat, and the number of each species found, into some compact form, and then write expanded notes elsewhere for the few species studied in detail. Other items that should go either in a journal or with the pertinent species list are the weather at the time (and possibly for a previous few days or weeks, if significant), the location and extent of each habitat through which you searched, time spent there, whether your effort was thorough or casual, etc.

For making such lists of species (but with minimum other comments), the small field cards or regional checklists which are sold by the National Audubon Society and by some local bird clubs and museums are helpful. A separate card or column should be used for each local field trip or for each region of similar habitats. The California Field Ornithologists offer a small booklet which is a complete state species list, with multiple blank columns. Such lists, however, require repeated scanning of many pages when all records of a particular species are desired.

A cross-ruled "roll-book" arrangement, with a row or more of space for each species and a column or more for each date, or for each location covered on more extended trips, does the two jobs of bringing together all the records on one species and those on all the species found at one locality or on one date. If you make your own page headings, a loose-leaf system with index pages giving species names, on margins that protrude beyond the actual entries on filler pages, is a time-saver. (To prevent confusion in case of misplaced filler pages, however, each one should be keyed to its correct place.)

Most desirable, however, is to have some system in which details beyond the dates, places, and numbers of birds, the weather, etc., can be readily accommodated. For a number of years I have used a system with more flexible space in which a species list for each major area covered on field trips away from my home area is in (or on cards with) the journal giving details of my field activities, habitats visited, etc. In another section of the notebook there is a page for each species on which any detailed notes are made, including those seen repeatedly in my home area. I frequently enter brief observations including only date, place, and numbers on these pages also, in columns that can be quickly scanned when desired. Additional notes, in paragraph style, are written on these same pages by disregarding the columns, and the cumbersome keeping of separate notebooks is thus avoided. A file of cards appropriately ruled with a few columns could be adapted to serve the same function — and might be easier to add to in the field.

Whatever the system used, it is important to be able quickly to check back to all previous notes pertaining to any one species from whatever locality, and similarly to be able to look up complete lists of all species found in a given area on successive visits to it.

STUDY PROJECTS AND REPORTING RESULTS OR OBSERVATIONS

When he has developed an adequate system of records, what can the amateur student of birds do with them? Perhaps more than in any other branch of biology, ornithological knowledge is continually being advanced by amateurs who publish the results of their studies. When these accounts are on previously little-known species, places, types of behavior, ecological relationships, etc., and the results are well reported, including a concise discussion of their significance, they are welcome material for publication in scientific journals (see suggested references).

Even such small items of distributional nature as unusual species in an area, or usual ones out of season, and particularly any documented observations of definite increase or decrease in populations, are welcomed by the regional editors of *American Birds* for their quarterly summary (there are two regions in California). If rarities are involved, the observation should always be supported by details of how the identification was made; it will be best verified if the bird is collected by a qualified person, is adequately photographed, or is seen by other qualified observers, all of whom agree upon its identity.

Among the more important contributions to ornithology that can be made by amateurs are critical observations of behavior, censuses over measured areas of habitat, and systematic gathering of data on nesting success and on actual flyways used by migrants. In recent years the Special Wildlife Investigations Section (nongame species) of the California Department of Fish and Game has been coordinating many such efforts and supporting some financially. Volunteers at Point Reyes Bird Observatory (see organizations list preceding the Appendix) have also contributed greatly. Yet there are

Cogswell 1966   Journal   p. 14

Feb 2   S + SW Johnson Landing, Hayward, Calif.
11:30-   From 1141-1208 hrs. I counted all birds on 100-acre plot of
12:20   intertidal mudflat (measured & staked here in 1965)
            using 9x binocs. + 20-30x scope from levee E of s. half
            of 2090 × 2090 ft. plot. The n. half of east edge of plot
            is bordered by narrow, eroding belt of pickleweed
            salt marsh. Johnson Ldg. is a 700 ft. peninsula of
            pickleweed marsh extending onto tideflat, with a low,
            rocky border that is included in plot (the marsh is not).
            The San Mateo toll station is on a longer peninsula
            1/3 mi. S of the plot. Otherwise the tideflat extends for
            miles to S and N. It is about 2500 ft. wide at a "0" tide
            level here, with 7 miles of shallow bay W of this to ship
            channel.

WEATHER: Sky 9/10 mod. low overcast with cirrostratus above.
            Wind W, very light. Temp. 59-60°F.
BAY: surface calm, no waves. Tide receding from H.W. of
            8.1 ft. at 0850; bits of mudflat exposed by 1200.
COUNT   OF BIRDS ON PLOT (all on water except as noted):
            Eared Grebe   1
            Western Grebe   16
            Pintail   5 (2 on rocks of J. Ldg.)
            Am. Wigeon   85 (20 at marsh edge)
            Canvasback   121
                    [PART OF LIST OMITTED]
            Dunlin   35
            Glaucous-winged Gull   2 (1 on rocks, J. Ldg.)
            Calif. Gull   1 (on mud)
                    TOTAL   457 birds, of 18+ species.
GEN.   OBSERVATIONS: Shorebird departures, chiefly from
            salt ponds E of s. 1/2 of plot: many small flocks at 11:44;
            + at 12:00± a mass exodus of Marbled Godwits + Willets,
            altho 350 of these remained on dike roost as I left.
            Other species noted in general area were:

Cogswell 1966		*American Wigeon*	p. 1
			Anas americana
DATE	LOCALITY (detail in Journ)	NO.	AGE, SEX, BEHAVIOR, HABITATS USED, REMARKS.
Jan. 29	N West Butte	2000	est, most of them foraging on new, short, green grass of w. slope of Sutter Buttes and flying to join that flock. Some 50-100 were with Pintails, etc, in grain fields (stubble) of Butte Sink to W.
"	Gray Lodge "a"	300	est. Unusually few on quick trip around visitors' loop road in marshy pond area.
Feb. 2	S Johns. Ldg.	85	in plot = 65 or bay near se. boundary; +20 resting on top of eroding marsh bank.
Mar. 19	Oliver S.P. vic	160±	incl. 100± on diked salt marsh.
Oct. 9	out. Pt. Reyes	⑧	at farm pond in grassland
Nov. 5	" " "	30±	" " " "
Nov. 13	Gray Lodge "a"	1,000±	SEPARATE SHEETS FOR EACH SPECIES ACCOUNT

Black Oystercatcher.

Haematopus bachmani

Apr. 23	S. Farall. S.	4	incl. ③ in "piping" flight.
May 29	Pt. Lobos S.R.	6+	in Bird Ids. area, s. pt. of Reserve. Up to 4 at a time in courting activity near + on ne. corner of largest id., at and above h.w. line. A bit later one was watched on a med-gentle slope of rocks below the big 'flat'. Here it was making lunges, both while afoot and by short flights, at 2 or 3 of the Brown Pelicans resting there. Several times it actually struck a pelican on the back or wings. Some 10-15 mins. later the pelicans were gone, and an (this?) oystercatcher was squatting flat on a gently sloping spot just above the encounter site—no doubt now on its 'NEST', probably with eggs, since it sat for a long time thereafter.
July 30	Yankee Pt.	2	
Aug. 27	Abalone Pt.	1	?? (dark bird of this size + shape) flew se. over ocean

profound gaps in our knowledge for many common species and many sections of the state.

Some specific suggestions for projects that do not require anything more than binoculars or spotting scope, a notebook, and alert observing time are:

1. Behavior of a single species: patterns in feeding, resting, pairing, social or aggressive-submissive repertoire – in relation to seasons, time of day, light intensity, weather, etc.

2. Utilization of a particular area: variation in species and numbers on a small lake, portion of a bay, shore, field, etc.; area surveyed can be adjusted to time available, but should usually be at least twenty acres (larger for wide-ranging birds).

3. Roosts: numbers of birds and times and directions of arrival and departure, through a season or a year.

4. Shorebirds: species and numbers involved in use of high-tide roosts and nearby tideflats (night and day if possible).

5. Gulls: numbers at major feeding areas, times, weather conditions, routes to and from roosts, etc.

6. Visible migrants: actual counts of those passing vantage points during extended sea-watches along the coast or inland along rivers or strings of marshes and ponds or in passes between mountains.

7. Nest success: number of nests, of eggs laid, and of young fledged in a given area of habitat for any species where nest contents can be inspected repeatedly without undue disturbance of the adults or endangering the young; care should be taken not to create a scent trail to ground nests that predators are likely to follow.

If one has other means at one's command, such as a banding permit, motion picture equipment, sound recorder with good microphone and playback, scuba gear, or merely a boat suitable for use on shallow inland or bay waters or the open ocean, the possibilities for study projects on water birds are greatly extended.

BINOCULARS AND TELESCOPES

The one bit of expensive equipment needed by every serious bird student is a prism binocular of medium to good quality. To be really adequate for the frequently distant birds treated in this volume, it should be of at least 7X35 or 8X40 size, which is most adaptable to different situations. It should have "coated" glass surfaces throughout, which greatly improves the light-gathering power. A binocular of 9X or 10X magnification (or some of the newer zoom models) is better yet for distant birds, but many are heavier. Still larger models are too cumbersome to use easily with fast-moving birds at closer range. Most people prefer a binocular with central focusing wheel for study of birds. High-quality American and European brands with these specifications may cost well over two hundred dollars, but will give a lifetime of service if cared for. Some of the less expensive imported brands (mostly of Japanese manufacture) are good also, but should be thoroughly checked for adequate operation at close and distant ranges and in both bright and dim light. Some are backed by good guarantees, others not.

For adequate viewing at longer ranges such as the minima to which many water birds will readily allow approach, a telescope is also needed. Prism or "spotting scopes" are usually used because of the greater viewing area they provide, and very good ones are now available with internal focusing mechanisms that keep out dirt. With an eyepiece of 20 to 25X an observer's identifying range is doubled or tripled as compared to that with a binocular alone. Additional, higher-powered eyepieces may be included (or a zoom scope acquired), but

11

heat waves above sunny fields or shores often reduce the quality of image seen with them. In any event, a tripod with a pan head to which the scope is attached sturdily enough not to wobble in a moderate wind is almost an essential. Mountings to attach the scope to a car window are also available.

The beginner should be cautioned, however, against too much dependence on magnifying instruments. Using either binocular or telescope is, of course, like looking through a tube, and much happens in other directions that may thereby go unnoticed. It is far more effective to search for birds (all but the most distant ones) with the unaided eye, and then put up the binocular. With a little practice you can get very proficient at doing this without taking your eyes off the bird, or at least keeping them fixed on nearby "landmarks" in the environment by relation to which the bird can then be found.

PHOTOGRAPHING BIRDS

So much is possible in this activity, and so great a variety of equipment is available, that space allows little comment here. Smith (1959) suggests the basic equipment needed for nature photography in general and gives helpful references. Unless one works in a blind or with a remote control on a camera aimed at a spot where a bird is likely to return (e.g., a nest), telephoto lenses of various sizes are needed for really effective photography of birds in most situations. To make satisfactory still pictures of moving birds or of those whose stopping places cannot be predicted, a single-lens reflex camera is a must because it permits viewing and focusing until the instant the picture is taken. A motion picture camera with through-the-lens viewing is also best, for the same reasons; but effective results are obtained by many people with less expensive types.

In certain protected areas where ducks, gulls, coots, and other water birds are fed and become relatively tame (or even beg food from people), closeup photography is possible without special precautions or expensive lenses — lakes in city parks such as Lake Merritt in Oakland, Stow Lake in San Francisco, El Estero in Monterey, MacArthur Park in Los

Angeles, are examples. Distant masses of water birds such as the geese and ducks on the wildlife refuges of the Sacramento and San Joaquin valleys, gulls about harbors and dumps, and shorebirds on favored high-tide roosts are also readily photographed with simple cameras.

BIRD BANDING AND ITS SIGNIFICANCE

In order to legally capture, mark, and release any wild bird protected by law, a permit is required from the California Department of Fish and Game and (for all but a few resident species) also from the U.S. Fish and Wildlife Service. While most users of this book will not become bird-banders, because of the associated restrictions, all who are interested in birds should know about the program and how to report a band they may find or have brought to them from a dead bird.

Individually numbered aluminum bands are placed on the legs of many hundreds of thousands of birds annually in North America — over 62,000 in California in one recent year. When properly applied, the band is like a ring on one's finger and causes no interference with the bird's normal activities or health. All bands used on water birds (and most land birds) in North America are issued by the U.S. Fish and Wildlife Service, which also keeps the master file of all bandings and returns and recoveries, except for some records on nonmigratory game birds kept by individual states.

If a banded bird is subsequently captured or found dead

13

and the number is carefully read (since there are no duplicates), the data contribute to knowledge of the range of individual birds on migration or in dispersal from hatching sites, on accuracy of return to former nesting or wintering stations, on longevity, and (if the same bird is recaptured repeatedly) on plumage changes with age and season, constancy of mates from year to year, etc. Among the more spectacular migration records of banded water birds in California are those of many of our wintering Snow Geese, which have been found thereby to summer in eastern Siberia. Some of the ducks of the Pacific Flyway have also been found wintering in the Hawaiian Islands and the smaller islands of the central Pacific southwest of them. When (as with waterfowl shot by hunters) large numbers are banded and later recovered from a given population of any one species, wildlife biologists have a means of estimating the population size without time-consuming censuses over a broad area.

When any banded bird is found, the date and place, along with the name and address of the finder, should be reported to the Bird-Banding Laboratory, U.S. Fish and Wildlife Service, Washington, D.C., or Laurel, Maryland, 20811. If the bird is alive, the whole number on the band should be carefully read and reported, the band left on the bird, and the bird released, since it might yield another record in the future. If the bird is dead, remove the band, flatten it out and tape it to a card, and send it along with the report. Even if the band is so worn that no numbers seem legible, the laboratory staff can often read them with special etching techniques. Both the person making such a report and the one who banded the bird are subsequent-

ly notified as to when, where, and by whom the bird was banded and the details of the later recapture.

In addition to metal leg bands, some banders carry on research using other markers such as colored plastic leg bands or bright-colored plastic wing tags (attached near the base of the wing and not interfering with flight), or tags on the back or attached to the nostrils. Authorization for any such marker or for dyes on the feathers, which are also used for temporary studies, must come from the U.S. Fish and Wildlife Service Banding Laboratory. Such markings give the additional advantage of establishing the location or time of banding, and sometimes the individual bird, without recapturing it. Any numbers or symbols on tags, the colors involved, and whether on right or left, should be noted in making a report. Identity of the species is usually not necessary.

CONSERVATION AND HUNTING

It was not too many years ago that concern for diminishing wildlife resources first resulted in laws to control the seasons and the number of birds, as well as the kinds, that could be taken by hunters. The Migratory Bird Treaty (between the United States and Canada, and later with Mexico), and the enabling act of Congress passed in 1918, are the basis upon which the federal government establishes hunting seasons and bag limits on migratory game birds each year and prohibits the taking of other native species defined as migratory, insectivorous, or endangered. Hunting regulations, with special exceptions to allow control of some species that become problems locally, are worked out in cooperation with individual state fish and game departments.

In the pursuit of waterfowl, coots, and upland game birds for sport in California alone, millions of dollars are spent annually by hunters on guns, ammunition, special clothing, and other equipment, as well as for licenses and transportation, food, lodging, and other away-from-home expenses. Millions are also expended by both state and federal governments in continual study of gamebird populations and in management of special refuges and controlled shooting areas, in order to build and maintain populations at levels that offer a harvestable surplus. The major National and State Wildlife Refuges and Management Areas in California, including the unique new San Francisco Bay National Wildlife Refuge, are shown on the map in the Appendix. Many private gun clubs throughout the state also manage their land so as to attract far more birds than they attempt to shoot. For ducks and geese, especially in the coastal regions and elsewhere away from the large government-managed areas, such private clubs have been of great overall value because of the maintenance of ponds and marshes in otherwise dry country.

When the take is controlled and proper management of habitats is extended to provide adequate food and cover around the year, the species populations, despite the large number of birds killed annually in the hunting season, are far larger than they would be if the refuges and hunting areas now managed primarily for game were to be converted to other uses. California's great valley, the Central Valley, is one of the truly major waterfowl concentration areas every winter. In the early 1960s it was estimated that some 70 percent of the approximately four to seven million ducks that winter in the state, and a half million geese, were found there each January. On the whole Pacific Flyway, however, the most important single stopping-place for migrating waterfowl is the Tule Lake-Lower Klamath Basin of northeastern California, through which nearly all of the dabbling ducks and geese funnel when enroute to the Central Valley or southward. These and most of the other refuges have regular visitors' routes that can be driven or walked in order to see these massive concentrations, although there may be temporary restrictions during the hunting season in order to protect the birds.

For diving ducks and many herons, some grebes, and most shorebirds and terns, the tideflats and shallows of San Francisco Bay, despite loss of large acreages to dredging and filling, are still the biggest block of suitable habitat on the coast. Hence the new refuge established there is of extreme importance. Also important in ratio to their lesser size are Humboldt Bay, Bodega Lagoon, Tomales Bay and Drake's Estero, and

Morro, Upper Newport, and San Diego bays. Still smaller but good units at such places as Moss Landing, Goleta, Seal Beach, Sunset Beach, Bolsa Chica, and Oceanside-Carlsbad serve as important homes for these birds, especially in transit.

Large segments of such invaluable tidal and wetland habitats, both inland and coastally, must be preserved if we are to have any sizable water bird populations in the future, for hunting either with shotgun or with binocular and camera. With no closed season on the latter, many a former scattergun devotee has learned that the rewards of an exciting day in the open, pitting his skill against that of wary birds, can be gained from a view or a picture rather than a bit of (expensive) meat for the table afterward.

A brief note on the collection of bird specimens for scientific purposes is also in order. Private collections of bird skins or eggs, formerly amassed as a hobby, are now appropriately forbidden; but there are many aspects of ornithology that can be studied only or best with a specimen in hand, and others (such as initial evolutionary changes) that require the comparison of specimens from many places. Many field identifications have been corrected or substantiated by the specimen deposited in some museum. The number of birds taken for such purposes is far too few to cause any dangerous reduction of population on any permanent basis, even locally. Students seriously interested in particular problems requiring use of specimens will usually be accommodated in solving them at a research museum, if they prepare themselves to undertake such work and make application to the director (see list of organizations preceding the Appendix).

However, no amount of control of the direct killing of birds, either for sport or for science, will support a population if the food and the nesting and retreat cover of the habitat do not suffice. Man's greatest effect upon most species of water birds in California today is his alteration of their ancestral types of habitats through such activities as filling and dredging shallow waters; "reclamation" of marshes, both fresh and salt; diversion of water from its natural drainage channels; creation of lakes by construction of dams; irrigation and other agricultural practices; broadcast application of pesticides on areas of great wildlife value, including uplands which drain into the wetlands, with the resulting pollution of waters; and (perhaps most of all in recent years) by the generally poor planning of urban development.

Critical declines in various groups of fish-eating and raptorial birds have occurred in the last ten to twenty years, some no doubt caused by disruption of nest sites by human activity. However, accumulation of the persistent chlorinated hydrocarbon type of pesticide (DDT, aldrin, dieldrin, chlordane, etc.) through food chains has been demonstrated as a major factor in several cases (see species account of Brown Pelican). There is widespread thinning of eggshells and disruption of nesting cycles, even when the adult birds seem otherwise not directly affected. This is an important conservation problem, especially in marine habitats where the food chains are longer (more steps in "who eats whom"); and it has severe effect, even though the chemical can scarcely be detected except in the organisms themselves.

Preservation or, where necessary, restoration of suitable habitat relatively free of this and other types of contamination is the only way to conserve many of our bird species. Good practices have been applied to dabbling duck habitats for years, but little has yet been accomplished for the diving ducks and shorebirds and various fish-eaters of our coastal waters — some of the prime avian attractions in California. The new San Francisco Bay National Wildlife Refuge and the smaller salt-water, tideflat, and marsh preserves of the San Pablo Bay and Seal Beach national wildlife refuges and those at Bolsa Chica and San Diego's Mission Bay will hopefully change this picture. Offshore nesting islands should also be under firm protection, but so far only Anacapa and Santa Barbara islands (the two smallest) of those off southern California have National Monument status, and the Farallones off San Francisco are a National Wildlife Refuge. Southeast Farallon now has the largest population of breeding sea birds of any, and is manned by personnel of Point Reyes Bird Observatory the year around, giving more hope for its future.

BEHAVIOR OF WATER BIRDS

FOODS AND FOOD-GETTING

Birds of a given species are often quite limited in the kinds of food they eat, and their behavior in seeking and taking in that food is apt to be even more steretyped. This includes in some a particular manner of flight or approach not otherwise used. Most such actions are apparently largely inherited and so, while some birds may learn to feed in completely new ways, this is a rare event. Shifts in diet are more frequent, the foods being taken by the ancestral method but utilizing new supplies of food similar to old but dwindling ones.

Since very few birds store food for future use except as fat within their bodies, many of their waking hours are spent in their typical food-getting activities, collectively termed foraging. Throughout the species accounts, the usual diet (of at least some species in each family) is mentioned, as well as the characteristic ways of obtaining it. For fish-eaters, for instance, these foraging techniques run the gamut from swimming beneath the surface using the feet only (grebes), the wings only (alcids), or both (shearwaters); to scooping or grabbing fish as the bird swims on the surface (White Pelicans, some gulls), standing or stalking slowly in shallows and then making a sudden grab (herons), diving head first into the water from overhead flight (most terns) or a lookout perch (kingfishers), or grabbing from near the surface on a low swoop past it (frigatebirds, some gulls, some terns).

Variation is also evident, especially in the larger families of birds. The strainers (sieve plates) in the bills of ducks obviously adapt them for getting small food items from mud, and so they do most of the time. But the expanded nail on the tip also enables them to pick up large items or to pluck vegetation from the bottom or on the ground out of water. The degree to which different species in the waterfowl family specialize in these activities varies, however, as does the depth of water in which they can successfully forage (by tipping-up or by diving). These and their differing agility on land have apparently been major factors in the evolution and survival of the many species in this family.

Another large family, the probing shorebirds, has instead diversified primarily in body size and length of bill for probing. Most of their food is of small animals or parts of them — felt (and tasted?) by a sensitive and flexible tip of the bill in a substrate from which a human has to use a sieve or careful washing and picking to find anything alive. The turnstones in this family have become specialists in another way, however, as their name indicates.

In contrast to these families of species diversified for feeding compared to each other, most of the gulls have retained a diversity of behavior within each species that enables them to meet rapidly changing conditions better than other water birds. They can feed while walking, surface-swimming, in aerial swoops, or by plunging. As long as suitable foods are available by these means (and the foods too are varied), special efforts to preserve gull populations are not usually necessary.

Plant-eaters are relatively few among the water birds — chiefly the geese, cranes, and coots, while many ducks, rails, and a few others take large fractions of both plant and animal material as food.

Perhaps the most bizarre of all food-getting techniques, each successful in particular habitats, have been developed in the phalaropes, avocets, and dippers (see species accounts).

FLOCKING AND ROOSTING

Most species of water birds are gregarious, at least during seasons when they are not breeding. Even those that feed singly may group together for roosting or for migratory flights. The pelicans, cormorants, some grebes, herons, terns, and alcids even nest in colonies, in some species as closely packed as "bill reach" of neighbors allows; but many of these are less gregarious for other activities.

When foraging, the most social of our birds are the shearwaters, the White Pelican, and the Double-crested and Brandt's cormorants, all of which fly to and from feeding areas in flocks and often engage in cooperative fishing activities in massed flocks. More loosely organized feeding "aggregations" occur among many species that seek fish or other prey varying in location from day to day or even hour to hour. The way these flocks build up in a typical example is described for the Herring Gull (see species account).

Among the bottom-feeders, the swans, coots, ibises, and most species of ducks, geese, and probing shorebirds usually seek food while in even more integrated flocks. However, the community alertness to danger from predators may simply outweigh the disadvantage of competition for food within the flock, which is often also evident. The plovers illustrate well a borderline situation, their inter-individual distances as they feed being so great they can as well be considered solitary, even though when danger threatens (or when otherwise flying or resting) they join into compact flocks.

Most of the water birds mentioned above also fly in flocks to and from feeding and resting areas, as well as on longer migration flights. The arrangement of individuals in a flock differs among the various families and to some extent in individual species. Some, such as the loons, terns, and most gulls (on local flights) travel in elongate straggling array, while others, like many shorebirds and ducks, arrange themselves into compact but irregular flocks. A few species form into precise diagonal lines (echelons) or V's, seen especially in Canada Geese and sometimes other geese, some ducks, and some gulls and shorebirds when on long-distance flights.

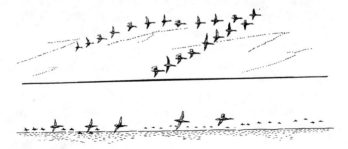

Detailed studies of the behavioral basis for such flock *structures* are just beginning.

For resting purposes, the pelagic forms (albatrosses, shearwaters, storm-petrels, jaegers, and some alcids) simply stay at sea or well offshore and sit on the water; but the extent of any movements or flocking involved for this purpose is poorly known (e.g., see Sooty Shearwater account). Loons, grebes, diving ducks, mergansers, and phalaropes that may feed closer to shore usually seek open water for resting and sleeping; and some geese, dabbling ducks, and gulls also do so temporarily, especially when disturbed elsewhere. For overnight or other long rest periods, however, these and most other water birds usually find places on land for sleeping where they will be undisturbed, flying long distances to reach them if need be. Thus certain rocky headlands or islets along the coast attract pelicans, cormorants, murres, and some gulls from miles of adjacent ocean; and remote sandy beaches are sleeping spots for terns, other gulls, and some shorebirds. Other shorebirds, most dabbling ducks, geese, and coots usually go to marshy areas or low islands, dikes, or peninsulas where surrounding waters make approach by land-based predators or humans difficult.

How do birds find what places are suitable for these roosts? The largest flocks, of course, may build up in the best places merely by virtue of birds that settle elsewhere being disturbed, then flying about the area, and finally being attracted to those already at rest in the undisturbed locations. But when multiple sites exist in an area, and the same one or few are used year after year, other factors are indicated. Probably there is considerable tradition involved, the older birds remembering the locations from previous years and the birds-of-the-year merely staying with the group. Again, there has been almost no basic research that provides real answers here. In only a few groups of birds, notably the swans, geese, and cranes, is there any good evidence that family groups stay together as a unit on migration and through the first winter.

The roosting needs of shorebirds that feed on intertidal mudflats are so special as to deserve separate mention. These

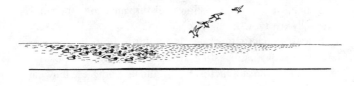

birds cannot occupy their feeding grounds when the tides reach levels of more than about four feet above mean lower low water (the average of the lower of the two daily low tides), as happens once or twice per day but at gradually changing times of day and night. Originally such birds moved for the duration of the higher tides into the upper parts of salt marshes or onto sandy points that remain above the water then, the surrounding waters protecting them from most disturbances by predators or man. Now, however, in many of our bays, most such areas are either altered to unsuitable dry land or deep water, or are so frequented by people that the flocks are repeatedly disturbed. About San Francisco Bay most of the shorebirds feeding on the tideflats now use the dikes and shallows of the extensive salt-evaporating ponds for this function of "waiting out" the high-tide periods. Possessed of a strong instinct to fly over water or along shores, these birds do not go any distance inland but may fly 5 or 6 miles along the bay shore at each ebb and flow of the tide, day or night, in going to and from their customary roosts. Flocks of mixed species of shorebirds totaling up to 60,000 birds can be found in the best of these places during the two to five hours of high water, and the preservation of such roosts is seen to be as important as preservation of the feeding areas for the welfare of these species. Watching these flocks passing from the roosts to feeding areas, or their return as the tide rises, is a fascinating experience. Even more striking to the casual observer, though, are the amazingly agile, close-order wheeling flocks of smaller shorebirds that show dark backs, then white bellies, in unison.

This behavior usually follows disturbance on the roosts, especially by a low-flying bird of prey.

Inland, the activity cycle of shorebirds probably consists of feeding by day and resting at night, since tidal exposure of feeding areas is not a factor. Coots, ducks, and geese may carry on a similar cycle there, but when disturbed repeatedly during the day (as during the hunting season) they tend to rest on safe areas in large flocks by day and feed in more scattered flocks at night. It is on the large waterfowl refuges (see map in the Appendix) that the massed concentrations of ducks and geese then are such great sights. They are so densely packed in areas closed to hunting that any food there is quickly eaten up. So there is a regular late afternoon exodus from such ponds toward feeding areas, often at fairly low heights. The return flight at dawn, however, is one of the truly spectacular sights of California's Central Valley — great skeins of waterfowl high in the sky, silhouetted against the sunrise. For a rival feeling of wildness in our low country, I would vote for the similar flights at various hours by flocks of Sandhill Cranes, perhaps even more appealing, despite smaller numbers, because of their larger size and louder voices.

MIGRATION

The term *migration* is sometimes applied to all movements of animals in numbers from one geographic location to another, but with respect to birds it customarily refers only to the movements between breeding areas and those used in non-breeding periods. Some degree of annual regularity and of movement over considerable distance is usually implied also. A few species, however, engage in equally long-distance movements on an erratic or "once every few years" basis — e.g., the flights northward into California of Wood Storks, Magnificient Frigatebirds, and Elegant Terns from more southern breeding areas, and the irregular influxes of Northern Fulmars, Black-legged Kittiwakes, etc., from northern regions in winter.

There is never a season in California when some migrants

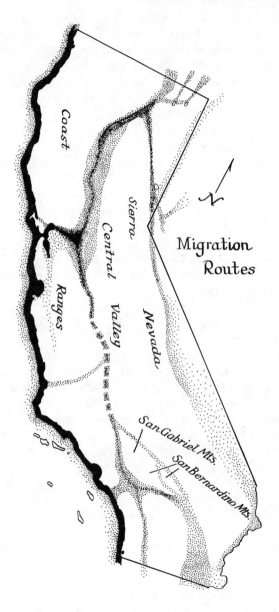

N

Coast

Ranges

Sierra

Central Valley

Nevada

Migration
Routes

San Gabriel Mts.

San Bernardino Mts.

are not passing. Our lowland and coastal areas are major wintering areas for a great variety of water birds. Some arrive in late summer while other species are still breeding here, and others do not reach the Central Valley or the coast until driven from the northern interior by truly wintry weather in December or January. The first of the northward migrants are detectable by January, e.g., Cinnamon Teal on the coast slope and White-fronted Geese along inner foothills; but the bulk of the spring departure of waterfowl comes in March, and that of some other water birds in April and May, after the summering terns have built up in numbers. The southward movement extends over a longer period, from July or August through the rest of the year, the particular span of dates depending on the species.

The Graphic Calendars in the Appendix summarize (by rows of asterisks) the dates of actual migration that have been reported for each species, as well as provide information on changes in numbers present in specific districts of the state; the main dates of arrival and departure can be inferred from this. You can thus make up a calendar of what to look for when.

Major routes of migration are fairly well known for the waterfowl, from the extensive banding programs of the U.S. and Canadian Wildlife Services and cooperating state agencies and a few private conservation organizations. The large concentrations of dabbling ducks and geese that winter in the Sacramento Valley and locally in the San Joaquin Valley are known to funnel into this area largely by way of the Tule Lake-Lower Klamath Basin mentioned above. Most of the birds that mass there in September, October, and November have come from breeding areas ranging from Utah to Saskatchewan to Alaska. When they go on southward, some fly to the east of the Sierra Nevada and on to the Colorado River or Imperial Valley, but most cross the low mountains southwest of the Tule Lake area (roughly along the Pit River) into the upper Sacramento Valley, either settling there for the winter or passing on to other wetland areas southward through the Merced-Los Banos-Dos Palos region. Considerably fewer continue migrating on into interior southern California. Large numbers of certain of the dabbling ducks (but few geese) also extend their migration routes on toward the coast to good wintering areas about San Francisco Bay, including certain of the "fresher" salt ponds and nearby gun clubs, or on to Morro Bay and the southern coastal lagoons.

A truly coastwise route is followed on migration through California by most diving ducks, although many of them reach the ocean from interior or northern Canada by overland routes to the north of us. Among the divers most commonly making such crossing farther south are Canvasbacks, Buffleheads, and Goldeneyes, some of which remain on fresh water in interior California all winter. The strictest coastwise migrants include the Brant, loons, and scoters, and (farther offshore) the alcids, jaegers, Red Phalarope, Sabine's Gull, and Arctic Tern. The loons and some others can be counted in impressive numbers from outer headlands at the right season (see Arctic Loon account). Gulls and terns also follow the coast for long distances, even the California and Ring-billed gulls passing north steadily in April, sometimes to beyond the Golden Gate, before taking still largely unknown routes to their interior breeding areas. The southward migration of at least some populations of these same species, in July and August, seems to take a more direct westward or southwestward route over the high Sierra Nevada to the Central Valley or on to the coast, more or less throughout the state; but tagged California Gulls from Wyoming also show up every fall in the Pacific Northwest before moving on south.

Among shorebirds, some of the southern breeding populations may make short movements to bayside or coastal localities from more interior breeding stations (e.g., American Avocets) and a few are nearly resident (e.g., Killdeer); but many reach our shores from breeding areas in Arctic regions. The earliest southbound flocks (frequently adults only) appear around the first of July and can hardly have bred. They probably consist of birds that failed to nest or had their nests interrupted. By a month or so later, however, post-breeding arrivals of both adults and birds-of-the-year are proceeding in force, after which many species begin their wing and tail molt. The Dunlin, latest fall arrival among the common shorebirds, is known to molt before leaving its Alaskan breeding grounds.

The most inveterate travelers among the shorebirds are the Golden Plover, famous for long overwater routes including flights from Alaska to many islands of the Pacific, and the Black-bellied Plover and Sanderling, both of which breed in high Arctic latitudes and range southward in winter to southern South America (some flocks remaining in California). Other transequatorial migrants include the phalaropes that go from the Arctic to oceanic wintering locations of the southern hemisphere, and some shearwaters that travel the whole of the Pacific Ocean, or at least its peripheries, from breeding stations in the Australian area, the same individual returning to the same nest site year after year. Experiments with albatrosses displaced (by jet plane) from nests at Midway to far corners of the Pacific, including some where the species does not occur, have shown that they can get home over long distances at speeds averaging a hundred or more miles per day.

These and other navigational performances remain one of the intriguing but as yet incompletely solved mysteries of ornithology. Further experiments on the actual orientation of migrants in cages and at release points have shown that various birds can find and maintain their proper direction for migration by reference to the sun, with appropriate correction for its movement across the sky from hour to hour. Some songbirds are even known to do so equally well under starry skies at night or in a planetarium, and the same is presumably true of the night-migrant water birds (grebes, rails, gallinules, and coots) as well as those that migrate either by day or night, such as the waterfowl and shorebirds.

Migration-watching can indeed be fascinating. One minor but interesting feature should also be mentioned. There are conspicuous postbreeding movements among the herons, ibises, pelicans, terns, and some gulls, away from all known nesting areas, carrying these birds at times far to the north of their breeding range and for some species to various high-altitude lakes in the Sierra, after which they finally depart for the south.

NESTING

All birds reproduce by means of eggs, and all but a few species incubate their eggs for definite periods of weeks until they hatch. After this, there is also a period when the young are dependent upon the parent for food and protection. The need for all this parental care ties adult birds to a definite place for the breeding period each year. The actual months involved in California begin as early as January for the Cassin's Auklet, whereas the Ashy Storm-petrel and a few other species are still raising young as late as October. Whatever their breeding period, the habitat needs of many species are most restricted then. These needs may differ sharply from those of related species, and even from those used by the same species at other seasons. A common shift is from fresh water or inland marshes (or northern tundra) to salt water when not breeding. Brief indications of the nesting habitats and usual type of nest and appearance of the eggs are given in the species accounts for each species breeding regularly in California.

Nests of many water birds are on the ground or in tussocks of marsh vegetation, since many species seek food or escape from danger in water or shore areas. Only the grebes and coots

regularly build floating nests, but some other marsh nesters may have nests that will float for a time if the water level rises temporarily — e.g., rails, American Avocets, Forster's and Black terns. The amount and kind of material used in making a nest varies from the Great Blue Heron's large platform of sticks to the seaweed mound of a Brandt's Cormorant, to the thin rim or lining of plants or debris used by a Spotted Sandpiper, to nothing at all in murres, many shorebirds, and some terns. Some sea birds dig prodigiously into the soil or secrete their eggs in deep crevices among rocks, and several kinds of ducks choose similarly protected sites in holes in trees. Most waterfowl also pluck their own down feathers to form a soft nest lining. Other than the hole-nesters, the only California water birds nesting in trees are most of the herons and egrets, usually in colonies.

Safe nest sites for water birds are often at a premium, especially next to waters with good food supplies. Hence it is advantageous for many species to nest in colonies in such areas. Islands, steep headlands, marshes, and remote dikes used by many water birds for nesting are traditional sites of great importance to their survival as populations. As such, these places need to be guarded against interference by humans during the nesting season if we are to continue to have these birds around. The Snowy Plover and Least Tern, both of which nested chiefly on open sandy beaches until recent years, exemplify how losses come about when people become prevalent throughout an otherwise adequate habitat.

The normal number of eggs per clutch varies greatly among species, but is often very regular within those with smaller clutches: one only in the storm-petrels, murres, puffins, and Cassin's Auklets; two (or one) in guillemots and other alcids, and in Elegant Terns; two to four in various grebes, herons, most gulls, terns, and shorebirds. Variable larger sets in the rails, coots, and ducks are normal. In general, the larger the set the greater the likelihood of death of the young before reaching adulthood.

In many birds with clutches of more than two eggs, incubation is not begun until the last or next-to-the-last egg is laid, so all young hatch in one to two days, whereupon the behavior of the adults shifts appropriately to one of bringing food to the young or guiding them to it. In some, however, incubation may begin after the first egg, with the result that the young are so much graduated in age that the smaller ones may starve because their older brothers and sisters get all the food – another means of nature adjusting functional clutch size to resources. Actual incubation periods among the water birds range from about 18 or 19 days for the smaller rails to some two months or so in the whole albatross-shearwater-petrel order. Where the information was readily available, the span of incubation (either known or as estimated from near relatives) is given in the accounts, following description of the eggs, for each species breeding in California.

The parent birds of a pair may share the duties of nest-building, incubation, and feeding of young more or less equally, as occurs in storm-petrels, cormorants, herons, rails, gulls, terns, alcids, and some shorebirds. In some of these – e.g., cormorants and herons – the male brings nest material which the female places in position, or the two do it together as a courtship and mating ritual. In some species the female does most or all of the incubation, as in American Bitterns, Ruddy Ducks, and geese, with the male continuing to guard the area and aiding in care of the young. Most ducks form into pairs during the winter or early spring, then migrate together to the female's home area. There the male stays with his mate only until she has incubation well under way, or sometimes until the young are just hatched. She then raises the young while the male retires to safer waters for his flightless molting period. In several shorebirds, notably the Killdeer and Spotted Sandpiper among California breeders, the male performs most of the incubation. However, it is only in the phalaropes among our birds that the usual role of the sexes is nearly fully reversed: the female is the more active in courtship and lays the eggs, but the male builds the nest, incubates the eggs, and raises the young without her help.

HABITATS AND BIOTIC DISTRICTS

The habitat or kind of environmental situation that birds of a particular species utilize is often much more restricted than beginners in bird study realize. Despite their mobility by flight, many birds are attracted to or held within very particular habitats by behavior patterns either inherited or acquired early in life through imprinting. These actual processes of habitat selection have been little studied as yet, but much is known of the habitats that most bird species seem to "prefer."

Thus, Mountain Plovers (a most *in*appropriately named species) are almost always found on broad expanses of gently rolling to flat grassland or fields with very short or no vegetation. The land itself, the insects on it, and the lack of tall vegetation are the essential features. A mudflat or lake or bay shore, where other shorebirds are so common, does not seem to attract them in the least. On migration they must of course cross such habitats at times, as well as the even more unsuitable rugged, often forested mountains. The fact that they never alight in such places (barring accidental groundings by storms or illness) merely testifies to the strength of their habitat-selection response.

Similarly, we find Black-legged Kittiwakes primarily offshore, and Heermann's Gulls along and just off the outer coast but seldom even halfway within San Francisco Bay, whereas the Ring-billed Gull of the same size is the primary gull species inland and quite uncommon on the outer coast. Close attention to habitat limitations and preferences of most birds will greatly foster the bird student's understanding of their lives and will be of considerable help also in identifying them.

In a book as limited in size as this one, full descriptions of the habitat preferences of each species cannot be given; this has been done very well by Grinnell and Miller (1944), the "official" but somewhat outdated distributional list of the birds of the state. Here a space-saving code for habitat preferences is provided, in the Appendix, for each species of regular occurrence in the state, along with a calendar in graph form showing the numbers and dates of known occurrence in the preferred habitats. By learning the essentials of this code and

In each key, pictures of families or subfamilies that are easily confused are arranged near each other. Each is numbered to correspond with the taxonomic sequence followed in the species accounts and in the Notes on the Picture Keys. A line by each picture indicates the scale, with small tick marks along this scale representing 6-inch (15 cm) intervals. When the size range in the family is more than one-third above or below the average bird illustrated, a ~ is placed by the line. See also the diagram Parts of a Bird, the list of Abbreviations and Terms, and the Size Guide which follow the Notes on the Picture Keys.

WHAT TO LOOK FOR ON THE BIRD AND IN THE PICTURE KEYS

1. General shape of major parts: body, neck, head (especially forehead), tail, and wings (wing-tips are visible on most birds even when at rest).

2. Bill: its length relative to diameter of head, continuing along the same axis; its shape, including any special tips or knobs or plates near the head.

3. Legs and feet: length and shape and, where possible, the color; whether the toes are separately lobed (few species), webbed (many), or fully webbed (pelicans, cormorants, and allies).

4. Feathering patterns: broad ones of light and dark are more helpful than colors, especially for family identification. (See also Techniques in Observing Birds, in the Introduction, on the importance of good lighting.)

5. Size, relative to some object of known length or another bird whose identity is known; without such comparison, judging size is notoriously inaccurate, especially on open flat terrain, on water, or overhead. (See Size Guide to lengths of some common birds, at the end of this section.)

6. Behavior, especially items mentioned in the notes accompanying the Picture Keys; others may be important for species identification within the family.

7. Special markings such as contrasting patches or stripes,

the wings and tail being particularly important and best seen in flight for ducks and shorebirds (for identification to sub-families, at least), and the wing-tips in gulls.

After using the Picture Keys and associated notes to select the probable family of a bird you want to identify, turn to that family in the species accounts — the number on the Picture Key corresponds to the sequence of families in the text. Compare the descriptions given under Recognition and the illustrations of the various species in the family. Usually you will find one that fits, and then you have identified the bird. However, if you failed to note some critical item on the bird before it departed (a common occurrence), then you may know only that it was one of several species. If none of the species in the family seems to fit your bird, go back to the Picture Keys and track down any other possibility. In any event, double-check by reading the full account — Habits, Range, and Occurrence in California, as well as Recognition. Some birds may not fit all categories but may still be correct — especially because of the gaps in our records of occurrence and knowledge of habitat use.

The really rare bird, completely out of season or usual locations for its species, is indeed less likely to be spotted than some beginners are prone to believe. In the event you think you have located such a bird, and it seems to be staying around, seek help or confirmation of the record. Some of the organizations listed at the end of this book (following the species accounts) can be of service in this.

NOTES ON THE PICTURE KEYS: RECOGNIZING THE WATER BIRD BY FAMILY

Picture Key A. *Birds on the Water*

1. **Loons.** Bill pointed, held horizontal; feet large and very far back. Swim fairly low; often peer leisurely below surface before diving; may dive and remain under water for long periods; pursue prey under water.

2. **Grebes.** No evident tail; neck moderate to long and slender; bill tapered to sharp point (or convex in one species);

PICTURE KEY A. BIRDS ON THE WATER

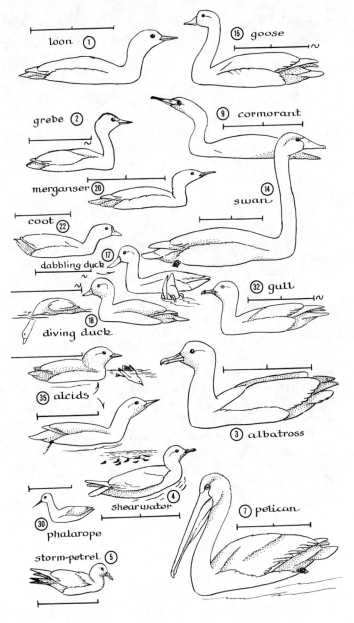

loon ①

⑮ goose

grebe ②

⑨ cormorant

merganser ⑳

⑭ swan

coot ㉒

dabbling duck ⑰

diving duck ⑱

㉜ gull

㉟ alcids

③ albatross

shearwater ④

phalarope ㉚

⑦ pelican

storm-petrel ⑤

(Numbers correspond to sequence of families in the text and Graphic Calendars) large feet, far back. Swim high or low, commonly dive, and some can submerge by sinking; pursue prey under water.

3. **Albatrosses.** Oceanic. Larger than any Gull (no. 32); note the elbows of long wings often apparent near base of tail. Swim high; may plunge head, or rarely dive a few feet.

4. **Shearwaters.** Normally oceanic. Shape gull-like (see no. 32), but neck appears shorter, the head held lower; feathering brown to gray and white; bill medium length, with hooked tip and nostrils as tubes basally. Swim medium high; may dive repeatedly for food using wings and feet, but usually seen resting, often in massed flocks.

5. **Storm-petrels.** Normally oceanic. Rather gull-like in shape (see no. 32), but smaller; colors mostly sooty-brown, but light gray in one; some with white rump; bill short, with hooked tip and tubular nostril openings on basal part. Swim high and often erratically, but usually not for long; dive rarely.

7. **Pelicans.** Large chunky body, neck upright with long beak sharply downward; beak has large pouch in which fish are caught, sometimes (one species regularly) by scooping from swimming position; swim quite high.

9. **Cormorants.** Feathering all dark (to whitish on belly in immature of one); bill with hooked tip; feet fully webbed. Swim low to very low (back sometimes at surface), the bill usually at upward angle, the moderately long tail near or at the surface; dive from surface and pursue prey. (See also species accounts on somewhat similar Boobies, no. 8.)

14. **Swans.** Very large, very long neck, feathering all white in California species. Swim high, the neck up when alert or curved to reach bottom for feeding, or may tip-up. (Compare with Geese, no. 15.)

15. **Geese.** Bill tapered from high base to semi-flattened terminal third with large nail on tip; feathering brown and/or white, often with some black. Swim high and may tip-up to feed. (Compare with Swans, no. 14.)

17. **Dabbling Ducks.** Body slenderer than most Diving Ducks (no. 18), neck of some is longer; beak typical duck type (gradually flattened toward tip, with thickened nail on tip and. sieve plates inside). Swim fairly high; tip-up to reach bottom

with bill, but dive only in extreme emergency and with difficulty. Males often contrastingly patterned, females not.

18. **Diving Ducks.** Body heavier for length than in Dabbling Ducks (no. 17); feather colors black, white, gray, brown in most (rusty in a few); bill typical of ducks (see note on 17 above), some with knobs at base, colorful in some. Swim medium to low, commonly dive to bottom for food, but in shallows may tip-up. (See also account on Ruddy Duck of the Stiff-tailed Ducks no. 19.)

20. **Mergansers.** Plumage often with much white or light gray; head crested (except males of one species); bill sub-cylindrical with coarse "saw-tooth" edges (visible only at close range). Swim moderately low, with bill horizontal; dive and pursue prey under water.

22. **Coots** (of family Rallidae). Beak tapered abruptly to blunt point and with shield on forehead; feet large, toes lobed. Swim high, often with head jerking; tip-up and dive at times for food. (See also species account of Common Gallinule, in same family.)

30. **Phalaropes.** Tiny version of Gulls (no. 32) as to head and body shape; bill straight, thin to needle-like; toes lobed or flanged. Swim very high; make quick jabs into water for food; often whirl repeatedly in place.

32. **Gulls.** Feathering white and gray plus some black (wing-tips of many), or mottled brown (first year), or these in mixed pattern; bill moderate to stout, with definite angle in outer half of lower profile. Swim high and relatively weakly, with tail and wing-tips· well above surface; may plunge head after food, but do not usually dive. (See also species accounts on Jaegers and Skua, no. 31.)

35. **Alcids.** No bright-colored feathering. Body short and chunky, neck length within range shown; beak two-thirds head length or less, usually convex above and slightly angled below (but see Puffin, with enlarged beak, in Picture Key B). Swim moderately high but dive quickly, *using wings* for underwater propulsion (note sketch of disappearing bird) and the feet for steering; often peer quickly below surface while sitting on it (bill-dipping).

Picture Key B. *Water Birds on Land or Perch**

Features noted in Picture Key A and visible here are not repeated.

7. and 8. **Pelicans** and **Boobies.** Fully webbed feet. Each has unique profile if beak and tail are seen well; booby is less upright.

9. **Cormorants.** Fully webbed feet. Upright stance; when moving on land, gait is a slow waddle or rapid hops aided by wings.

11. **Herons, Egrets, Bitterns.** Long neck "disappears" when head is drawn back; many have long plumes in breeding season; feathering variously patterned but not brightly colored. Usually walk or wade slowly to feed; many roost and nest in trees. (Compare with Cranes, no. 21.)

12. **Storks.** Heron-like shape (see no. 11), except for bill; bare head and neck in California species. (Compare with Cranes, no. 21.)

13. **Ibises.** Small heron shape (see no. 11), except long curved bill (compare with Curlews, no. 26). Some perch above ground, but California species rarely does. All or mostly dark feathering.

15. **Geese.** Legs longer and more central under body than in ducks (see nos. 17 and 18). Walk with but slight or no waddle, neck bending to graze or pick up food, or held stiffly erect in alarm. (**Tree** or **Whistling Ducks**, no. 16 — not shown here — are intermediate between Geese and Dabbling Ducks, see species account and fig. 18.)

17. **Dabbling Ducks.** Legs a little longer and more under body than Diving Ducks (no. 18), less so than Geese (no. 15). Walk with slight to definite waddle, but may run; many feed some distance from water at times.

---

*Families which are rarely seen on land or on a perch out of water in California are omitted. See Picture Keys A and C, and species accounts, for characteristics of birds in these families. Skimmer (no. 34), Kingfisher (no. 36), and Dipper (no. 37) families, each with one very distinctive species in California, are not shown on Picture Keys; see figs. 44 and 47.

PICTURE KEY B. WATER BIRDS ON LAND OR PERCH

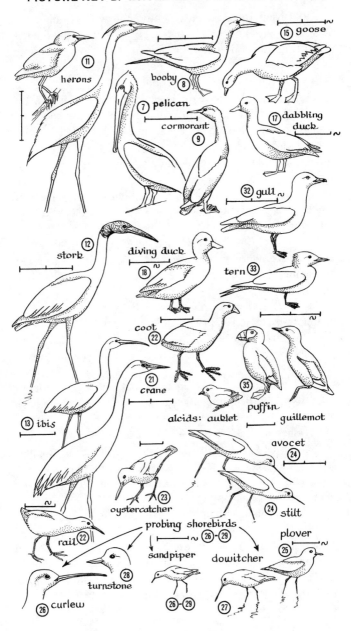

(Numbers correspond to sequence of families in the text and Graphic Calendars)

18. **Diving Ducks.** Heavier body, shorter legs, and larger feet than Dabbling Ducks (no. 17). Walk with pronounced waddle; normally do not feed on land, nor get far from water (except those that nest in tree holes). (**Mergansers,** no. 20 — see Key A - are similar except for bill.)

21. **Cranes.** Distinctive silhouette; compare with Herons (no. 11) and Storks (no. 12), noting bill shape and usual walking stance. They do not perch above terrain level.

22. **Rails.** Neck short to medium-long; feathering mostly brown or reddish-brown and gray with light and darker markings, a number with barred sides; bill from one-third to one and a half times head length, stout, blunt-ended; large feet. Usually walk or skulk amid vegetation, occasionally in open for short distance, and may swim briefly.

22. **Coots.** Chunky body and huge feet with flat lobes along toes. Stance varies; See species account for behavior. (Gallinules, in same family are intermediate between Rails and Coots.)

23. **Oystercatchers.** Bill shape and body size unique; our common species has all black feathering.

24. **Avocets** and **Stilts.** The bold black and white patterned plumage, very long legs, and slender pointed beak are distinctive.

25. **Plovers.** Legs medium to medium-long; feathering smooth or mottled brown or gray and white, with dark or black chest bands in some; bill shorter than head, with distinctive swollen area about one-third distance from tip. Normally feed by visual search method: running to a new spot, stopping and standing very still several seconds, bending to pick up any food item seen, running again to a new point, etc. Probe only in pursuit of escaping prey.

26.-29. **Probing Shorebirds** or "probers." The smaller species (from about robin size downward) in subfamilies **Tringinae** (no. 26) and **Calidridinae** (no. 29) are often called **Sandpipers;** but some of them, and most of the larger species, are given individual names: **Sanderlings, Turnstones, Snipe, Curlews, Tattlers, Dowitchers, Godwits, Willets,** etc. The smaller sandpipers (below about blackbird size) and other small shorebirds, if unidentified to species, may be referred to as "peeps." The beak is always slender, though it varies greatly in length in different species and is pointed in **Turnstones** (no. 28) only. Colors are mostly browns and grays to whitish below, often with darker mottling; some bright rusty or reddish-brown in breeding feathering. They typically feed by inserting bill into mud or sand substrate, finding prey by touch (or taste?) — but see species accounts for specialties.

32. **Gulls.** Most features are "medium," body held horizontal. Color (hue of gray) of mantle (upper wing surface and back), wing-tips, and bill and feet are important for species identification.

33. **Terns.** More slender than Gulls (no. 32) of comparable size, head of some has crest at rear. Feathering of most is pale gray and white, usually with black on crown in spring and early summer (body all black in one species). Wings long with pointed tips (over tail); tail usually forked, shallowly to very deeply; feet webbed but small, and legs very short. Gull-like stance; rarely walk any distance.

35. **Alcids.** Strictly marine; seen on land only in nesting areas or comparable nonbreeding resting spots, usually on rocks. Most squat on full tarsus, shuffle to walk; but Puffins stand on webbed toes only.

Picture Key C.   *Water Birds in Flight**

The sketches and the commentary, unless otherwise stated, pertain only to normal full-speed horizontal flight. Considerable difference is to be expected on takeoff (often commented on) and landing, or when the bird is engaged in aerial displays. Overall speed of flight of most water birds is medium to fast (40 to 70 miles per hour airspeed), and can be assumed to be in this range unless described as slow. Wingbeat rate is indi-

---

*Patterns seen only or chiefly in flight are described here. For features of body, neck, head, beak, and foot shape, and most aspects of significant color patterns, see Picture Keys A and B.

cated by small letters in the space beside each sketch (if in parentheses, indicates occasional only) as follows:

s = slow; not over 3 or 4 per second, and thus countable

m = medium; about 5 to 9 per second, borderline or uncountable

r = rapid; definitely too fast to count by eye

vr = very rapid

i = interrupted; flaps only irregularly, or in short bursts separated by periods of gliding or soaring

1. **Loons.** Distinctive droop of head and feet, the latter extending beyond the short tail; wings narrower and more pointed than Cormorants (no. 9) or Geese (no. 15). Takeoff only by long run on water surface (not possible from land).

2. **Grebes.** White patch in secondaries (except in Pied-billed). Takeoff by long run on water surface; helpless on land. Much less prone to fly in daytime than loons; migrate at night.

3. **Albatrosses.** Oceanic. Wings narrow, with wrist bend beyond midpoint (and another joint in "finger" area beyond this sometimes evident). Expert at dynamic soaring close to ocean swells, but occasional slow flaps except in high winds. Course is often in wide, meandering arcs alternating with low glides along wave front, and rarely more than 30 feet above water except on steep banks in high winds.

4. **Shearwaters.** Oceanic. Wings narrower and less arched than Gulls (no. 32), held stiffly extended on glides with little or no wrist bend apparent. Short bursts of wingbeats followed by short to long glide (according to wind speed), often very close to water, with higher turns banking into wind but usually not over 20 feet up. The course is thus often a succession of long arcs following along wave fronts or troughs, occasionally rising over crests. Flapping speed is inverse to body size for a given wind condition. Flocks are open or closely massed.

5. **Storm-petrels.** Oceanic. Dark brown to nearly black (or light to medium gray in one), with paler line along wing coverts; some have white on rump or upper tail coverts also. Bill held at downward angle; wrist bend usually prominent

except in slow fluttering. Brief gliding periods interspersed with flapping in most. Course is usually erratic (species differ) and low over water, even mostly in wave troughs.

6. **Tropicbirds.** Oceanic. The mostly white feathering, beak shape, and (in adults) long central tail feathers are distinctive. Wingbeats vary from deep to shallow, but the birds may soar on updraft over a ship. Course is often irregular; speed medium to slow.

7. **Pelicans.** Beak angle, neck in S-curve, and long broad wings are distinctive (but see Herons, no. 11). They commonly flap a few times, followed by long glide, but may flap steadily for long periods also. The White Pelican commonly soars in circles in updrafts when traveling over land; the Brown Pelican dives from the air. Takeoff by kicking feet in unison.

8. **Boobies.** Wing and especially tail shape is unique; bill held slightly downward. Only brief glides while traveling, but much wheeling and gliding (often higher) when foraging. Takeoff is cormorant-like (see note 9 below).

9. **Cormorants.** Neck straight forward or slightly above horizontal; wings fairly broad, bluntly pointed. Flight is mostly low, in lines or irregular echelons or bunches; except that Double-crested, when traveling over land, may soar in circles at times. Takeoff by kicking both feet together several times, or by drop from high perch.

10. **Frigatebirds.** The wing, tail, and bill shape are unique. These birds soar and glide, slow to fast, with only occasional flaps, high to low over water or shore; they never alight on water, but swoop to the surface or pursue other birds with powerful wingbeats.

11. **Herons, Egrets, Bitterns.** Wings long and broad, typically arched; long or medium neck folded (on other than very short flights) in S-curve so that head rests just ahead of wings, beak pointing straight forward, long legs trailing. Speed is slow and course is somewhat bouncy, with wingbeats most evident in bitterns; may make long glide on descent.

12. **Storks.** Heron-like (see no. 11), except that neck is extended. In rising air, may soar in circles at times.

13. **Ibises.** Appear goose-like (see no. 15) at a distance but

PICTURE KEY C. WATER BIRDS IN FLIGHT

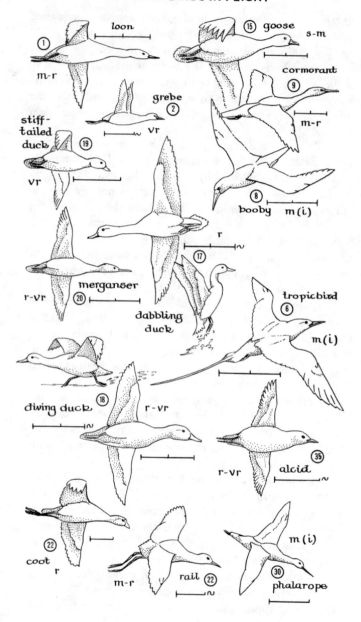

loon ① m-r

⑮ goose s-m

grebe ② vr

cormorant ⑨ m-r

stiff-tailed duck ⑲ vr

⑧ booby m(i)

r ~

merganser ⑳ r-vr

dabbling duck ⑰

tropicbird ⑥ m(i)

diving duck ⑱ r-vr ~

r-vr alcid ㉟ ~

coot ㉒ r

m-r rail ㉒ ~

m(i) ㉚ phalarope

PICTURE KEY C. WATER BIRDS IN FLIGHT

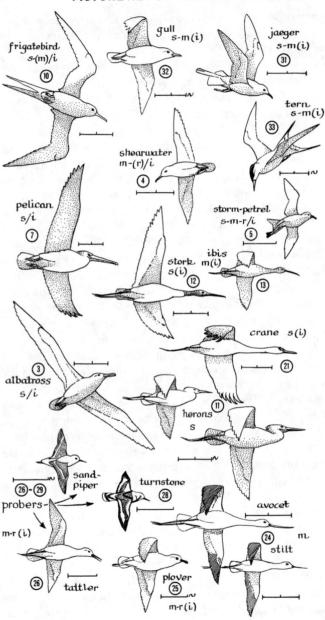

frigatebird
s-(m)/i
⑩

gull
s-m(i)
㉜

jaeger
s-m(i)
㉛

tern
s-m(i)
㉝

shearwater
m-(r)/i
④

storm-petrel
s-m-r/i
⑤

pelican
s/i
⑦

stork
s(i)
⑫

ibis
m(i)
⑬

crane s(i)
㉑

albatross
s/i
③

herons
s
⑪

㉖ - ㉙
sand-
piper

turnstone
㉘

avocet

probers-

m-r (i)

㉔
m
stilt

㉖   tattler

plover
㉕
m-r(i)

Numbers correspond to sequence of families in the text and Graphic Calendars) long curved bill and trailing legs are distinctive. Sometimes soar on thermals when traveling a distance. Flocks are bunched or in echelons and V's.

14. **Swans** (not pictured here) are similar to Geese (no. 15), but feathers are all white and neck is longer.

15. **Geese.** Neck forward, short tail at rear, and wings broader than in most ducks (nos. 17-20), and wingbeats, even of small geese, somewhat slower. Some species form into regular echelons or V's, others into curving lines for long-distance travel, often at high altitudes, the birds calling frequently. Birds approaching landing from high up may tumble (sideslip) violently. Takeoff by a short run or by a jump.

17. **Dabbling Ducks.** Straight-line bill-to-tail silhouette in most (but see Wood Duck account); wings longer and broader than Diving Ducks (no. 18), less so than Geese (no. 15). Contrasting color patches in wings of most, especially secondaries and their coverts (= "speculum"; see wing section on diagram Parts of a Bird, later in this chapter), which differ distinctively among the species. Takeoff usually without a run on surface, even nearly vertical a short way. Flocks are usually irregularly bunched, but some species fly in echelons or V's at times.

18. **Diving Ducks.** Straight-line silhouette or with bill held somewhat down; wings narrower and shorter relative to body size than in Dabblers (no. 18). Many have white on speculum and some on other parts, but most are not brightly colored (see, however, Redhead, Canvasback, and Harlequin Duck accounts). Takeoff requires a run on surface (short to long, according to species and wind and water condition). Flocks are variously bunched, those over the ocean commonly elongate.

19. **Stiff-tailed (Ruddy) Ducks.** Wings smaller in relation to body than other ducks (nos. 17, 18, 20). Grebe-like takeoff and flight (see note 2 above), and similarly loath to do so.

20. **Mergansers.** Prominent white in secondaries and coverts (larger in males); beak in line with neck. Takeoff requires run along surface (or drop from a height).

21. **Cranes.** Wings long and broad, but less arched than in Herons (no. 11). Long neck straight forward (or slightly down) and legs trailing beyond tail give distinctive silhouette, similar

only to Storks (no. 12). Wingbeats (on local flights at least) have distinctive quick upward flip of wing-tips and slower downstroke. Speed of flight medium, course and height variable; they fly high and in long echelons or irregular lines on migration, and often call in flight.

22. **Rails.** Wings quite convex, rounded, of solid color (except one rare species with patch of white). Flight observed is usually short and low, dropping into marsh suddenly. Flight appears weak, legs commonly drooping, but some migrate considerable distances at night.

22. **Coots.** Large feet trailing (drooping on short flights), distinctive when coupled with head and body shape; wings quite convex. Takeoff by short to long run. (Gallinules are intermediate between Rails and Coots.)

23. **Oystercatchers** (not pictured here; see Picture Key B for shape and size) are similar in flight to larger species of the tringine subfamily (no. 26) of Probing Shorebirds, but wingbeats are of medium frequency.

24. **Avocets** and **Stilts.** Similar to Probing Shorebirds (nos. 26-29), but very long legs and bold black and white patterns are even more striking.

25. **Plovers.** Similar to Probing Shorebirds (nos. 26-29) of comparable size, but larger head and short bill evident except at a distance, and maximum flock sizes are smaller.

26-29. **Probing Shorebirds** (see note to Picture Key B). Wings pointed, flaring at base toward edge of tail, the wrist bend usually apparent, with or without white stripe; extended neck in line with body, the beak held slightly down (less so when enroute for a distance). Slightly irregular wingbeats result in a slightly to definitely erratic course, especially after takeoff (see also Spotted Sandpiper account). Flocks are usually bunched, sometimes closely so; may wheel and turn in unison.

30. **Phalaropes.** Very like sandpipers (nos. 26, 29); two species with white line in wing, one without. Course is usually slightly erratic. Flocks are irregularly bunched; speed variable.

31. **Jaegers.** Gull-like (see no. 32), but wings more pointed and wrist bend usually more evident (but see Skua account).

Feathering patterns of light or dark phases (see species accounts); bill hooked at tip. Wingbeats shallow and slow to medium (i.e., gull-like) on ordinary flight, but in pursuit are deep and driving (falcon-like).

32. **Gulls.** Bill held nearly horizontal when traveling, variously angled in search of food or danger. Speed slow to medium. Wings usually quite arched on glides, which are frequent during local flights, and wingbeats are then shallow (unless against strong wind). On uprising air from ships, bridges, sea-cliffs, etc., may soar long distances; in still air or going overland, usually flap steadily. Flocks are open and irregular, except may form echelons or V's going long distances. Exact hue of gray or brown of back and wings (mantle), and wing-tip and tail patterns, are important for species identification.

33. **Terns.** Wings more pointed and (in most) narrower than in Gulls of comparable size; tail usually forked, deeply so in several. Flight speed medium (occasionally slow or fast); wingbeats vary according to wind condition. Most Terns forage at from 15 to 50 feet up, the bill held at a decided downward angle, with a stop and a steep downward plunge into the water; or at other times (some species regularly) by shallow swoops close along the water surface, or by erratic flying to catch insects in air. Course is bouncy in all but the largest few.

35. **Alcids.** Wings fairly long and narrow, roundish to pointed at ends; feet often visible at side of tail, especially on turns, when they are spread. Flight is commonly low over water, except at colonies. Takeoff by run on water surface or by dropping from high point. Flocks are usually irregular and loose; small groups are more compact.

TOPOGRAPHY OF THE BIRD BODY
AND CHANGES IN FEATHERING

The parts of the typical water-bird body (its topography) are shown on the accompanying diagram. Technical names used for such parts in the Recognition paragraphs of the species accounts will be found on that diagram, or are explained

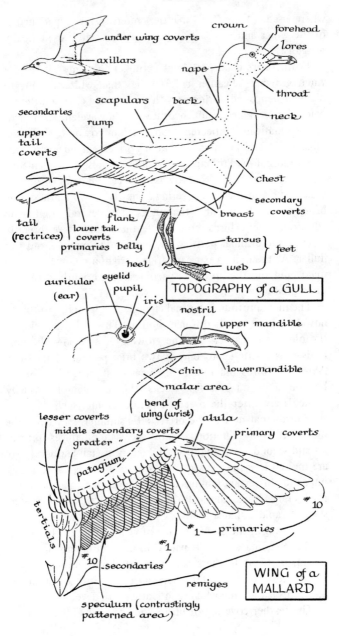

under wing coverts

axillars

crown
forehead
lores
nape
throat
neck

secondaries
scapulars
back
rump
upper tail coverts

secondary coverts
chest
breast

tail (rectrices)
lower tail coverts
primaries
flank
belly
heel

tarsus
web
feet

TOPOGRAPHY of a GULL

auricular (ear)
eyelid
pupil
iris
nostril
upper mandible

chin
lower mandible
malar area

bend of wing (wrist)
alula
lesser coverts
middle secondary coverts
greater "
primary coverts
patagium

tertials
#10
secondaries
#1
#1
primaries
#10

remiges

WING of a MALLARD

speculum (contrastingly patterned area)

where used. (*Note*: The "mantle" refers to the upper surface of the wings and the back, taken together, and is used mostly for gulls.)

Feathers are dead protein structures (alive only at the base when growing). Feather colors other than white are due to pigment deposited during growth; or in some cases a different colored appearance results because certain wavelengths of light are prismatically reflected from the microscopic structure. Most blues and bright reds, and any iridescent hues, are such structural colors. Prolonged exposure may also fade pigments, especially the yellowish browns, oranges, and nonstructural reds. Thus a bird with a covering of newly grown feathers may be considerably brighter than it is after several months. In addition, many birds produce feathers that have brightly colored subterminal parts, with the exposed margins and tips duller. As the dull parts wear off, the brighter colors below are uncovered and the bird thus becomes more brightly patterned or colored without molt.

The most profound seasonal changes in appearance of birds, however, come about through molting. This is the shedding of the old, worn feathers and the growing of new ones. In most birds it is restricted to one or two brief periods of the year. Almost all birds older than one year molt all feathers of the body in a period of a few weeks (the annual molt), usually immediately after the breeding period. Some, such as many shorebirds that nest in the brief arctic summer, may delay the molt until after migration. In most species the remiges (primaries and secondaries), and often the rectrices (tail feathers) also, are molted in a precise sequence so that flight is not seriously impaired. In certain families of water birds (e.g., grebes, waterfowl, cranes, rails, and at least some alcids), these feathers are shed so rapidly that a flightless period of a few weeks ensues. Any molting at the opposite time of year, before the beginning of the breeding season, is partial in most birds, not involving the major flight feathers, and may be absent entirely. Yet if the species shows brighter colors in the breeding season due to molt, this is the time when such feathers are developed.

The feather covering of birds, and the molts by which it is

altered, have been referred to for years by certain terms that were in several respects unsatisfactory. A new and carefully defined system has been proposed and is being used in the *Handbook of North American Birds* (Palmer 1962, 1976). The terms used here in the species accounts are consistent with this system, although some of its complexities have been avoided. These terms are defined, along with others, following the table of the sequence of molts, plumages, and feathering:

Molt	Plumage (Yields)	Feathering	Usual Season(s) Worn
- - -	natal down	natal down	at or soon after hatching
postnatal (not a a true molt)	juvenal	juvenal	few weeks or months in and just after breeding period in which hatched
1st pre-basic (partial)	1st basic	1st nonbreeding, or 1st "winter" (includes retained juvenal and new 1st basic feathers)	late summer or fall through 1st winter
1st pre-alternate (partial)	1st alternate	1st breeding (nuptial), or "summer" (includes some juvenal and 1st basic)	spring and summer before to just after age 1 yr. (but see duck accounts)
2nd pre-basic	2nd basic	2nd nonbreeding (all feathers new), or 2nd "winter"	late summer or fall through 2nd winter (but see duck accounts)
2nd pre-alternate (partial)	2nd alternate	2nd breeding (nuptial) or "summer" (includes remiges and usually other basic feathers)	spring and summer to just after age 2 yrs.

Subsequent molts, plumages, and feathering are as in the second year. However, the age at which the bird actually reaches sexual maturity often influences whether the plumage is the definitive or "adult" type. This age varies from the spring following the year of hatching in many species (particularly smaller ones) to as late as the bird's third spring (approaching age 4), as in the larger gulls, or even its sixth or seventh, as in albatrosses.

ABBREVIATIONS AND TERMS USED
IN SPECIES ACCOUNTS AND ILLUSTRATIONS

1. *For measurements,* given at beginning of Recognition section:

ca.	=	approximately
cm.	=	centimeters
ft.	=	feet
g.	=	grams
in.	=	inches
kg.	=	kilograms
L.	=	length, the "straightened out" distance from tip of bill to tip of tail
lbs.	=	pounds
m.	=	meters
WS.	=	wing spread, with wings fully extended
Wt.	=	weight

2. *For relative numbers,* in Range and Occurrence in California sections:

Abundant = 1000 or more (with Very Abundant in some cases, to indicate over 10,000)
Very Common = 250-999
Common = 50-249
Fairly Common = 10-49
Uncommon = less than 10, but frequently more than 3
Rare = 3 or less if Regular, but may be up to 9 if very Irregular

The above designations are equivalent to the number of lines (5 to 1, both Uncommon and Rare as 1) on the Graphic Calendars in the Appendix. Each category is based on the number of individuals that have usually been reported by

SPECIES ACCOUNTS

1. FAMILY GAVIIDAE (LOONS)

(Picture Keys A and C)

Common Loon (*Gavia immer*) (fig. 1)

Recognition. L. 28-35 in. (71-89 cm.); WS. 52-58 in. (132-147 cm.) Larger and with heavier beak than other loons (except Yellow-billed). In nonbreeding feathering all loons are dull gray above and white below (but see Red-throated account). In breeding feathering, beginning between Feb. and May, the Common Loon has black sides and back with numerous white spots, and a black head and neck with a partial collar of white.

Habits. Despite its name, this species is sometimes the least "common" of the three regular loons along the California coast, but it is the only one normally found on inland lakes and is more frequent far within bays than the others. Loons are powerful divers and may remain submerged as long as two minutes. A good place to watch them feeding is at the narrow mouth of an estuary where the tidal current brings a rich supply of fish. *Food*: in salt water, chiefly sand dabs, herring, surf-perch, sculpins, rockfish, and the like — mostly of little economic value.

Range. Winters primarily along coast from the Aleutians to Mexico and on coasts of e. U.S. and w. Europe. Breeds nly in Canada and extreme n. U.S., a few south (only merly?) to ne. Calif.

Occurrence in California. On ocean near shore and in bays ommon to Fairly Common Sept.-May (few stragglers gh summer are mostly immatures, and often retain noning feathering); Common (Irreg. Very Common) in Nov.

competent observers in from one-half to one day, in the preferred habitat of that species, during a general search for various birds.

3. *For regularity of occurrence,* in the same sections:

Regular(ly) = found every year, or at least most years, in the location or season or habitat referred to; implied when no other term is used

Irreg. = found irregularly, with distinct fluctuations in numbers, locations, or dates of occurrence from year to year, the fluctuations evident and repeated within a few years

Occ. = found occasionally, with longer-term fluctuations, timing of which is unpredictable, in numbers, locations, dates, or habitats

Local(ly) = occurring in some parts of the suitable habitat but absent from other seemingly similar habitat within the range as given

Casual(ly) = has occurred a number of times in area or season specified, usually as single individuals, but data are insufficient to disclose any pattern

Vagrant = a bird occurring out of the normal (regular or irregular) range of its species

Accidental (used only in Appendix) = has occurred, but unlikely to occur again in same area, because far from normal range

4. *For locations,* in Range and Occurrence in California sections:

c. = central
e. = eastern
n. = northern
s. = southern
w. = western

These abbreviations are also used in normal combinations such as ne., cw., ce., sc., etc. See the map (in the Appendix) for approximate portions of California implied by such designations.

cismontane = west of the Cascades, Sierra Nevada, and major s. Calif. mountain ranges
Co. = County
Eurasia = entire landmass of Europe and Asia, or this less the tropical southeast portion of Asia
I., Is. = Island, Islands

Mt., Mtn., Mts. = Mount, Mountain, Mountains
Pt. = Point
R. = River
Res. = Reservoir

For states of the United States (U.S.): standard commonly used abbreviations

For provinces of Canada: certain abbreviations in common use, as B.C. for British Columbia, Sask. for Saskatchewan, Ont. for Ontario, Que. for Quebec; others are spelled out

Baja Calif. for Baja California, Mexico

5. *For different feathering, sex, and ages,* terms in Recognition section and abbreviations used on illustrations:

♀ = female   ♂ = male
ad = adult: having the feather pattern or color characteristic of sexually mature birds
br = breeding feathering (including the alternate plumage): the feathering normally worn by breeding adults; essentially the same as nuptial, spring, or summer "plumage" referred to in many books (but see species account of Mallard for dabbling duck specialties)
dny = downy young: a recently hatched bird with a sparse to dense coat of soft fluffy feathers
eclipse plumage: a long-used term for the dull-colored feathers (basic plumage) worn by male ducks following their annual (prebasic) molt – i.e., midsummer to or through fall
feathering: the entire covering of feathers worn at any one or all seasons; may be composed of one to several plumages
im = immature: a general term for non-adult birds of any age after the downy young and differing from the adult; most frequently used for "first winter," and sometimes so specified
jv = juvenal: the first plumage (and feathering) subsequent to the downy young; worn at time of first flight but normally partly molted again within a few weeks
juvenile: the bird itself (wearing juvenal plumage); at first, often of less than adult size, but most birds are nearly or quite the size of the adult by the time they can fly
nbr = nonbreeding feathering (including basic plumage; or commonly, after the first year, equivalent to it): in most

species, normally worn in fall and winter (the "winter plumage" of many bird books), but also seen on many nonbreeding stragglers at other seasons (see also "eclipse plumage")
plumage = all the feathers that grow in one molting period, whether the entire feathering or only part of it is involved
subadult: having a feathering and other features nearly like the adult, but not yet sexually mature

Size Guide

Average length of some common birds

Centimeters	Inches	
14½	5¾	Least Sandpiper, Junco
16	6¼	House Sparrow
16½	6½	Western Sandpiper
18½	7¼	Northern Phalarope
21½	8½	Dunlin
23	9	Brewer's Blackbird
25½	10	Killdeer, Robin
30½	12	Scrub Jay, Mourning Dove
38	15	Forster's Tern (but 6½ inches = tail)
33-41	13-16	small ducks (teal, Bufflehead, Ruddy Duck)
46	18	Common Crow
50	19½	Ring-billed Gull
43-51	17-20	medium ducks (many species)
53-63	21-25	large ducks (e.g., Mallard)
61	24	small geese (e.g., *minima* or Cackl subspecies of Canada Goose)
63-66	25-26	Western Gull (a typical large gul'
71	28	medium geese (e.g., White-fron
81-86	32-34	large geese, (e.g., large or "hon Canada); large cormorant
99	39	Great Egret
124	49	Great Blue Heron
135	53	Whistling Swan
152	60	White Pelican

FIG. 1

Arctic Loon br

Common Loon br

nbr

Red-throated Loon nbr br

and mid-May migrations, and then farther offshore as well. Rare in same seasons on large fish-stocked lakes in valleys and foothills throughout the state (Uncommon some years in se. desert area and Salton Sea, but not through summer there). At Tahoe and other sizable mountain lakes Uncommon or Rare and usually Irreg. mostly Apr.-May and Oct.-Dec. A few formerly bred at lakes in Mt. Lassen area in June-July.

Note: The **Yellow-billed Loon** (*Gavia adamsii*) breeds in the Arctic and winters normally south to B.C. in fair numbers. It is slightly larger than the Common Loon and similar to it except for the apparent upturned shape of the *yellowish* beak in adults and the paler side of the head and neck (except for a dark auricular patch) in nonbreeding feathering. The top ridge of the bill (culmen) is never dark beyond the middle, whereas dark extends near or to the tip in the Common Loon. Further details on identification of these two species are given by Binford and Remsen (1974: *Western Birds,* 5:111-126). Since 1965 one to five Yellow-billed Loons have been seen along the Calif. coast nearly every year mid-Oct.-early May (once June) — once to within Mexican waters but mostly from Monterey northward, where some have stayed for months in one area (see Appendix).

1. LOONS

Arctic Loon (*Gavia arctica*) (fig. 1)

Recognition. L. 23-28 in. (58-71 cm.); WS. 44-48 in. (112-122 cm.). Large duck or small goose size; beak straight and relatively more slender than in Common Loon. Non-breeding feathering similar to Common Loon but adults usually darker above. Breeding feathering with black and white *bars* in four zones on back, the hind head and neck light gray, and a blackish patch on the throat.

Habits. This loon is similar in its feeding habits to the others. Although all loons are usually silent in winter, soft or guttural calls may be given by groups when actively feeding. In migration along and off the coast north of Pt. Conception, the Arctic is by far the most numerous of the loons. As seen from vantage points such as Pt. Reyes, Pt. Pinos, Pt. Arguello, or the Channel Is., they pass steadily by in little groups or long, loose flocks, mostly within 50 feet of the water. Although only those closest to the observer can be identified to species, they can be recognized as loons by the characteristic silhouette up to a mile or more away with good binoculars. At the peak of the fall flight in mid to late Nov. up to 1000 or more loons per hour can be counted thus from one point — one of the most visible of our bird migrations. At the south, small numbers pass to and from the Gulf of California over the deserts, a few sometimes remaining for weeks along the Colorado R. or on the Salton Sea. Occasionally on such overland flights loons or grebes alight on shiny pavements (mistaking them for water) or come down because of exhaustion. They are then helpless, for they can barely walk and cannot take flight except from a large water surface.

Range. Winters along the seacoast from s. Alaska to s. Baja Calif., and in Europe and Asia. Breeds in n. Canada, Alaska, and n. Eurasia.

Occurrence in California. On ocean near shore Fairly Common to Common, late Oct. and late Dec.-Mar.; Very Common or Abundant in Nov.-early Dec. and late Apr.-May migrations, and many also farther offshore then. Stragglers frequent to July, and early arrivals in Sept.; Rare in Aug. Casual (usually single birds) on various inland lakes including Salton Sea, Oct.-mid-May.

Red-throated Loon (*Gavia stellata*) (fig. 1)

Recognition. L. 24-27 in. (61-68 cm.); WS. 42-45 in. (107-114 cm.). About the size of the Arctic Loon and with similarly slender beak which is, however, distinctive in appearing *upturned.* Nonbreeding feathering also differs from the other loons in having small flecks of white within the gray of the back feathers, visible only at somewhat closer range than the beak shape (immatures of the other species may show pale gray feather *edges* in fall and winter). The hind neck and head are also usually paler gray in winter Red-throats. In breeding feathering (March or later) there is a reddish-brown patch on the foreneck, and the medium gray of the rest of the neck and head has fine black and white streaks.

Habits. Similar to other loons when in California waters, but this species commonly feeds close to surf and thus is seen quite often by human "beachcombers." It has apparently declined in numbers since 1968, but before that was usually commoner than the Arctic Loon well within bays and estuaries, though rarest of all inland.

Range. Winters in the Aleutians and from the coast of B.C., to nw. Mexico; also on coasts of e. U.S., Europe, and e. Asia. Breeds in circumpolar arctic and subarctic areas, south near both coasts to Vancouver I. and Nova Scotia.

Occurrence in California. On ocean near shore Uncommon to Fairly Common (formerly Common) Oct. and Dec.-May, and in greater numbers migrating past outer headlands in Nov.; a few stragglers also recorded through summer. Casual on coastal lakes and inland, esp. Salton Sea, mostly Oct.-Apr.

2. FAMILY PODICIPEDIDAE (GREBES)

(Picture Keys A and C)

Red-necked Grebe (*Podiceps grisegena*) (fig. 2)

Recognition and Habits. L. 17-22 in. (43-56 cm.); WS. 30-32 in. (76-79 cm.). A large grebe with long neck, but not as long nor as slender as that of the still larger Western Grebe; the Red-necked Grebe's beak is shorter, heavier, and with yellow only near the base. In breeding feathering the neck is bright reddish brown, in nonbreeding mostly dull gray to gray-brown. The cheeks are white or partly so, in winter often appearing as a white crescent. Look for this uncommon grebe on bays and coves of the outer coast, but expect only one or a few even in the most favorable spots such as Tomales Bay, Berkeley harbor, or Moss Landing.

Range and California Occurrence. Breeds in subarctic regions south to n. U.S., including in some years Upper Klamath L., Ore. (with eggs in late May, young in June). Winters in Eurasia and along North American coasts from sw. Alaska and se. Canada to Fla. and s. Calif. In c. and n. Calif., Rare to Uncommon but Regular Oct.-Apr. (recorded also Aug.-Sept., and May-mid-June). In s. Calif., Rare and Irreg. Oct.-Mar. Also found near offshore islands and about seventy times statewide in cismontane lowlands and Sierra foothills, mostly Oct.-Apr.

Horned Grebe (*Podiceps auritus*) (plate 1)

Recognition. L. 12½-15 in. (32-38 cm.); WS. ca. 24 in. (61 cm.). A small grebe very similar in nonbreeding feathering to the Eared Grebe, but dark cap ends abruptly at larger white cheek area, often almost meeting white of other side at nape. The foreneck usually is pure white or nearly so. Breeding feathering (acquired gradually mid-March to May) is distinctive. Beak straight, and whole head and beak somewhat heavier than in Eared Grebe.

Habits. This species has become increasingly common in California coastal areas since the 1930s, now often outnumbering the Eared Grebe on large open bays where both may assemble in loose flocks at favorable feeding spots. Like all grebes and loons, they obtain their almost strictly animal food by pursuit beneath the water surface, using only their powerful feet for propulsion. Sometimes their erratic darting after small fish can be seen from above on a pier. Grebes also swallow so many feathers from their own dense plumage that their stomachs are often half full of such. It is thought that this feather mass, in which fish bones and scales become embedded, serves to protect the stomach lining until digestion

FIG. 2

jv. Coot

Red-necked Grebe

nbr

nbr

Western Grebe

is complete. *Food*: many kinds of small fish such as anchovies, darters, carp, sculpins, perch, etc.; also a variety of crustaceans and, in fresh water, insects.

Range. Winters along coast from the Aleutians to s. Calif. and Irreg. inland in c. and s. Calif.; also on coasts of e. U.S., Europe, and e. Asia. Breeds from nw. and nc. U.S. to Alaska and across nc. Eurasia.

Occurrence in California. Fairly Common to Locally or Occ. Common on bays and lagoons, and somewhat fewer on ocean near shore (including near offshore islands), Oct.-Apr.; fewer yet and Irreg. Sept. and May; Rare through summer. Irreg. Uncommon to Rare on lakes at lower altitudes Oct.-May, Casual Sept.-Apr. on mountain lakes and in ne. Calif.

Eared Grebe (*Podiceps nigricollis*) (plate 1)

Recognition. L. 12-14 in. (30-36 cm). A small grebe similar in nonbreeding feathering to the Horned Grebe, but side of head (and usually foreneck also) partly dingy gray, or cap not sharply defined. The beak appears thinner and slightly upturned and the rear of the body is often held higher out of the water than in the Horned Grebe. In breeding feathering the Eared is a blackish bird with distinctive crest and straw-colored "ear" plumes, from which it is named.

2. GREBES

Habits. Formerly much commoner in California than the Horned Grebe, and still so inland and on salt ponds, this species now remains to breed in the state only occasionally west of the major mountain divides. They are colonial nesters, constructing a flimsy but anchored platform of aquatic plants or debris floating on water up to 6 feet deep. Often the tip of the *nest* is barely out of the water, and the 3-5 dull white or brownish *eggs* may not be entirely above it. Both parents incubate, for a total of 21 days. When leaving the nest, an adult grebe usually covers the eggs with some of the nest material. In spring and early summer, pairs engage in mutual displays, standing upright facing each other and turning the head from side to side, showing off the plumes. Occasionally there are short "races," but much less spectacular than those of the Western Grebe. *Voice:* a mellow *koo-r-r-eep,* given from late winter through summer; also a shorter *quer-ip* and other varied notes, some rather harsh.

Range. In America, winters from B.C., along coast, and inland from Sacramento Valley and n. Utah, south to n. South America; breeds over most of Mexico, w. U.S., and sw. Canada, but not near the coast north of s. Calif. Also found in Eurasia and Africa.

Occurrence in California. Abundant on salt ponds, Common to Fairly Common on bays, fewer on ocean near shore (including islands), and Uncommon to Fairly Common on lakes of valleys and foothills from Sept. (first arrivals mid-Aug.) to mid-May, with stragglers Irreg. in summer. Very Abundant Aug.-Oct. at Mono L. and from Nov. (Sept.-Oct. some years) through May at Salton Sea. Locally Uncommon to Common on lakes in s. Calif. mountains and Sierra Nevada and the ne. plateau, late Mar. or Apr. to Nov.-Dec., some Irreg. through winter on open water. Breeds at marshy lakes primarily June-Aug.; most colonies at low altitudes now depleted, but still numerous in Modoc Co.

Note: The **Least Grebe** (*Podiceps dominicus*) is a dusky slate to brown species with a body about the size of a jay (L. 9-10½ in., or 23-27 cm.). It can be further distinguished from the Pied-billed Grebe (fig. 3), the only one with which it might be confused, by its dark bill, yellow to nearly red eyes, and large white patch in the extended wing. The Least Grebe has nested at least twice and occurs Irreg. in very small numbers along the lower Colorado R. (see Appendix). The species ranges from s. Ariz. and s. Texas through Mexico to most of Central and South America.

Western Grebe (*Aechmophorus occidentalis*) (fig. 2)

Recognition. L. 22-29 in. (56-73 cm.). The largest grebe, further distinguished by long, very slender neck (sometimes hidden, when sleeping). Coloration all year is white, gray, and black. Beak slender, yellow.

Habits. These grebes often gather in winter in large "rafts" near the kelp beds off ocean beaches and on large bays. In these groups many birds may be asleep, long neck laid flat on the back and beak on the chest. When actively feeding, they may spend up to a minute or more below the surface in pursuit of fish, which they apparently spear with the beak closed, at least at times. The Western, like all grebes, migrates chiefly at night and is therefore rarely seen in flight. *Food*: primarily fish; fewer crustaceans and other invertebrates than other grebes. *Courtship*: before and at nest-building time, pairs engage in mutual diving, neck-swaying, "dancing" with water weeds held in bill, and a most spectacular "race" in which the birds rush rapidly in standing-up posture for up to 30 yards, their necks arched up and heads down. More than two occasionally race as a group, especially on lakes in spring migration. *Nest*: an anchored bulky mound of cattails, tules, etc., usually floating amid marsh plants by open water. *Eggs*: 3-4, pale buff or bluegreen, often stained or faded; incubation about 23 days, shared by two parents. *Voice*: a loud shrill *krrik,* or *krik-ker-reeeek* in late winter and on breeding areas, where a long continued *reek-ker-reek-ker-reek* . . . in a see-saw rhythm is also common.

Range. Winters on coastal waters from se. Alaska to wc. Mexico, and inland from c. Calif. and Nev. southward. Breeds on lakes from s. B.C., n. Alberta, and Minn. south to Colo. and ne. and ce. Calif., Irreg. to cw. and s. Calif.

Occurrence in California. On ocean near shore and bays, Common to Abundant Oct.-May, Uncommon to Fairly Common through summer. Fewer at same seasons on large lakes both near coast and inland (esp. Salton Sea), but Fairly Common to Common at nesting colonies Feb.-Sept. (Occ. to Nov.) on marshy lakes in Central Valley and sw. Calif. Recent successful colonies (some years) on Sacramento Valley refuges, near Corcoran, and at Sweetwater Reservoir, San Diego Co. Former colony at Clear L. wiped out by pesticides in 1950s, and large ones at Tulare L. and Buena Vista L. by drainage of these shallow lakes for agriculture. Common on the Modoc plateau and Locally south to Topaz L. Mar. or Apr. to Sept. on larger lakes, nesting May-Aug. on suitably marshy ones, with stragglers noted through Dec.

Pied-billed Grebe (*Podilymbus podiceps*) (fig. 3)

Recognition. L. 12-15 in. (31-38 cm.); WS. 22-24 in. (56-61 cm.). A small grebe with blunt, heavy bill with a curved upper profile. This, with a neck almost as thick as the head, gives it a "snake-like" look. General color brownish, the white belly usually not visible. Only the spring and summer adults have a black throat and a black band across the whitish bill (therefore "pied"). Juveniles have irregular light streaks on the sides of the head.

Habits. Least gregarious of our grebes, the Pied-billed is also the most widely distributed in California. Even in winter it prefers ponds or coves near shores of lakes, but it may take refuge on deeper waters when disturbed. Although the pair defends an area of marshy pond or lake up to 150 feet across, some find suitable territories in the larger irrigation ditches. When surprised by a person at a fair distance, the Pied-billed often quietly submerges without diving and comes up amid plants at another place; or it may expose only the beak and then go back under. *Food:* a great variety of small fish, crustaceans, insects, and water snails. *Nest:* a mass of dead plants in water or beside tall marsh; anchored or resting on the bottom. *Eggs:* usually 4-7, pale greenish, often stained brown or faded, both parents sharing incubation of about 23 days.

FIG. 3

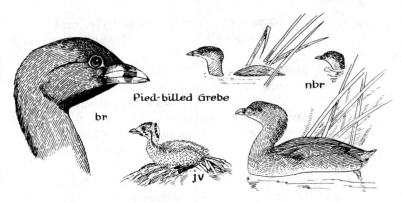

Pied-billed Grebe

br

nbr

jv

Downy young: striped black and white, ride on parent's back (as do the young of all grebes). *Voice:* males in breeding areas give a long, mellow "whinnying" that often becomes a series of louder *kowp* notes; very variable, sometimes drawn out to a wail.

Range. From across c. and s. Canada to s. Argentina. Moves south in winter from areas of U.S. and Canada where waters freeze; resident elsewhere.

Occurrence in California. Fairly Common all year on lower altitude ponds and large ditches throughout the state, nesting Mar.-Oct. (even Dec.-Feb. known in s. Calif.) on marshy borders of such (Locally Common in best habitat then). Also Fairly Common to Uncommon (Rare in winter) on mountain lakes, and nest more Locally. Fairly Common on tidal channels, lagoons, and to a lesser extent on open bays (Rare on ocean) Aug.-Apr. or early May.

3. FAMILY DIOMEDEIDAE (ALBATROSSES)

(Picture Keys A and C)

Black-footed Albatross (*Diomedea nigripes*) (fig. 4)

Recognition. L. 27-32 in. (68-81 cm.); WS. 6.3-7.1 ft. (2.1-2.3 m.). Distinguishable from other albatrosses by its medium brown plumage and dark bill. A white area surrounds the base of the bill; the undertail coverts and often some of the belly are paler brown to white in some birds (presumably the oldest), which may also show white above the base of the tail.

Habits. This "black gooney" of north Pacific sailors is the chief ship-following albatross in that area, and the only one now ranging commonly in waters close to the California coast. Expert at fast gliding and steep banks on broad turns, they fly most easily when winds are strong and steady. When taking flight they "taxi" by running with wings outspread for some distance. In *feeding,* albatrosses sit on the water or dive with difficulty, using semi-opened wings. It is thought that they may capture most of their prey, such as squid and fish, at night in this manner. Near ships in the daytime they scavenge animal matter of any sort and seem especially avid for fats, being readily enticed close to small boats with suet or a mixture of bacon fat and a cereal "carrier" that will float. Despite the casual sea observer's impressions, there is strong evidence that individual albatrosses of this species do not ordinarily follow one ship for more than a couple of hours or so. Like all the other "tubenoses" (shearwaters and storm-petrels), albatrosses never come to shore except to mate and nest; they *must* drink salt water, because of the active excretion of body salt via glands above the eyes that drain through the nostrils. *Voice:* a somewhat raspy squeal when in feeding competition; otherwise usually silent off our shores.

Range. Almost the whole n. Pacific Ocean. Do not nest until seven or more years old, so nonbreeders are widespread at all seasons. Breeds Oct.-July in colonies totaling about 50,000 pairs on w. islands of Hawaiian chain; smaller colonies on other c. Pacific islands.

Occurrence in California. Uncommon on open ocean, chiefly well offshore where Locally Fairly Common (Occ. Common) late Apr.-July, but more Irreg. to quite Rare Nov.-Feb. Irreg. at any season to within a few miles of mainland, but seldom visible from c. Calif. coastal headlands.

Note: The **Laysan Albatross** (*Diomedea immutabilis*) also nests in Hawaii and ranges the n. Pacific, but is rarely seen within 100 miles of Calif. shores (about thirty-five records, of one to four birds each; see Appendix). This species is white-

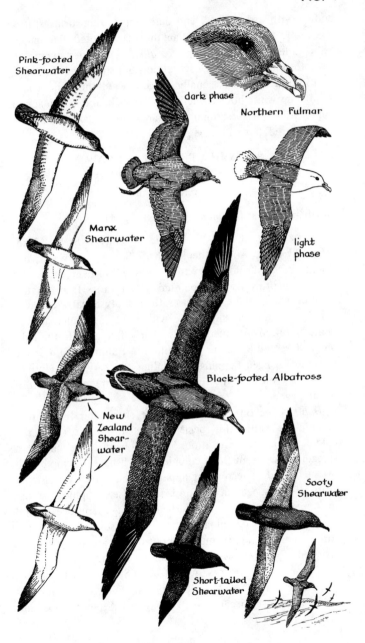

FIG. 4

Pink-footed
Shearwater

dark phase

Northern Fulmar

Manx
Shearwater

light
phase

New
Zealand
Shear-
water

Black-footed Albatross

Sooty
Shearwater

Short-tailed
Shearwater

69

bodied with a dark back and dark upper wing surfaces; it is less prone to follow ships than the Black-footed. The larger **Short-tailed Albatross** (*Diomedea albatrus*), formerly common close to our shores, is nearly extinct, but still breeds in the w. Pacific and may occasionally range toward the Calif. coast. Adults are white-bodied, including the back, and have blackish upper wing surfaces; immatures are brown, darker than the Black-footed, but with pale pink bill and feet.

4. FAMILY PROCELLARIIDAE
(SHEARWATERS, PETRELS)
(Picture Keys A and C)

Northern Fulmar (*Fulmarus glacialis*) (fig. 4)

Recognition. L. 17-20 in. (43-51 cm.); WS. 40-42 in. (102-107 cm.). About the length of Sooty Shearwater, but thicker of head and body, the *yellowish* bill being especially heavy; wing-tips also less pointed. Color varies from a fairly dark brownish phase (but paler than Sooty Shearwater) to a light phase that is gray above with head and underparts all white; mottled intermediates are fairly frequent. In the light phase it resembles some gulls, but the Fulmar's flight is the usual stiff-winged flapping and gliding of shearwaters, close to the surface most of the time. Color phases are independent of sex and age.

Habits. The only member of the shearwater family that nests in the Arctic, this species forages far out to sea from the colonies. Fulmar "migrations" are thought to be chiefly a spreading out from these colonies, which have varying breeding success, thus accounting in part for the year-to-year irregularity of this species in California waters (see Graphic Calendar in Appendix). *Food*: small fish and pelagic crustaceans and mollusks, or dead animal matter of any sort on the ocean; formerly noted in great masses about whale carcasses that were stripped of marketable portions and left at sea.

Range. Breeds from s. Alaska to e. Siberia and Kurile Is.; also in ne. Canada and nw. Europe. In winter or as stragglers Irreg. south over oceans to s. Calif., ne. U.S., w. Europe, and Japan.

Occurrence in California. Usually found Oct.-Mar. or Apr. on ocean and mostly offshore, but Irreg. to near outer coast at many places, esp. near headlands. Numbers vary markedly from Uncommon to Abundant in different years, with smaller numbers persisting into or even through the summer following peak winters; usually far fewer off s. Calif. Irreg. and Rare Dec.-Feb. in outer parts of large bays.

Pink-footed Shearwater *(Puffinus creatopus)* (fig. 4)

Recognition. L. 19-20 in. (48-51 cm.); WS. ca. 45 in. (114 cm.). A large shearwater with wingbeats noticeably slower than in the Sooty, and with whitish underparts and irregular dark blotches on sides and on the largely white underwing coverts. The dark-tipped pale bill and all-pink feet are also noticeable at close range.

Habits. This species commonly associates with the large flocks of Sooty Shearwaters passing along or feeding and resting off the California coast, but it usually makes up less than one percent of the total flocks. However, Pink-foots also occur in small groups of their own or singly, and may be locally more numerous. Feeding and migration habits are similar to those of the Sooty Shearwater (see account).

Range. Breeds on islands off Chile, migrating north to spend our summer (their winter) off w. North America at least as far as B.C.

Occurrence in California. On ocean, mostly well offshore, Fairly Common or Irreg. Common Apr.-Oct.; Occ. Very Common or Abundant late Apr.-May and Aug.-mid-Oct. Fewer and Irreg. in Mar. and Nov.-Dec., and only scattered records in Jan.-Feb. Relatively less numerous near shore off s. Calif. than off c. and n. coast at all seasons.

Note: Some ornithologists consider the **Flesh-footed Shearwater** *(Puffinus carneipes)* to belong to the same species as the Pink-footed. It is very similar in size, shape, and bill and foot color, but with the feathering all dark brown. The Flesh-footed breeds near Australia and New Zealand and ranges to the w. Pacific and Indian oceans and in small numbers to the c. and e. Pacific. It has been found in or near continental shelf waters off Calif. about fifty times, mostly off Monterey but also since 1958 off s. Calif. and in 1972 off Eureka (see Appendix).

New Zealand Shearwater (*Puffinus bulleri*) (fig. 4)

Recognition. L. 15-16½ in. (38-42 cm.); WS. ca. 40 in. (102 cm.). About the size and speed of wingbeat of the Sooty (or slightly slower), but underparts including the whole underwing are white. The cap and tail are dark, and the upper tail coverts and most of the back medium-light gray. The wings are browner, with a conspicuous dusky band in a broad ⌒⌐ pattern from primaries to bend of wing and along inner coverts to and across the lower back. (The much paler immature Kittiwake and some smaller gulls — see figs. 40 and 41 — and a few central Pacific petrels are the only other species with such a pattern.)

Habits. Formerly considered rare near California, this species has been found in ever greater numbers and quite regularly on fall boat trips off Monterey Bay and northward since 1954. It mixes with Sooty Shearwaters or flies in open flocks of its own, displaying a somewhat greater tendency to alter height above the water than the other shearwaters.

Range. Breeds in n. New Zealand area. Migrates through Pacific near South America and in w. to e. North Pacific at mid latitudes.

Occurrence in California. Rare to Uncommon mid-Aug.-mid-Nov. (Irreg., Fairly Common or Common late Sept.-Oct.) on oceans from Monterey area northward, mostly well off-shore but Occ. can be seen from headlands. Very Rare in inner continental shelf waters off s. Calif.

Sooty Shearwater (*Puffinus griseus*) (fig. 4)

Recognition. L. 17-19 in. (43-48 cm.); WS. 41-43 in. (104-109 cm.). The Sooty is the shearwater "standard" for the beginner. It is about the size of a Ring-billed Gull, with feathering all dark brown except for silvery gray in variable amount on the underwing coverts (often not visible unless the bird banks away from the observer in good light). The bill and feet are black.

Habits. The Sooty is the only species of shearwater occurring in massed flocks of many thousands off California, and is also the most regularly found even where these do not form. Sometimes there are truly spectacular flights, close enough to shore, even in and out of the Golden Gate, to attract attention from the general public. Locations vary in different years and often from day to day. Such flocks defy any accurate censusing, but when carefully estimated on a time basis going past a fixed point, literally millions may be seen over a period of hours. The oft-suggested explanation of movement in a big circle is not borne out by observations on many recent boat trips, on some of which massed flocks sitting on the water have been carefully estimated at 500,000 or more. (Birds in one such concentration southeast of the Farallones on April 30, 1955, were feeding on Euphausid crustaceans which colored the water orange-red for miles; but just hours later the shearwaters had mostly dispersed.)

The factors influencing the size and location of these massed flocks are little known, even as to variety of food taken. Elsewhere, stomach samples have shown the *diet* to include squids, crab larvae, and small fish such as anchovies. To capture such prey the shearwaters dive, when necessary even from the air; but in the streaming flocks in rapid flight they often seem to be uninterested in food. For a species that essentially circumnavigates the entire Pacific every year, the urge to move on is probably high. Studies in the central Pacific indicate that their migration there is probably nonstop for days. *Voice*: a nasal *mraah* or *graah-ah* when in feeding competition.

Range. Breeds on islands near Australia, New Zealand, and s. South America, ranging thence north through both the Atlantic and Pacific oceans, chiefly around periphery of main basins, to s. Alaska, Greenland, and Norway. In early fall large numbers migrate directly southwest across the c. Pacific.

Occurrence in California. Common to Very Abundant Mar.-Oct., and Uncommon or sometimes Fairly Common through winter on ocean, mostly well offshore but Irreg. close inshore (less frequent in s. Calif.) and Rare into narrow-mouthed bays. They sometimes "crash" into land in dense fog.

4. SHEARWATERS, PETRELS

Short-tailed Shearwater (*Puffinus tenuirostris*) (fig. 4)

Recognition and Habits. L. 15-16 in. (38-41 cm.); WS. 38-39 in. (96-99 cm.). Slightly smaller than Sooty Shearwater (not noticeable except by direct comparison) and similar in color except for underwing coverts, which are light brown (must be seen in good light to be certain of the *absence* of silvery gray, some Sooties having very little). As indicated by the newly adopted name for this species (formerly called Slender-billed Shearwater in America), it has a shorter tail — but this too is not noticeable afield. Under the same wind conditions, its wingbeat is somewhat faster than that of a Sooty. In behavior, too, it is similar. It is thus a difficult species to pick out until after most of the Sooties are gone, in late Nov. or Dec.

Range. Breeds in se. Australian area, migrating north through w. Pacific and Bering Sea, then back south along e. Pacific to off Mexico; some at least thence southwest across c. Pacific with Sooty Shearwaters.

Occurrence in California. Apparently now Rare to Uncommon, though formerly Fairly Common, late Aug.-mid Feb., Occ. Common mid-Dec. and Casual Apr.-May, on ocean mostly well offshore.

Manx Shearwater (*Puffinus puffinus*) (fig. 4)

Recognition and Habits. L. 14-15 in. (36-38 cm.); WS. 33-35 in. (84-89 cm.). A small white-bellied shearwater with dark bill and underside of wing-tips, partly pink feet, and (in our subspecies) dark undertail coverts. It flaps more quickly and frequently than the Sooty. In southern California, where it is more common, the Manx occurs in denser flocks and separate from other shearwaters, especially Oct. to Dec. when they feed on young squid. At times the Manx also mixes with other species of shearwaters, especially when farther offshore. The breeding and homing of the Atlantic subspecies have been much studied in Great Britain.

Range. Breeds on islands off w. Mexico (subspecies *opisthomelas*) and migrates north to s. Calif., Irreg. farther north, Occ.

as far as B.C.; other subspecies in c. and sw. Pacific, Atlantic, and Mediterranean.

Occurrence in California. Uncommon to Fairly Common Aug.-mid-Dec., and Irreg. Common or even Locally Abundant late Oct.-Jan., on ocean off s. Calif., with peak numbers often only a mile or two from shore; Rare to Uncommon and Irreg. Feb.-July. Now fewer and less regular at similar seasons north to Monterey area, though formerly Very Common there in Jan., May, and July of some years.

5. FAMILY HYDROBATIDAE (STORM-PETRELS)

(Picture Keys A and C)

Fork-tailed Storm-petrel (*Oceanodroma furcata*) (fig. 5)

Recognition. L. 8 in. (20 cm.); WS. 18 in. (46 cm.). Unique among our storm petrels in its medium *gray* upperparts, appearing light gray in bright light. Wings somewhat browner; blackish area around eye; underparts whitish. Tail shallowly forked, as is that of most storm-petrels.

Habits. Occurring off California shores away from its northern nesting grounds, this species, like most storm-petrels, *feeds* by plucking small crustaceans or fish from the water surface as it flutters along over it, or by "skimming" oil that floats out from larger dead animals. Storm-petrels may also alight to feed where food is abundant, and may even dive a short way with great effort. Compared to other storm-petrels, the Fork-tailed seems to fly with somewhat shallower wingbeats and with longer periods of fliding interspersed, often low in the wave troughs. The incomplete and "bumpy" nature of the Graphic Calendar (in Appendix), which is taken from actual observations, attests the irregularity of distribution of this species when it is not nesting.

Range. Breeds on islands off nw. Calif. and north to Alaska, west to islands off e. Siberia. Nonbreeders appear at greater distances from colonies, primarily in fall.

Occurrence in California. Uncommon to locally Fairly Common on ocean, chiefly well offshore except near nesting colonies (Locally Common on islets near Crescent City and Trinidad) and Occ. near shore elsewhere. Sometimes Common south to Monterey area, even into San Francisco Bay a few times. In s. Calif., records from May to Dec., plus one in early Mar. Pairs at colonies Jan.-Feb., nesting late Mar.-June (a few on through fall).

Leach's Storm-petrel (*Oceanodroma leucorhoa*) (fig. 5)

Recognition and Habits. L. 8-9 in. (20-23 cm.); WS. ca. 19 in. (48 cm.). Similar to Ashy Storm-petrel but (in our area) somewhat darker sooty brown, almost blackish in appearance, and usually with white rump. As in most storm-petrels, there is a lighter brown band diagonally along the wing coverts. The flight of the Leach's is said to be more "fluttery" in nature than that of the Ashy, with many abrupt changes in direction. Its main foraging range is apparently several hundred miles offshore, even to waters northeast of Hawaii, although none breed there. Each adult spends several days at sea, then comes back to take its stint on the *nest* when the single white *egg* is being incubated for the 41-42 days it requires. The young is then fed in the nest (usually nightly) for about 8 weeks. On nesting colonies on small islands off Humboldt Co., hundreds have been banded from late Feb. through Sept. Fewer nest on Southeast Farallon I. off San Francisco, where they lay in crevices similar to those used by the Ashy. Where there is enough soil, this species may excavate a burrow up to 3 feet long. Apparently no colonies exist on the southern California Channel Is., but another subspecies nests on the Los Coronados Is. southwest of San Diego. In this and another subspecies that breeds in Mexico and wanders to southern California waters, the rump is often only partly white at the sides, or not at all.

Range. E., n., and w. parts of North Pacific Ocean, breeding on islands off continents south to tip of Baja Calif. and s. Japan; also in n. Atlantic Ocean.

Occurrence in California. Rare as forager within 20-30 miles of shore, but sometimes Fairly Common late Mar.-Sept. well beyond this range, particularly off s. Calif. in late Aug.-Sept.

FIG. 5

Leach's Storm-Petrel

Ashy Storm-Petrel

Fork-tailed Storm-Petrel

Black Storm-Petrel

(dark-rumped subspecies). Present in nesting colonies Feb.-Sept. (few in Jan. and Oct.), with eggs noted late May-mid-July. Calif. colonies on islets off Crescent City and Trinidad, the Farallones, and Los Coronados Is.

Ashy Storm-petrel (*Oceanodroma homochroa*) (fig. 5)

Recognition. L. 8 in. (20 cm.); WS. 18 in. (46 cm.). This species and the Black Storm-petrel are the only all-dark-rumped storm-petrels occurring off central and northern California. In southern waters, the dark-rumped subspecies of the Leach's Storm-petrel requires great care to distinguish (see account). Like most storm-petrels, the Ashy shows a somewhat paler band diagonally from the bend to the inner trailing edge of the upper wing surface, the rest of the feathering being sooty brown except for a paler band along the underwing coverts and a characteristic blue-gray cast on foreparts that is evident only at short range. Tail is shallowly forked, flight rather "average" for a storm-petrel — neither particularly erratic nor languid.

Habits. Although Ashy Storm-petrels breed commonly on Southeast Farallon I., it is doubtful that they are now the most numerous sea bird there, as Dawson reported in 1923. They can, however, be quite numerous without being often detected at sea. They pass to and from their nesting islands only at night, thus avoiding attack by gulls; and the Farallon birds apparently forage between the islands and the mainland much less often than they do farther offshore. Thus they are often missed on the one-day charter boat trips off central California, even though this species is the most frequently encountered of any of the family there. When in forward flight, Ashy Storm-petrels usually follow each few wingbeats with a short glide, and may make frequent upward turns. At times they settle on the water, but usually only briefly. They usually build no actual *nest,* the single *egg* (white, or with irregularly arranged red-brown speckling at large end) being laid in a shallow to deep crevice amid rocks, or in a burrow beneath a rock. The incubation period (as in all tubenoses) is very long; and the *young,* covered by fluffy gray down, also develops slowly. It is fed by regurgitation, probably only every few days by parents returning from great distances.

Range and California Occurrence. Breeds on Southeast Farallon I. (where Common, at least formerly), more sparsely on islands off s. Calif. (and Occ. nw. Baja Calif.). Pairs at colonies Feb. onward, eggs primarily May-July, but some so late that young are found even to Jan. Ranges from nearby parts of e. Pacific to unknown distances. Common to sometimes Abundant on Monterey Bay Sept.-Nov; lower numbers off s. Calif. increase in fall, probably by migration.

Black Storm-petrel (*Oceanodroma melania*) (fig. 5)

Recognition and Habits. L. 9-9½ in. (23-24 cm.); WS. ca. 20 in. (51 cm.). A large storm-petrel, all dark except for pale band across upper wing surfaces, as in the Ashy, but underwing darker and head and body all dark sooty brown without any gray tone on foreparts. Tail somewhat more deeply forked and legs longer than in Ashy, but best distinguished from it in field by its slower, more languid, wingbeat and consequently more buoyant flight. In southern California, at least, this species forages closer inshore than other storm-petrels; most of the central California records are from outer Monterey Bay. In both areas it occasionally is found in massed flocks, sometimes mixed with Ashy Storm-petrels.

Range and California Occurrence. Breeds on islands off both w. and e. sides of Baja Calif. north to Los Coronados Is. near San Diego; Fairly Common as foragers May-early Nov. (sometimes Apr., Dec.-Jan.) through continental shelf waters about Channel Is., and as post-breeding visitors (chiefly July-early Nov.) north through Monterey Bay (sometimes Very Common to Abundant in Sept.) and Irreg. to Pt. Reyes.

Note: The smallest storm-petrel is the **Least Storm-petrel** (*Halocyptena microsoma*) — L. 5½-6 in. (14-15 cm.); WS. 13 in. (33 cm.) — thus truly only sparrow-sized. Its feathering is all dark brown except for a paler brown band on the greater wing coverts. It has a characteristic wedge-shaped tail. These storm-petrels are said to fly erratically and usually very close to the water. Breeding on islands off Baja Calif. and first collected in U.S. waters in 1927, they have recently been found Irreg. in extreme s. Calif. waters, mostly well offshore late July-mid-Oct. In Sept. 1969, 500 were found off Morro Bay; a few have been noted off Monterey Co., and one off Humboldt Co. (see Appendix).

Note: The **Wilson's Storm-petrel** (*Oceanites oceanicus*) is a small dark petrel with a conspicuous white rump, *square* tail, long legs, and feet with yellow webs (feet extend beyond tail in flight). It is primarily a Southern Hemisphere species that summers regularly in the n. Atlantic, but very Irreg. into the n. Pacific. Single individuals (or once, three) have been seen in Calif. waters at least twenty times Aug.-Nov., mostly on Monterey Bay.

6. FAMILY PHAETHONTIDAE (TROPICBIRDS)

(Picture Key C)

Red-billed Tropicbird (*Phaethon aethereus*)

Recognition and Habits. L. 18-20 in. (46-51 cm.), plus another 18-22 in. (46-56 cm.) tail streamers; WS. ca. 44 in.

(112 cm.). Tropicbirds are gull-sized, mostly white-feathered sea birds with stout pointed beaks and (in the adult) greatly elongated central tail feathers. They fly with strong flaps in pigeon-like manner, or with short periods of gliding and soaring at variable heights above the water, and capture fish by plunging head first into the ocean, like a big tern. The Red-billed Tropicbird is the only species of the three in the world that normally reaches California. It is distinguished in adult feathering by the combination of red bill and white tail streamers, plus considerable black barring on the upperparts. Immatures have a yellowish bill and even more black barring, and a black band across the nape.

Range and California Occurrence. Tropical e. Pacific Ocean north regularly to Baja Calif., and Rare and Irreg. into s. Calif. waters; also in Caribbean and Atlantic and Indian Oceans. Calif. records chiefly off San Diego May-early Oct. (mostly late July-mid-Sept.), but a few individuals noted north to off Monterey and one even off Ore.

7. FAMILY PELECANIDAE (PELICANS)

(Picture Keys A, B, and C)

White Pelican (*Pelecanus erythrorhynchos*) (fig. 6)

Recognition. L. 53-68 in. (135-173 cm.); WS. 8-9½ ft. (2.6-3.1 m.). Readily told from Brown Pelican by its feathering — all white (or "off-white" in young) with black wingtips — and by yellow bill and feet. In the breeding season there is a sparse yellowish crest on the nape, and a thin plate develops above the middle of the upper mandible. The White Pelican is likely to be confused only with the Snow Goose (fig. 17) and swans (fig. 13), and then only at great distances.

Habits. Primarily an inland bird in spring, the White Pelican is one of the most spectacular of water birds, both there and on the shallow bays and salt evaporators where they also congregate following breeding. They often fish by swimming in companies, a long arc of the big birds slowly closing in on schools of fish in shallow water and scooping them up in pouches used as dip nets. Flocks "commuting" between feed-

FIG. 6

Brown Pelican — ad — 1st. yr.

White Pelican

ing areas may be seen circle-soaring upward in rising thermals of air and then peeling off as they reach the top, to glide downward to another thermal or to flap ponderously onward in their direction of travel. Large flocks of 2000 or more engage in similar behavior on their longer migration passages. When far off and circling in sunlight, the birds in such a flock appear first glistening white and then dull gray. *Food*: fish, mostly of little market or sport value, such as carp, catfish, chub, suckers, minnows, and some perch and occasional bass; also occasionally salamanders and crayfish. *Nest*: of mounded dirt and debris, or practically none; always in colonies, usually on barren islands or remote dikes in lakes. *Eggs*: usually 2, dull white; incubation by both parents, probably about 30 days. *Young*: naked at first and fed in nest; with white down all over, after 3-4 weeks; they finally wander around colony in masses known as "pods" for several more weeks before they fly.

Range. Breeds in colonies locally from B.C., n. Alberta, and Sask. south to Minn., Utah, w. Nev., and ne. Calif.; also Occ. or formerly in interior c. and se. Calif. Migrates over much of w. U.S. except the Coast Ranges north of c. Calif. Winters from Marin Co. and n. Sacramento Valley south through w. Mexico, a few in mild winters in ne. Calif.

Occurrence in California. Common to Very Common July-Dec. (fewer Occ. through winter) on primary salt ponds about San Francisco Bay; Locally Uncommon to Common on major lakes and marshes of Central Valley and coast slope of c. and s. Calif. at same seasons, also Sept.-Mar. at lakes of s. desert areas.

The drainage of larger lakes in the Central Valley for agriculture reduced the species to only occasional breeding there, although colonies of several thousand existed at Buena Vista and Tulare lakes as late as the early 1950s. Islands in the Salton Sea where smaller numbers nested (Mar.-July) up to 1957 have been inundated by rising water level there. Still breeds in some years at major lakes on ne. plateau, where Common to Very Common Apr.-July (sometimes Mar.-Sept. or longer). Migrating flocks noted overland, especially s. Calif. mountains and passes, every month but June, though primarily Mar.-May and Aug.-early Nov. Regular but Uncommon visitor Apr.-Sept. to Tahoe and Occ. other Sierra Nevada and s. Calif. mountain lakes.

Brown Pelican (*Pelecanus occidentalis*) (fig. 6; plate 2)

Recognition. L. 45-54 in. (114-137 cm.); WS. 6½-7½ ft. (2.1-2.5 m.). Distinguished by pelican shape and general brownish-gray tone of the feathers. Adults are often quite silvery above and have a conspicuous white or cream area on the head and neck, reduced by rich reddish brown in the breeding season. The immature has an unpatterned gray-brown head and a whitish belly through at least the first winter.

Habits. The fishing method of this familiar coastal marine bird is to fly along at 20-40 feet or so (exceptionally to 60 feet) above the surface until accessible prey is sighted. Then the big bird makes a quick downward plunge, extending its neck and holding its wings far back as it hits the surface or even disappears below it. When fish are caught, the pelican sits on the surface to drain water from its pouch before swallowing. It is then that gulls, especially the Heermann's Gull, try to rob it of the catch. When not feeding, Brown Pelicans are often seen flying close to the water surface, or even practically touching it with wing-tips, on the windward side of long swells, where the "cushioning" effect of air gives them additional lift.

Colonies of Brown Pelicans in California produced hundreds to thousands of young each year on Anacapa I. and islets near other Channel Is., and in some years at a northern outpost of the species at Pt. Lobos in Monterey Co. Until the 1960s, that is; a rapid decline at that time was traced to accumulation of chlorinated hydrocarbon pesticides (DDT, etc.) in their bodies, resulting in sterility in some and in thin eggshells which broke under those birds that did attempt to incubate. Apparently only 3 young were raised in the entire state in 1970, a year in which researchers made a special effort to find all; since then there has been some increase, and the population is still augmented each summer and fall by both adults and immatures moving northward from Mexico. Although one major source of the DDT contamination has been allayed, however, it still accumulates in fish from their supporting food chain in the ocean, and hence in our pelicans. The outlook is thus for a still endangered breeding population for some years. Nonbreeding flocks still gather in large numbers at various rocky headlands and at particular estuaries such as at Moss Landing, where they find easy fishing in the outflow of water from the condensers of a power plant. *Nest*: a scanty mound of sticks or trash on rocky to low brushy slopes of islands. *Eggs*: 3 (or 2), dull white, incubation period estimated 30 days. Care and habits of *young* similar to White Pelican (see account); first fly at about 9 weeks.

Range. Breeds on coasts from s. (formerly c.) Calif. to Chile, and from N.C. through Caribbean to ne. South America. Nonbreeders more commonly to c. Calif. and sparingly north to B.C. and Nova Scotia. Casual inland in major valleys.

Occurrence in California. Common to Very Common (formerly Locally Abundant) late June-Nov., and Fairly Common to Uncommon rest of year, on ocean chiefly near shore and in outer parts of bays. They favor rocky coasts for resting, but at times use tideflats, sandy beaches, and harbor structures. Nesting Mar.-early Aug. Regular though usually Rare to

Uncommon on Salton Sea and lower Colorado R. Aug.-mid-Oct. (recorded Apr.-Nov.), and Casual inland elsewhere.

8. FAMILY SULIDAE (BOOBIES, GANNETS)

(Picture Keys B and C)

Blue-footed Booby (*Sula nebouxii*) (fig. 7)

Recognition and Habits. L. 30-33 in. (76-84 cm.); WS. ca. 55 in. (140 cm.). The body of a booby is about the size of a large cormorant or gull, the neck is moderately long, and the tail is longer than in most diving birds and prominently wedge-shaped. The beak is a little longer than the thickish head and is stout and straight, tapering to a blunt point. This species is distinguished afield by its dull brown to gray-brown coloration with irregular white patches on the mantle (also on rump in adult), the brown fading below to a whitish belly; at close range the many fine white streaks on the head are prominent and there is a small bare-skin throat pouch. The bright blue feet are displayed during courtship; as in the related pelicans and cormorants, the webs extend to all four toes. Boobies normally fly at heights of 20 to 50 feet above the water, singly or in flocks, and make slanting or nearly vertical dives to well below the surface after fish, often coming up ten or more feet away.

Range and California Occurrence. Entire Gulf of California, and from c. Baja Calif. on Pacific side south to Galapagos Is. and Peru. On lower Colorado R. and esp. Salton Sea, Irreg. Rare to Uncommon July or Aug.-mid-Oct. (up to forty at a time known) fewer and Occ. Nov.-Apr. Rare on Pacific side north of Mexico, but recorded as far north as Ventura and twice to wc. coast (see Appendix).

Brown Booby (*Sula leucogaster*) (fig. 7)

Recognition and Habits. L. 26-29 in. (56-74 cm.); WS. 54-59 in. (137-150 cm.). Adult Brown Boobies are easily distinguished by their solid dark brown coloration with square-cornered white belly, light greenish to bluish beak, and

FIG. 7

Brown Booby

Blue-footed Booby

yellow to pale greenish feet. Juveniles are duller brown, paler below, but in two shades foreshadowing the adult pattern. In behavior this species is similar to the Blue-footed (see account), but more often fishes far from shore.

Range and California Occurrence. Worldwide on tropical oceans, breeding north to Gulf of California. Casual July-Dec. at Salton Sea and on Colorado R.; individuals recorded several times offshore from San Diego and Santa Barbara Cos., and once in Marin Co.

9. FAMILY PHALACROCORACIDAE (CORMORANTS)

(Picture Keys A, B, and C)

Double-crested Cormorant (*Phalacrocorax auritus*) (fig. 8; plate 2)

Recognition. L. 32-36 in. (81-91 cm.); WS. 50-54 in. (127-137 cm.). About the size and shape of Brandt's Cormorant with head similarly thicker than neck; distinguished from it by yellow pouch on chin (lower bill also often extensively yellow). Adults are brownish-black with green gloss, in spring with short, curled crests on head, white-tipped in some birds. Immatures are medium brown, paler on the belly than the Brandt's, some being quite whitish below.

Habits. This is the only one of the three species of cormorants in California that occurs regularly on fresh water, and it is also the commonest on largely landlocked bays such as Humboldt, San Francisco, and San Diego. Prior to 1955 several thousand could be seen roosting on the power line over San Francisco Bay between southeast Richmond and Brooks I., but this site was then largely abandoned, and none of comparable size has been reported since. When taking flight, cormorants must either kick along the surface of the water (they use both feet in unison) or lose considerable altitude starting from a perch. The attraction of the power line wires, despite their instability in a wind, was apparently this insurance of adequate altitude for takeoff.

Where abundant, this species often engages in group fishing, as does the Brandt's (see account). *Food*: sculpins, smelt, river and bay perch, catfish, flounders, carp, suckers – a great variety of fish mostly of little economic value; also some crustaceans and rarely amphibians. *Nest*: a platform or mound 2 feet in diameter, of sticks and trash, placed amid spreading tree branches or, in island or coastal colonies, on wide rock ledges or slopes; each pair defends a small adjacent area for perching. *Eggs*: 2-7 (usually 3-4), pale blue; incubated 25-29 days. *Young* soon get dense black down, are dependent on parents for up to 9 weeks although they fly well at 7 weeks.

Range. Breeds on Pacific coast in sw. Alaska and from Vancouver I. south to c. Mexico; inland from s. Wash. to se. Calif., se. Idaho, n. Utah, and plains and Great Lakes areas of n. U.S. and s. Canada; also on and near Atlantic and Gulf coasts from Newfoundland to Tex., and in Cuba. Winters in same coastal areas (except ne. U.S.) and interiorly in Calif. and cs. U.S.

Occurrence in California. Fairly Common to Locally Very Common Aug.-May (and fewer through summer) on large bays, salt ponds, and lagoons along entire coast; also on rocky to sandy outer coast chiefly from San Luis Obispo Co. south; quite rare in summer on c. Calif. outer coast except Locally near few small nesting colonies – e.g., Southeast Farallon I. and s. Marin Co. On inland lakes and large streams, Locally and Irreg. Rare to Fairly Common in Central Valley and sw. Calif. lowlands in the same period. Breeds late Apr.-Aug. on s.

FIG. 8

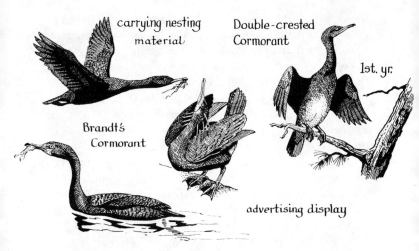

carrying nesting material

Double-crested Cormorant

1st. yr.

Brandt's Cormorant

advertising display

coast and perhaps still Locally in trees inland, but from Jan.-July near Salton Sea and on Colorado R. Recorded Mar.-Oct. on lakes in mountains and ne. Calif. plateau, sometimes Locally Fairly Common to Common Apr.-July around nesting colonies in latter district.

Brandt's Cormorant (*Phalacrocorax penicillatus*) (fig. 8; plate 2)

Recognition. L. 33-35 in. (84-89 cm.); WS. 48-52 in. (122-132 cm.). Larger and with proportionately heavier head and neck than the Pelagic Cormorant; about like the Double-crested in these features, but distinguished from it by *buffy band of feathers* behind the pouch, which is bright light blue in breeding season and dull gray at other times. At a distance the buffy band is sometimes mistaken for a yellowish pouch by beginners, but the light color never extends to the bill as in the Double-crested. Breeding-season adult Brandt's Cormorants have slender white plumes scattered variably along the head and neck and are glossed purplish on the upperparts. Immatures are dark brown, somewhat paler below, but appear blackish at a distance or in dull light.

Habits. All around the Monterey Peninsula, at Pt. Reyes, by the corners of the Golden Gate, and elsewhere along the rocky coast of northern California, the irregular lines of cormorants flying to and from feeding areas are mostly of this species. More locally to the south, as at Morro Rock and at La Jolla, this may also be true, but elsewhere on the south coast it is exceeded by the Double-crested. When large schools of fish are located, the cormorants converge on them in a big arc, the advancing front of which is under the water or surfacing while the laggards fly over the main group to alight and dive prompt-ly where the prey is more accessible. Like all cormorants, this species is often seen perched with its wings spread wide, apparently to dry them after an extended period of foraging.

The breeding activities of Brandt's Cormorants are readily studied at La Jolla and at Pt. Lobos on the Seventeen-Mile Drive near Monterey. Each pair defends its own nest and area within "bill-reach" in the closely packed colonies. There is an elaborate series of displays, well described by Williams (1942: *The Condor,* 44:85-104) and summarized in Palmer (1962), the most spectacular being the advertising stance of the male (fig. 8). *Nest*: a low mound of seaweeds obtained by diving and tugging, or by stealing from neighbors (sometimes low-growing land plants are used). *Eggs*: 3-6, pale blue to chalky white; quickly eaten by gulls if left unattended, so human intruders and large predators should be restricted from colon-ies; incubation by both parents, alternately, but period un-known. *Young* soon have dark gray down; are fed by regurgita-tion.

Range. Resident from Vancouver I. south along outer coast and outer parts of bays to tip of Baja Calif. and through Gulf of California. Small numbers "commute" between mainland and islands up to at least 30 miles offshore.

Occurrence in California. Very Common to Locally Abun-dant all year along outer coast, especially near rocky areas south to Morro Bay area, less numerous and more Local south to La Jolla. Common in outer parts of large bays (e.g., in San Francisco Bay to Pt. Richmond) but only Occ. inner parts or on small estuaries or lagoons. Breeds Mar.-Aug. (Jan.-July at La Jolla) on rocky headlands or islets of mainland coast or islands well offshore. In some years found well upstream on nw. Calif. rivers during salmon runs, Aug.-Oct.

Pelagic Cormorant (*Phalacrocorax pelagicus*) (plate 2)

Recognition. L. 25-27 in. (63-68 cm.); WS. ca. 40 in. (102 cm.). The duck-sized body (rather than goose-size of other cormorants) and very slender neck, with the head scarcely of larger diameter, distinguish this species. Breeding-season adults have very glossy feathering, and from Jan. or Feb. to May or June most have a white patch on the flanks and scattered white filaments on the neck and shoulders. The pouch and lower "face" are dark red, but often inconspicuous. Immatures are dark brown with some gloss (violet and greenish, like the adults, but duller).

Habits. Actually somewhat less prone to fly far offshore when breeding than the Brandt's, despite its name of "pelagic," this species is primarily found along the more rugged areas of the rocky coast, including straits such as the Golden Gate. It is less gregarious than the two larger cormorants and chooses narrower ledges for nesting, usually on the steepest sea cliffs. *Nest*: a scanty mound or ring of seaweeds or other plants or shoreline trash. *Eggs*: 3-5 (occasionally 6-7), pale bluish to chalky; incubation by both sexes, alternately. The *food* of the Pelagic Cormorant as indicated by stomachs examined includes up to ten percent or more crustaceans in addition to the usual small rocky-shore fish.

Range. Breeds from arctic Alaska and extreme e. Siberia south to nw. Baja Calif. and n. Japan. Winters from s. Alaska to s. (?) Baja Calif. and se. China, the Calif. populations being essentially resident, although individuals wander when not nesting.

Occurrence in California. Fairly Common all year (Locally Common or Very Common some years near largest nesting colonies) along rocky coasts of mainland and offshore islands south to San Luis Obispo Co. and s. Calif. islands; fewer and more Local along mainland s. coast and in outermost parts of bays. Breeds on steep rocky coasts or islets, Apr.-Aug. (or Sept.).

10. FAMILY FREGATIDAE (FRIGATEBIRDS)

(Picture Key C)

Magnificent Frigatebird (*Fregata magnificens*)

Recognition and Habits. L. 37-45 in. (94-114 cm.); WS. 7-8 ft. (2.3-2.6 m.). Frigatebirds (also called Man-o'-War Birds) are large, long-winged sea birds with long and deeply forked tails and long, heavy hooked beaks. They glide and soar super-efficiently, "adjustments" of the wing and tail serving to carry them even into a brisk wind without flapping. When necessary, as in takeoffs and chases, they do flap deeply. The general impression is of great overgrown swallows, except for the long and rapacious beak. Frigatebirds make their living almost entirely by pursuing or diving on terns, boobies, pelicans, or other fish-catchers, forcing them to disgorge their prey which the frigatebird then catches adroitly in the air or from the water. They do scavenge from the ocean surface also, but only on the wing — for their feet, though webbed, are so small as to be useless except for perching on the bushes or trees of their nesting or roosting areas. The adult male Magnificent Frigatebird is all black except for a large red throat pouch inflated like a balloon in the early phase of breeding. The female is blackish with a white breast but dark throat. The lack of a brown band on the upper wing surface of the adult male, the darker throat of the female, and the lack of rufous coloring on the whitish head and foreparts of the young distinguish this species from the Great Frigatebird, which also occurs in the east tropical Pacific and might reach the California coast.

Range and California Occurrence. A wide-ranging bird that might be seen over almost any habitat near the coast, though usually over the ocean within 100 miles of land. Resident from Baja Calif., ne. Mexico, and s. Fla. south to n. and e. South America. Post-breeding stragglers (usually single birds) occur in some years on Pacific coast through s. Calif. (and in Imperial and lower Colorado R. valleys), very Rare farther north (see Appendix).

11. FAMILY ARDEIDAE (HERONS, EGRETS, BITTERNS)
(Picture Keys B and C)

Great Blue Heron (*Ardea herodias*) (frontispiece; plate 4)

Recognition. L. 46-54 in. (117-137 cm.); WS. to 7 ft. (2.3 m.). By its great size and long-legged, long-necked shape and gray coloration, this species is easily told from all other birds in California, except the Sandhill Crane (fig. 26). Even though commonly referred to as "cranes" by persons unacquainted with true cranes, herons are readily distinguished from them by the features shown in Picture Keys B and C. In addition, most herons are less gregarious when feeding. Adult Great Blue Herons have a largely white head with a black stripe extending into a plume to the rear, and whitish spear-like tips to the lower neck and chest feathers. First-year birds lack these plumes and are more uniformly gray or brownish gray about the head and neck.

Habits. Standing about 4 feet tall when in normal alert posture, a Great Heron attracts the attention of persons not otherwise interested in bird study. Its size seems even greater when it spreads very broad, long wings and utters deep rasping croaks as it flaps ponderously away, then slowly glides to a new vantage point. When foraging, Great Blues are usually solitary or at least well spaced-out along lake or stream or across a tide flat or field. Here they stalk slowly along or hold a stiff posture with neck forward as the prey is sighted — until a quick lunge is made for it with the beak. *Food*: nearly 75 percent is fish, mostly of species not valued by man; the remainder includes crustaceans, frogs, salamanders, lizards, snakes, large aquatic insects, and small rodents (which are often obtained in open fields in California's valleys), and even occasional birds. *Nest*: similar to and often in mixed colonies with Great Egrets (see that account and frontispiece), but also in more numerous small single-species colonies, these some-times in *Eucalyptus* or native upland trees. Where undisturbed, they may even nest in shrubs. *Eggs*: 3-7 (usually 4), pale bluish green to olive; incubated about 28 days. *Young* are fed in nest by regurgitation; first fly at about 8 weeks.

Range. North America from se. Alaska and s. Canada to s. Mexico and Jamaica; in winter retires from n. interior region (although may persist until most water freezes) and some reach n. South America.

Occurrence in California. Fairly Common all year about shallow bays and marshes near entire coast, and on marshes and lakeshores of cismontane and se. lowlands; fewer Irreg. along streams, in open fields, and to outer rocky coast and islands. Mostly near many scattered breeding colonies Feb. (or Jan.)-June or July, and Common near largest colonies then. Locally Common July-Oct. (fewer other months) on fish-bearing salt evaporators. Regular but usually Uncommon May or June to Oct., and Irreg. other months, about lakes and marshes in forest belt of mountains (most winter records in s. Calif.). Probably Fairly Common in similar seasonal pattern at ne. lakes and marshes, where some also breed and are thus Locally Common.

Green Heron (*Butorides striatus*) (plate 8)

Recognition. L. 20-22 in. (51-56 cm.); WS. 24-26 in. (61-66 cm.). About the size of a crow and appearing somewhat like one when in flight, but the wings are held more arched (and, of course, long beak and legs are evident except when far away). Bluish-green and reddish-brown adult coloration is distinctive (see plate 8). Immatures are streaked with brown on neck and chest, but show enough greenish tinge on wings and back to be told from the larger, stockier American Bittern and immature Black-crowned Night Heron (see fig. 10). Only the Least Bittern among the family is smaller.

Habits. Green Herons are among the most solitary of their family, except for the marsh-inhabiting bitterns. The preferred foraging area of the Green is a wooded or log-strewn shore, or they may fish from a perch on a snag or tree a foot or so above deeper water (and even dive therefrom on occasion). With trees bordering streams and ponds now so generally disturbed by human activities from central California southward, this species is greatly reduced in much of the state. Fair numbers can still be found near the few riparian woods areas remaining by permanent streams and ponds. *Voice*: a characteristic *skyow(k)* of alarm, often given as it flies away. *Nest*: of sticks, amid outer or upper dense branches of trees (rarely on ground). *Eggs*: 3-6, pale green or blue-green; incubation about 20 days. *Young*: are fed in or near nest for about 3 weeks.

Range. Breeds from w. Wash. and Ore., ne. Calif., sw. Utah, across most of c. and e. U.S. and south to Argentina. Winters from cw. (Occ. nw.) Calif. and Sacramento Valley and from s. U.S. elsewhere to n. South America.

Occurrence in California. Locally Fairly Common in n. Coast Ranges in summer, and Uncommon and more Irreg. all year on and near tree-bordered streamsides, ponds, and lakes of cismontane lowlands south to San Diego Co.; a few reach coastal marshes in spring and fall. Breeds Apr. (or late Mar.) to July. Recorded on Colorado R. and in Imperial Valley every month except Mar. and July-Aug., with one nesting record. In ne. Calif., reported Apr.-early Oct.; and in Aug. in Sierra Nevada and s. Calif. mountains.

Little Blue Heron (*Florida caerulea*) (fig. 9)

Recognition, Habits, and Range. L. 25-29 in. (63-74 cm.); WS. 38-41 in. (96-104 cm.). This small heron ranges normally from New England and the southern Great Lakes and Texas to South America, and north on the Pacific Coast to Sonora. Observations of supposed stragglers in California were reported for years, but most of these were in the white immature feathering, without adequate details to verify that they were distinguished from the immature Snowy Egret. In adult feathering, Little Blues are unmistakable — smooth blue-gray except for maroon head and neck. The beak is gray to bluish basally, with the terminal third black. The legs are dark greenish to black, paler in the immature. First-year birds have all white feathers except for dark tips on the primaries, but they have a bluish-gray bill basally, without the yellow lores of the Snowy Egret.

Occurrence in California. So many records are now at hand

(including at least two immatures collected and one adult photographed) from coastal bays, lagoons, and nearby ponds that there is no doubt that this species occurs here repeatedly. They are usually seen feeding in salt-marsh channels and in open shallows of nearby bay shores or ponds. Most records (two to four maximum at one location) are from July through Jan. in Orange Co. and the San Francisco Bay area (few also Jan.-May). Recently a pair of adults appeared in eastern Marin Co., where they were suspected of nesting with other herons.

Cattle Egret (*Bubulcus ibis*) (fig. 9)

Recognition. L. 19-21 in. (48-53 cm.); WS. 36-38 in. (91-97 cm.). Body about the size of a Snowy Egret, though somewhat stockier. The neck is shorter and the head and beak thicker than the Snowy's. Feathering white, or with buff on crown, foreneck, and mantle in spring and summer. Beak, lores, and iris yellow (or orangish-red when breeding). Legs and feet blackish or dull greenish yellow (to orangish when breeding).

Habits. This is one of the birds that feeds closely around (or even atop) large mammals in Africa, and they prefer to do likewise about cows in America, often walking along within a foot or two of the hooves or head. From such a position they make quick lunges at grasshoppers, crickets, beetles, or other large insects disturbed by the grazing animal, and on ocacasion take lizards or mice. At other times they also forage by themselves in open grassland, especially where the flow of irrigation water (or a grass fire) similarly drives prey into view. In Florida they have also become a common feeder close to the bulldozers on garbage dumps – a habitat also widely available in California. Nesting can be expected in any pro-tected groves of dense-canopied trees or tall shrubs near good feeding areas. *Nest*: of twigs, 10-18 inches across, with shallow cup on top. *Eggs*: 4-5 (occasionally 6), light blue; incubated 21 days or more.

Range and California Occurrence. A species of tropical, subtropical, and south temperate parts of the Old World. It invaded South America before 1930, was in Florida by the early 1940s, and from there has spread north and west, finally

FIG. 9

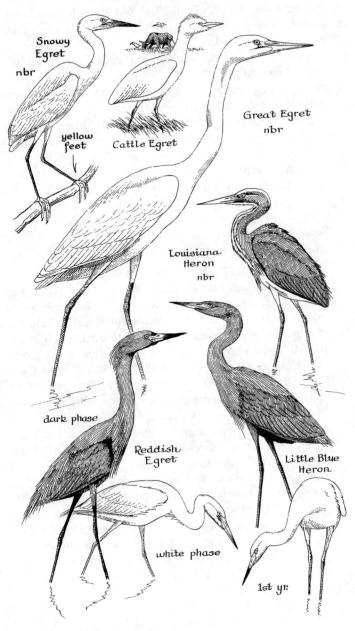

Snowy Egret nbr

yellow feet

Cattle Egret

Great Egret nbr

Louisiana Heron nbr

dark phase

Reddish Egret

Little Blue Heron

white phase

1st yr.

reaching Calif. in 1962. By the late '60s it was Regular in the Imperial Valley; by 1974 it had been recorded nine times north to Humboldt Co. (and on into Ore.). In cismontane c. and s. Calif. it is now Regular and Uncommon to Locally Fairly Common Oct.-Mar. (Occ. Aug.-May), frequenting pastures and irrigated fields and to nearby trees for roosting; also to be expected at garbage dumps (as in Fla.). Hundreds began nesting (Apr.-July) in 1970 at s. end of Salton Sea, with thousands subsequently as post-breeding visitors in the Imperial Valley. After such a remarkable spread, this species can be expected to increase further throughout Calif. lowlands. In e. U.S. it breeds north at least to N.J.

Reddish Egret (*Dichromanassa rufescens*) (fig. 9)

Recognition and Habits. L. 27-32 in. (68-81 cm.); WS. 42-46 in. (107-117 cm.). A medium-sized heron with two distinct color phases (at all ages): a *dark* one, with plumage all slaty gray except for reddish-brown head and neck; and a *white* one, with plumage all white except grayish toward tips of primaries. Some are intermediates with patches of light and dark. Bill pale pinkish or lavender basally (lores same color), with black terminal third to half. Legs dark blue to black. Other distinctive features are the shaggy appearance of the head and neck and the proclivity for rapid dashing about when feeding. Although the dark phase predominates in Texas, most of the California records are of the white phase. Like many herons, this one is a colonial nester, and feeds on small fish, amphibians, and crustaceans.

Range and California Occurrence. Resident from s. Fla., Tex. coast, and both sides of Baja Calif. (at about lat. 30°) south to Central America and West Indies. Stragglers occur north to se. and sc. states of U.S. and in se. and coastal s. Calif. (Casual to Monterey Co.), chiefly individual immatures in Orange, San Diego, and Imperial Cos. July-Nov. (a few records Dec.-June). Coastal birds usually forage on tideflats, marshes, or lagoons.

Great Egret (*Casmerodius albus*) (frontispiece; fig. 9)

Recognition. L. 37-41 in. (94-104 cm.); WS. 42-46 in. (107-117 cm.). The heron characteristics (see Picture Keys B and C) and all-white feathering easily distinguish this species from all but the smaller Cattle and Snowy Egrets (and a few others very rare in California). From all these except some Cattle Egrets, the *all-yellow bill* and black legs and feet identify the Great Egret — or, as it has been called until recently, the "Common" Egret. Adults have long, loosely branched feathers (aigrettes) extending from the back to beyond the tail in the breeding season, and a more orange-yellow bill then also.

Habits. Although egrets are the most conspicuous of herons when foraging (much like the Great Blue Heron; see account), it is at their nesting colonies that the variety of their behavior is evident. Various displays indicate aggression toward neighbors, advertising for a mate, courtship, etc., the latter types with the aigrettes lifted and spread. These activities pass into a quiet period when eggs are incubated; but when young are being fed in the nest, the parents make regular commuting trips to and from distant feeding grounds. At Audubon Canyon Ranch, near Bolinas, a mixed colony of Great Blues and Great Egrets nest in redwood trees and forage largely in the nearby shallow lagoon; but some fly the ten miles or so to Tomales Bay. This ranch is a sanctuary, and there is a trail to a vantage point from which one can view the nests without disturbing the birds. (The schedule of times available should be requested.) Other colonies of Great Egrets about San Francisco Bay have declined or disappeared, as have a number in the Central Valley. Several may still exist in the Central Valley, and at least in some years at Salton Sea and at Havasu L., on the Colorado R. All heronries in areas other than remote ones are subject to much disturbance from man. Even though few nests are actually robbed any more, these birds are prone to desert nests with eggs if highly alarmed. After the great decimation of these and other birds by plume hunters of the late 1800s, no egrets were found in California for many years. All herons and egrets are now protected by stringent federal law, but even ground-level intruders into colonies should be held to few or none if their populations are to be maintained. *Nest*: a flattish to bulky platform of sticks (see frontispiece), usually well up in trees over, or rarely in, marshes near water. *Eggs*: usually 3-4 (occasionally 5-6), pale blue-green; incubated about 24 days. *Young* first fly at about 6 weeks.

Range. In America breeds from nw. Calif., e. Ore., and Idaho, east to N.J. and south throughout Middle and South America. Winters from n. Calif., w. Nev., and the se. U.S. southward. Also occurs in most warmer parts of the Old World, retiring from s. Europe and c. Asia in winter.

Occurrence in California. Fairly Common all year, or Locally Common July-Nov., about coastal lagoons, marshes, salt evaporators, and bayshores; fewer in most inland valleys about marshes and lakes, Irreg. onto open fields or along streams. Also Common Mar.-July, near the largest nesting colonies at both bayshore (e.g., Humboldt, Bolinas, San Francisco bays) and Central Valley locations (south from upper Sacramento Valley). Former colonies on the lower Colorado R. and southern end of Salton Sea now abandoned, but the species is still Fairly Common there all year. Uncommon or Locally Fairly Common Mar.-Aug. on ne. plateau, nesting May-July. Rare and Irreg. at Sierra Nevada and s. Calif. mtn. lakes mostly July-Sept.

Snowy Egret (*Egretta thula*) (frontispiece; fig. 9)

Recognition. L. 22-26 in. (56-66 cm.); WS. 36-42 in. (91-107 cm.). The Snowy Egret is the commonest all-white-plumaged heron smaller than the Great Egret in California (but see note on increase and distinction of Cattle Egret, in that account). When judgement of size cannot be depended upon, as is so often the case, the black legs, yellow feet, and black bill with yellow lores further distinguish the adult Snowy from any other heron here. In immatures, the bill and legs are duller and may have gray to light green areas, especially up the back of the legs, such birds sometimes being mistaken for immature Little Blue Herons (see account). However, they are still distinguished by yellowish lores and toes. From Feb. or so into the summer, adult Snowies have long, recurved loose-webbed plumes on the back and shorter ones on the foreneck and as a crest on the nape.

Habits. The Snowy Egret is both the most active in its foraging habits and the most social on the nesting grounds of any of our herons. Although a Snowy may stalk prey like other species, it is also often seen to engage in such specialties as stirring the water with one foot, spreading its wings in semiparasol fashion over the area searched, and leaping a few times in quick succession after active prey. *Food*: a variety of small fish, crayfish and other crustaceans (including a type of "shrimp" that damages growing rice), and large insects in abundance. *Nest*: a shallow platform of sticks, usually rather low in trees (see frontispiece), or of tules, etc., on the ground in marshes; nearly always in dense colonies of good size except at margins of the range. *Eggs*: usually 3-4, pale blue-green; incubation period (18+ days) not reliably established. *Young* are fed in nest for at least 3 weeks.

Range. Breeds from c. Calif., ne. Nev., Utah, Okla., and se. N.Y. south through most of South America. After breeding, occurs north to Wash. and across n. U.S. Winters from coastal c. Calif. (a few in the Central Valley) through s. Calif., sw. Ariz., s. Tex., and Fla. southward.

Occurrence in California. Common all year, except Fairly Common Dec.-Feb., about bayshores and coastal lagoons, marshes and ponds, and more Locally through the Central and Imperial valleys, including to flooded fields; fewer inland in cismontane s. Calif., and Uncommon at other inland low altitude locations. Known recent nesting colonies near Redwood City, San Rafael, Pittsburg, Los Banos, Bishop, and s. end of Salton Sea. In nw. and ne. Calif., Rare and Irreg. May-Oct., to Fairly Common some years (Aug.-early Sept.). Vagrants also noted at lakes in the Sierra Nevada and s. mountains June-Oct.; Rare to offshore islands.

11. HERONS, EGRETS, BITTERNS

Louisiana Heron (*Hydranassa tricolor*) (fig. 9)

Recognition and Habits. L. 24-28 in. (61-71 cm.); WS. 36-38 in. (91-97 cm.). A rather slender, long-necked heron, largely slate-gray with conspicuous white belly and rump and a narrow median light line on the foreneck. In breeding feathering the adult shows maroon at the base of the neck and buffy plumes from the lower back, with a short white plume on the hind-head. The legs and basal half of the bill are grayish. Immatures have the same pattern, with brown neck and chest and some brown in the wings, plus light greenish beak and legs. This species is a fairly active forager in bay shallows or pools of river backwaters, taking a variety of small fish and crustaceans as food, plus insects when away from salt water. In lower latitudes it is said to prefer the coastal mangrove swamps.

Range and California Occurrence. Resident from c. Baja Calif. and nw. Mexico, Tex., La., Ala., and Md. south to cw. and ne. South America. After breeding, wanders north to various parts of sw. and c. U.S. In sw. Calif., Uncommon late Aug.-May (most Regular late Dec.-Feb.) on coastal lagoons and tideflats, chiefly in San Diego Co. but Occ. north to Orange Co. and Casual to Santa Barbara Co. Records of vagrants at Salton Sea and on lower Colorado R. span most months.

Black-crowned Night Heron (*Nycticorax nycticorax*) (fig. 10)

Recognition. L. 23-28 in. (58-71 cm.); WS. 42-48 in. (107-122 cm.). A stockily built heron of moderate size with a thicker neck and head than any except the similarly shaped American Bittern. Adults are easily identified by whitish to pale gray underparts, neck, and head except for black crown, from which in breeding feathering a few slender white feathers project over the greenish-black back; wings gray, legs light green (or pinkish when breeding), bill black or with some greenish at base. Immatures through first year are streaked brown on whitish below and brown (duller than the bittern) above, with whitish streakings. (They also lack the blackish coloring on the primaries and side of the throat that a bittern has.)

FIG. 10

Black-crowned
Night Heron

ad

1st yr

Least
Bittern

American
Bittern

Habits. Foraging chiefly at night or in twilight, this species is unique among our herons. Its usual call-note, a short, flattish *quokk,* often given as they fly, is distinctive. As to both feeding and nesting habitat, this is the most adaptable of all herons, some colonies having persisted in the heart of large cities where their food was obtained from debris-laden harbor areas or park lakes after the daytime human activity had subsided. For unknown reasons, these herons are much reduced in some urban areas but still breed regularly in others. Near San Francisco Bay since 1968, the largest colony (to about 400 pairs) has been on a semi-marshy shrubby island. Black-crowns are quite gregarious except when actually searching for food, and many come regularly to favorite dense tree canopies or masses of tall tules or cattails to spend the day. The high winter numbers shown on the Graphic Calendar (in Appendix) are from a few such favored roosts in Central Valley marshes. Major disturbance of such a roost causes all its birds to take flight and mill about; but this provides a clue to their real numbers that is otherwise difficult to obtain. *Food*: more than half consists of fish, largely of no commercial value; many crustaceans, and in freshwater areas, large insects and occasional small mammals. *Nest*: of twigs and/or marsh plants, rather frail until added to in successive seasons, always in colonies and variably placed high in trees, in shrubs, or on ground. *Eggs*: 2-6, pale blue-green; incubated about 25 days. *Young* fly at about 6 weeks; fed by adults for a time afterward.

Range. Breeds from c. Wash., e. Wyo., and Sask. south and east throughout U.S. and coastal parts of Mexico. Winters from s. Ore., n. Utah, and se. U.S. to Panama. Also resident in South America and found in most of the Old World.

Occurrence in California. Common to Locally Very Common Oct.-Jan. at roosts in Central Valley marshes, and Fairly Common all year about lakes, marshes, and larger streams throughout most of the lower altitudes of the state, frequently roosting and nesting in dense trees; feeding also about harbors and rocks (or even kelp beds of outer coastline), Irreg. to offshore islands; Uncommon in nw. Calif. and Rare in ne. Calif. in midwinter. Breeds Feb.-July, Very Common or Abundant at best colonies. Also Common at ne. lakes Apr.-Aug., nesting commonly there; and Uncommon Apr.-Oct. at lakes in s. Calif. mts. (has bred at Big Bear Lake).

Note: The **Yellow-crowned Night Heron** (*Nyctanassa violacea*) is about the size and shape of the Black-crowned, but the adult has a lavender-gray body, blackish face, and yellow to white crown and broad "ear stripe." Immatures are very similar to the Black-crowned but have longer legs, smaller white spots above, and a shorter, thicker bill. Normally this is a species of se. U.S. to South America, breeding north on the Pacific Coast only to c. Baja Calif. But at least six individuals have reached s. Calif. (one inland) in Mar.-June and late Oct.-mid-Nov. of various years since 1950, and one exceptionally regular pioneer returned yearly (dates range from Apr. to mid-Nov.) from 1968 to 1973 at San Rafael, where it foraged along the bay shore.

Least Bittern (*Ixobrychus exilis*) (fig. 10)

Recognition. L. 13-15 in. (33-38 cm.); WS. ca. 19 in. (48 cm.). Smallest of the heron family, body size about that of a robin. In summer more often heard than seen; but if it shows itself, the chestnut to tan neck (streaked below in immatures), blackish crown and dark back with pair of white lines, and especially the large buffy area of upper wing coverts, are distinctive.

Habit. A very secretive bird, even more so than some of the rails, the Least Bittern is no doubt considerably more numerous and certainly more regular in California than indicated by the Graphic Calendar based on published records (see Appendix). Listening and watching quietly where you can look over the tops of a marsh known to be inhabited by them will often yield more views than attempting to drive them into flight. Much of their activity includes clambering about amid the dense cattails and tall bulrushes (tules), with rare flights over them. *Voice*: a series of up to five or six low-pitched *coo* notes, sometimes with weaker beginning and ending; also more guttural *tut-tut-tut* calls, especially in alarm. *Nest*: a rather flimsy platform or cup of marsh plants, usually placed over water a foot or more deep. *Eggs:* 3-6, smooth pale blue or greenish; incubated 17-19 days by both sexes. *Young* may leave nest as early as 5 days of age, usually at about 10 days.

Range and California Occurrence. Breeds over most of e. U.S. and from s. and e. Ore. south to Nicaragua. Rare to Uncommon (though probably Regular) in the larger cattail and tule marshes of the Central Valley of Calif., and Locally in sw., se. (and ne.?) Calif. (and probably elsewhere in much of w. U.S.). Calif. nestings reported mid-Apr.-early July. Winters from nc. (Rarely nw.) Calif., nw. Mexico, s. Tex., and s. Fla. to Brazil.

American Bittern (*Botaurus lentiginosus*) (fig. 10)

Recognition. L. 24-34 in. (61-86 cm.); WS. 37-50 in. (94-127 cm.). A stocky medium-sized bird with streaked brown feathering similar to an immature Night Heron, but with a buffier tone and lacking prominent white streaks or spots on the mantle. The American Bittern is slightly more slender also, and has a distinctly slenderer bill. In all but the juvenal plumage, the throat is white with conspicuous black at sides, and in flight the blackish wing-tips are also helpful.

Habits. More secretive than the true herons but much less so than the Least Bittern, this species is usually flushed quite readily from marshes along roadsides or dikes when one walks near. It ordinarily flies only a short way, and in somewhat more gangly attitude, with legs drooping more than a typical heron. Then it drops back into the tall marsh. Bitterns very rarely perch much above ground level. Occasionally one that is approached slowly may "freeze" with its bill pointing vertically and peer at the intruder with both eyes from under its chin. Amid marsh plants, of course, this immobile stance and the streaked neck and chest provide excellent camouflage.

Food: a variety of fish, amphibians, snakes, crustaceans, snails, and mammals or birds. *Voice*: male on breeding grounds gives a mellow, ventriloquial *oong-ka-chunk* up to eight times at few-second intervals; also a dry, nonresonant version, and a hoarse croak or nasal *haink* when disturbed. *Nest*: a platform of dead marsh plants, sticks, and/or leaves, usually in shallow water or sometimes floating. *Eggs*: 4-5, smooth buff to light olive brown; incubated 24-29 days by female. *Young* are fed in nest for about 2 weeks.

Range. Breeds from c. B.C. throughout most of s. Canada and U.S. (only Locally in s. U.S.). Winters near the coasts from sw. B.C. and Md., interiorly from n. Utah and Ohio, south to Cuba and Central America.

Occurrence in California. Fairly Common Oct.-Mar. and Uncommon to Rare rest of year in marshes (of both tall dense and more open sorts) throughout Central Valley; somewhat fewer elsewhere in cismontane lowlands. Eggs or young reported Apr.-July; also occurs as nonbreeder in Imperial Valley and in bayside salt marshes late Aug.-May, and at s. end of Lake Tahoe and in ne. marshes May-Aug., where it probably also breeds.

12. FAMILY CICONIIDAE (STORKS)

(Picture Keys B and C)

Wood Stork (*Mycteria americana*) (fig. 11)

Recognition. L. 36-47 in. (91-120 cm.); WS. ca. 65 in. (165 cm.). A mostly white-plumaged bird about the size of the

FIG. 11

Wood Stork

Great Egret, but with all the long flight feathers of the wings and the short tail black. In other white water birds with black on the wings, it is restricted to primaries or their tips. This stork, the only one in America, has the heavy beak typical of its family — high at the base and somewhat downcurved in the terminal half — yet was called "Wood Ibis" for years. The head and neck are bare and scaly in adults, and dark gray like the beak and legs. Immatures have dingy white to pale gray feathering and may show yellowish on the beak.

Habits. These gregarious storks breed in wooded wetlands of varied sorts; but when in California they favor the open muddy shallows of lagoons and brackish marshes, or fields undergoing heavy irrigation. Here they both pick and probe shallowly for any animal food available. When not feeding, and particularly for overnight roosting, they usually gather on high perches in trees. In flight, unlike herons, storks keep the neck extended in front. When rising thermals are frequent, they may soar upward in circles without flapping their long wings. *Voice*: a hoarse croak, usually given only when disturbed.

Range and California Occurrence. Resident from nw. Mexico, Tex., and se. U.S. south to Argentina. Variable postbreeding movements northward across much of (chiefly s.) U.S. In Calif. found Regularly July-Sept. only in the Imperial Valley in irrigated fields, marshes, and lakeshores, but great yearly variation in numbers even there. They may be present in s. Calif. coastal lagoons and marshy lakes for several years in sequence, then absent for years. Vagrants have been recorded

recently north to Morro Bay and inland to Woodland (Aug.-
Dec.) and Modoc Co. (early Aug.).

13. FAMILY THRESKIORNITHIDAE (IBISES, SPOONBILLS)

(Picture Keys B and C)

White-faced Ibis (*Plegadis chihi*) (fig. 12)

Recognition. L. 19-25 in. (48-63 cm.); WS. ca. 35 in.
(89 cm.). A dark bird the size of a Snowy Egret but with long
down-curved bill which is thicker at the base than in curlews
(see Picture Key B and fig. 31). In good light the feathers of
adults appear dark purplish to purplish-red with green or
bronze sheen, especially on wings. A narrow white zone of
feathers edges the base of the bill in breeding season only.
Immatures are duller and may have whitish streaks on the head
(as do some winter adults).

Habits. When feeding either in shallow water or on fields
undergoing irrigation, these birds often probe deep into the
mud with their five-inch-long bills after earthworms and burrowing insects; or they may feed in shallow water or at the
surface like herons. Flocks of ibises are unorganized when
feeding, but gather into straggling lines when flying any distance. Their general appearance when in the air far off is
similar to small dark geese; but of course the long bill and the
legs extending beyond the tail distinguish them when closer.
This most northern of the ibises is now only an irregular
breeder in the marshes of the San Joaquin and Imperial valleys
(last nesting records in 1940s and and around 1960, respec-
tively), even though it breeds commonly in northern Utah.
Fragmentary records indicate recent breeding in Modoc Co.
Nest: a large cup of dead tules or cattails, lined with grass,
placed amid or on mounds or tall marsh plants or exceptionally low in trees. *Eggs*: 3-5, pale green or blue; incubated for
21 days, mostly by female. *Young* are fed in or near nest for
about 5 weeks.

Range. Breeds from n. Utah and probably ne. Calif. (and
Occ. in Central Valley of Calif., e. Ore., Neb., and Minn.) south

FIG. 12

White-faced Ibis

br

Roseate Spoonbill

to sc. Mexico. Winters from c. Calif. south to sc. Mexico and s. Tex. Also found in South America south of the tropics. (Other very similar "Glossy Ibises" are found in parts of the Old World and in e. U.S., the White-faced being considered only a subspecies of the Glossy by some.)

Occurrence in California. Uncommon to Common all year, but varying yearly, in Imperial Valley fields, low marshes, and shallow lakes; Locally Very Common to Abundant there June-Sept. some years. Also Uncommon in San Joaquin Valley and lower Sacramento Valley (Locally Common near Los Banos and Gustine) Aug.-Apr. and Occ. through summer. Reported at shallow lakes in ne. Calif. Apr.-Aug. (Locally Abundant in w. Nev.). Formerly Common as breeder near Los Banos, with eggs May-July, and Occ. still breeds in Imperial Valley. Nonbreeders Irreg. chiefly July-Dec., at s. Calif. coast slope ponds; Rare north to near Monterey.

Roseate Spoonbill (*Ajaia ajaja*) (fig. 12)

Recognition. L. 28-33 in. (71-84 cm.); WS. 48-53 in. (122-135 cm.). This spectacular bird is named from its 6- to 7-inch-long greenish-yellow bill, each half of which is flattened toward the tip into a 2-inch-wide spatula. It is of medium size and leg length, the adult feathering being bright pink except for white neck and back and a dark red band on the upper wing coverts. The immature is all white except for touches of pink posteriorly.

Habits. Spoonbills feed primarily upon small fish and invertebrates they catch adroitly between the seemingly clumsy bill-tips, either from the water itself or by "feeling" along in the muddy bottom. their pattern of flight and other traits, including nesting in colonies, are rather like those of ibises and some herons (see accounts and frontispiece).

Range and California Occurrence. Found normally along the Gulf Coast of U.S. and through Mexico (north to Sinaloa on Pacific side) and south to Argentina. Spreads Irreg. north, Casual to c. Calif., and some have occurred repeatedly July-Oct. since 1950 around Salton Sea and Imperial-Laguna Dam area of Colorado R. Many of the records are of small groups of immatures.

Note: **Flamingos**, the very long-legged, very long-necked birds of the family Phoenicopteridae, have been reported frequently since 1958 from various locations in coastal California — usually one or two that stay in good feeding areas such as shallow salt-evaporating ponds sometimes for months, the records just about spanning the year. Probably all are escapees from among the many in zoos and other places where these birds are exhibited; some have been so tame as to be obviously so. The nearest normally occurring species in the wild state, the **American Flamingo** (*Phoenicopterus ruber*) breeds in the Caribbean area to Yucatan, Mexico, and occurs Irreg. north to s. Florida.

14.-20. FAMILY ANATIDAE (WATERFOWL)

14. Subfamily CYGNINAE (Swans)

(Picture Key A)

Whistling Swan (*Olor columbianus*) (fig. 13)

Recognition. L. 47-58 in. (119-147 cm.); WS. ca. 7 ft. (2.3 m.); Wt. 10½-18½ lbs. (4.5-9 kg.). Almost everyone knows a swan, at least the Mute Swans commonly kept on

FIG. 13

Whistling Swan — usually yellow spot

Introduced Mute Swan

Trumpeter Swan

park lakes. From these the Whistling Swan, our only native species of regular occurrence, can be distinguished by the absence of a knob or sharp "forehead" at the base of the bill. Adults have an all-black bill, sometimes with a yellow spot in front of the eye which distinguishes them from the larger Trumpeter Swan. Many Whistlers lack this spot, however, and should not be called Trumpeters unless identified by position of the nostril or by voice and/or size. In their first winter, young Whistlers often have a dull pinkish bill with black borders and are washed with dingy brownish gray on the head and foreparts (as are the young of other swans). Adults are entirely white-feathered in all these species, though sometimes stained rusty on head and neck.

Habits. The maximum wintering populations of this next-to-largest waterfowl in western America are in California's Central Valley. In favored feeding areas such as the winter-flooded asparagus or corn fields west of Lodi and Stockton, single flocks of 5000 or more are often seen. There they grub in the mud for the peppery fruits and waste grain not utilized

14. SWANS

by the farmers. Elsewhere they also feed on the bulbous roots of sedges, succulent grasses, and a variety of seeds, plus water snails and occasionally some amphibians. When on water deep enough to require it, they reach the bottom by immersing the long neck or even tipping-up like a duck (fig. 13). Once settled for feeding, a flock of swans often allows a closer approach than geese and ducks do, tending at first to swim away with necks held stiffly vertical (unlike the arched neck of the Mute Swan). If pressed further, they take off by an amazingly short run for so big a bird, and seldom mix with other waterfowl in flight. *Voice*: resonant, somewhat nasal calls, goose-like but higher pitched and less abrupt (scarcely "whistle"-like, however); often given in flight.

Range. Breeds in w. and n. Alaska and far n. Canada. Winters from B.C. to n. Baja Calif. (but mostly s. Ore. through the San Joaquin Valley) and in middle Atlantic states.

Occurrence in California. Fairly Common to Locally Abundant late Oct.-early Mar. on wet fields, marshy ponds, and lakes of Central Valley, and decidedly fewer but frequent on other lakes at same season in nw. to wc. lowlands; fewer yet along coast and in s. Calif., except Irreg. up to 50 or so in Imperial Valley. At Modoc Plateau lakes, Abundant in Mar. and mid-Oct.-Nov. and variably Very Common to Uncommon through winter when and where unfrozen water is available; also Irreg. Rare or Uncommon Nov.-Jan. on large lakes in Sierra Nevada and Jan.-Feb. in s. mountains.

Trumpeter Swan (*Olor buccinator*) (fig. 13)

Recognition and Habits. L. 52-62 in. (132-157 cm.); W.S. 7-8 (10?) ft. (2.3-2.6 to 3.3 m.); Wt. 20-38 lbs. (9.1-17.2 kg.). Although the Trumpeter is distinctly larger than a Whistling Swan, the surest field identification is by *voice*, which is deeper in pitch and much more sonorous. At close range the Trumpeter's black bill has a reddish cutting border (as does that of some Whistling Swans!), and there is never a yellow spot on the lores. The gradual increase of this species in the interior northwest, after once being reduced to endangered species level, apparently accounts for its reappearance since 1960 in central California. Trumpeters are perhaps more prone than the Whistling Swan to occur coastally in winter here; but if mixed in a large flock of that species, only very careful study is apt to pick them out.

Range and California Occurrrence. Breeds from s. Alaska to ec. Ore. (Malheur Natl. Wildlife Refuge) and nw. Wyo., wintering on any open water and coastally in that region, Rare and Irreg. south to c. (formerly s.) Calif. Recent Dec.-Mar. records on coastal lakes and bays and inland Nov.-Mar. from nw. and ne. Calif. to e. San Luis Obispo Co.

Note: **The Mute Swan** (*Cygnus olor*) (fig. 13) sometimes escapes from captivity or from park ponds and lives for a time in the wild (at least six records of one or two such birds in cw. and sw. Calif.). In ne. U.S. they have even become established as ongoing populations. Adults are easily told by the orangish or reddish beak with a black base that has a knob by the forehead. Immatures have a rudiment of the knob as an abrupt "forehead" by the wide black border of the pinkish bill.

15. SUBFAMILY ANSERINAE (GEESE)

(Picture Keys A, B, and C)

Canada Goose (*Branta canadensis*) (fig. 14)

Recognition. Five different subspecies in three sizes:

"Honkers" (2 subspecies) L. 31-40 in. (79-102 cm.); WS. 5-6½ ft. (1.6-2.1 m.); Wt. 6-13½ lbs. (2.7-6.1 kg.).
"Lesser" Canada (2 subspecies) L. 25-30 in. (63-76 cm.); WS. 52-62 in. (132-157 cm.); Wt. 4-6½ lbs. (1.8-3.0 kg.).
Cackling (subspecies *minima*) L. 22-27 in. (56-68 cm.); WS. 43-57 in. (109-145 cm.); Wt. 2½-4 lbs. (1.1-1.8 kg.).

A dark goose with largely brownish body and wings, except for blackish primaries; white lower belly and tail coverts (upper ones as white band), blackish tail, and a *black neck and head* with *white bib-like patch* on cheeks and throat. Of the three size groups occuring in California (fig. 14), the little Cackling Canada (*minima*) is darker brown below and has a bill more abruptly tapered than the larger subspecies. There is a

FIG. 14

Canada Goose

Honker

Lesser

Cackling

ad

White-fronted Goose 1styr

large dark subspecies which occurs primarily on the northwest coast. The others, all rather light brown on chest, are scarcely distinguishable except by direct comparison of size. The Lessers are about the size of the Snow (fig. 15) and White-fronted (fig. 14) Geese, but can be told from them of course by their darker necks (and by voice) even when so far away their other features cannot be seen.

Habits. Geese are strongly social birds, often gathering into flocks of thousands for migration and wintering. Even in these large flocks, however, the family groups (a pair with their young-of-the-year) can be distinguished with careful watching, for they stay intact through all maneuvers, feeding, etc. Unlike ducks, geese pair for life, and the young migrate with their parents and return with them to the breeding area, where the old gander finally drives them off. The nest and vicinity, and where possible a sizable area nearby, are defended vigorously by the male. The spectacular threat displays, noisy honking and all, can be seen in Feb. at Lake Merritt, Oakland, where a small breeding population is now established (from former wing-clipped birds). These and other scattered breeding records on lakes near San Francisco Bay (second Graphic Calendar in Appendix) are for the large Honkers, which also form the bulk of the wintering flocks toward the coast. In the Central Valley, however, they are far outnumbered by the small Cackling Canadas. Where each is plentiful, these geese commonly segregate by size to a great extent.

FIG. 15

Snow Goose

blue phase

white phase

1st yr

Ross' Goose

Emperor Goose

1st yr

Voice: loud resonant *honk* (or *ahnk*) notes uttered frequently but irregularly, mostly when in flight; the Lessers give a similar but higher-pitched and less resonant call, the Cackling a mere *ank* or *lek-luk*. *Nest*: a mound of grasses, usually with stick foundation; on ground of islands or in marshes, sometimes in trees (or on elevated platforms provided in management areas). *Eggs*: 4-7 (even 10 occasionally), creamy white; incubated 28-30 days by female. *Downy Young*: yellow with medium dusky wash on crown, most of back, wings, and in a point on flanks – a pattern typical of many waterfowl. (e.g., see downy young of Mallard in plate 5.).

Range. Breeds throughout Canada and Alaska – the Cackling in w. Alaska; Lessers from c. and n. Alaska to Aleutians and n. Hudson Bay; Honkers farther south in Canada and south to ne. and cw. Calif., Utah, and across n. U.S. to e. coast. Winters from se. Alaska and B.C. over most of U.S. (except n. interior) and south into Mexico; also to Japan. The Cackling winters predominantly in the Central Valley of Calif.

Occurrence in California. Common to Abundant Oct.-Apr. in marshy lakes, moist grasslands, and fields of the Central Valley; somewhat fewer in coastal valleys and hills and s. coastal plain; and Irreg. Uncommon on bayshores and s. mountain lakes. Uncommon to Fairly Common through summer at few reservoirs in c. (and nw.?) coast counties, where they breed Feb.-June. On ne. plateau lakes and grasslands, Common May-Aug. (when breeding), Abundant in main migrations

Feb.-Apr. and Sept.-Nov., and variably Very Common to Rare through winter, dependent upon unfrozen water. Considerably fewer at same seasons (and breeding) at lakes in n. Sierra Nevada and Cascades.

Brant (*Branta bernicla*)* (fig. 16)

Recognition. L. 22-26 in. (56-66 cm.); WS. 44-50 in. (112-127 cm.); Wt. 2.7-3.8 lbs. (1.2-1.7 kg.). A small goose with black head, neck, and chest, relieved by *partial collar of white streaks* on foreneck. Bill and feet black; body and wings similar to a dark Canada Goose but whiter on the sides (showing below dark wings when on water) and, in most of those in the west, blackish on much of the belly (subspecies *nigricans* or "Black Brant"). (A light-bellied form from the east Canadian area is very rare in California.)

Habits. These almost exclusively marine geese formerly gathered during migration and winter residence in flocks of up to 10,000 or more on certain favored bays all along the coast, with much smaller numbers elsewhere. In mid and late winter, up to a few thousand can still be seen at the relatively undisturbed places, and a fair number irregularly on smaller estuaries and bays, especially those with sandy to sandy-mud substrate. Pollution and disturbance long ago made San Francisco Bay unsuitable as a major wintering area for them, as did dredging of Mission Bay at San Diego, the eelgrass on which they prefer to feed being practically exterminated there. When disturbed from such feeding grounds on bays, the flocks gather on the open ocean beyond the surf or may rest on remote sandbars or beaches. At times, some also feed on rocky benches exposed by the tide on the outer coast. During the main migration periods, numbers can be seen flying steadily over the ocean paralleling the coast, mostly in long irregular lines or skeins close to the water or with frequent changes up to a hundred feet or so above it. Some of those from the Gulf of California come northward to the Salton Sea in spring, but

*Includes *Branta nigricans* of the 1957 American Ornithologists Union *Checklist of North American Birds*, as merged by the A.O.U. Committee in 1975.

FIG. 16

Black Brant

apparently do not return in fall by that route. Compared to other waterfowl that breed in the far north, some Brant remain at favored wintering areas quite late; but the numerous summer records in the state are undoubtedly of nonbreeding individuals, perhaps mostly immatures. *Voice*: low-pitched, mellow *gronk*, repeated a few times or rolled into a guttural *g-g-gr-gr-r-r-onk*.

Range. Breeds in arctic areas around the world, the subspecies *nigricans* or "Black Brant" from ne. Asia across n. Alaska to cn. Canada. This subspecies winters on salt water from nw. B.C. to c. Baja Calif., and in Asia south to Japan and China; the light-bellied subspecies winters primarily on seacoasts from Mass. to N.C. and in Europe and North Africa. Casual in migration or winter inland.

Occurrence in California. Common to Locally Abundant Oct.-May on large shallow bays with eelgrass beds (chiefly Humboldt, Tomales, Morro, and San Diego bays and Drake's Estero) and nearby ocean waters; fewer at same seasons on smaller and mud-bottomed estuaries. Stragglers remain at favored coastal sites well into June or July most years, Occ. all summer. At Salton Sea, Regular, sometimes Common, as migrant Apr.-early May; stragglers Occ. there Jan.-Mar. and June-Nov. and in the Central Valley Oct.-Apr.; Casual on the ne. plateau Oct.-Nov. (The light-bellied subspecies, the "Atlantic Brant," is also Casual, all but one of the sixteen records as of 1975 from the coast, Humboldt to San Diego cos.)

Note: The **Red-breasted Goose** (*Branta ruficollis*), an Asiatic species, has been noted six times in Calif., all presumed escaped from captivity (see Appendix).

Emperor Goose (*Philacte canagica*) (fig. 15; plate 3)

Recognition and Habits. L. 26-28 in. (66-71 cm.); WS. 48-56 in. (122-142 cm.); Wt. 5-7 lbs. (2.3-3.2 kg.). A bluish-gray goose with scale-like effect produced by black, then white, tips of feathers, less conspicuous in immatures. Adults have a white head and hindneck and a largely black foreneck, whereas immatures are very dark over the whole foreparts (compare blue phase of Snow Goose in fig. 15). Bill pinkish; legs and feet yellow or orange, unlike all other geese except White-fronted, which has different color pattern (see plate 3). The Emperor Goose is mostly a marine species in winter, but about a third of the nearly sixty records in California are from interior points (where hunters are more numerous, of course).

Range and California Occurrence. Breeds near and on coasts of far ne. Siberia and w. Alaska. Winters primarily in Aleutians to Kurile Is.; Rare and Irreg. Oct.-Apr. south to c. or s. Calif. Small groups have wintered at such places as the Carmel R. mouth, Santa Cruz, Morro Bay, and Seal Beach. Individuals (or families?) also apparently mix with large goose concentrations of other species in the Central Valley and Modoc Co. refuges.

White-fronted Goose (*Anser albifrons*) (fig. 14; plate 3)

Recognition. L. 26-30 in. (66-76 cm.); WS. 53-62 in. (135-158 cm.); Wt. 4-7½ lbs. (1.8-3.4 kg.). (In one rare subspecies, L. to 34 in. (86 cm.) and WS. to 65½ in. (167 cm.). A medium-sized dark goose with neck and head the same brown color as body and wing coverts. Adults have a white forehead (the "front") and irregular black blotches on the belly ("speckle-belly" of hunters), pinkish to orangish bill, and yellow feet. Immatures lack the white front and black blotches, and their bills and feet are duller but similar to the adult. A large white area below and a smaller one above the base of the dark tail (as in the Canada Goose, fig. 14), which is also white-tipped (unlike the Canada), are helpful from the rear.

Habits. Although far less common toward the coast, the White-fronted Goose is as numerous as the Canada in the Central Valley, and locally more so. Mixed flocks of White-fronted and either Canadas or Snow Geese may be seen feeding on the fallen grain in stubble or disked fields, or grazing where short green grass is available. All these geese also feed at times by tipping-up in open marshes and shallow ponds or lakes, but ponds in protected areas are of greater importance to them for retreat. When not disturbed, most of their feeding is done in the early morning and late afternoon; but during the hunting season they commonly feed at night and go to protected ponds for the day. Like other geese, they also migrate both by day and night, often high up whence their occasional calls draw attention to the passage. Even on such long flights, flocks of White-fronts are usually in irregular lines; they may form distinct V's but these tend to be less stable than those of the Canadas. The Jan.-Feb. migration (second Graphic Calendar in Appendix) brings these birds from the Imperial Valley through San Gorgonio Pass and westward over or near the mountains of Los Angeles Co., then presumably on to the Central Valley. From there, major departure is in the later migration period (first Graphic Calendar). *Voice:* loud, moderately resonant notes, as *kah-la-luk,* sometimes shortened to a *ka(n)k* similar to but lower in pitch and less nasal than Snow Goose; also a high-pitched "tootling" at times.

Range. Breeds in w., c., and n. Alaska, w. arctic Canada, w. Greenland, and across arctic Eurasia. Winters from sw. B.C. through w. Wash., Ore., and Calif. (chiefly interior valleys) to coastal Tex. and La. and s. Mexico, and in the Old World to n. Africa, India, and Japan. The rare large subspecies (*gambelli*) is known chiefly from the upper Sacramento Valley and marshes near Suisun Bay.

Occurrence in California. Abundant Oct.-Mar., Fairly Common from early Sept. and to early May, in Central Valley on moist to wet grasslands, some cultivated fields, and lakes and marshes; Uncommon to Fairly Common in similar habitats of c. and s. coast ranges; fewer yet at coastal lakes and marshes through winter. In Imperial Valley and Colorado R. areas, variably Fairly to Very Common Sept.-Feb., with stragglers to early May. In ne. Calif., Abundant Oct.-Nov. and late Feb.-Apr.; Irreg. through winter.

Snow Goose (*Chen caerulescens*), including the "Blue Goose" (figs. 15, 17, plate 3)

Recognition. L. 26-31 in. (66-79 cm.); WS. 53-61 in. (135-155 cm.); Wt. 4-7 lbs. (1.8-3.2 kg.). Except for the smaller Ross' Goose (see that account), the typical Snow Goose in California is the only goose with plumage mostly white, only the primaries being black and their coverts light gray. White Pelicans (fig. 6) and Wood Storks (fig. 11) are the only other large white birds with black wing-tips likely to occur in California. The dark phase ("Blue Goose") is bluish to brownish gray with white head and upper neck, rump, lower belly, and tail coverts, and white-edged inner wing coverts. Because birds with these contrasting types of feathering are now known to interbreed commonly north of Hudson Bay, and many intermediates exist, both are now included in the same species. There is even some indication that the dark phase is increasing in frequency and spreading westward, but so far not over five dark ones out of many thousands are usually found in California. Adult Snow Geese have pink feet and bills. These parts are duller, to gray or tan, in immatures, which also have a pale grayish tone to wing coverts, secondaries, and much of the body in the white phase − but the general impression is still of a white bird. Immature dark-phase birds are smooth gray-brown with white belly and rump and some white on the inner wing area. A blackish "grinning patch" along the margin of the bill of the Snow Goose (visible only at close range) is an absolute distinction from the Ross' Goose (fig. 17).

Habits. Much more gregarious than our other wintering geese, this species is usually encountered in large flocks (30,000 to 80,000 or more in best areas) or not at all. They also tend to stay on or near marsh-bordered waters more than the White-fronted and Canada Geese, and are especially concentrated in the hunting season on closed areas of the major

FIG. 17

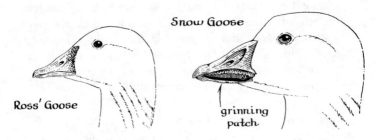

Snow Goose

Ross' Goose

grinning patch

waterfowl refuges and management areas of the Tule and Lower Klamath lakes and the Central Valley from the Sacramento (near Willows) and Gray Lodge (near Gridley) to Merced and Los Banos. Watching the dawn and late afternoon flights of geese and ducks to and from such a protected resting area is truly one of the greatest experiences for outdoor enthusiasts in California. Most of the birds fly high over areas where guns are fired; but enough are lower, or fly low between feeding areas, so that California still provides some of the best goose-hunting in the whole nation. This will prevail, of course, only so long as the careful management of these favored wintering grounds is continued and the breeding grounds and migration stopping-places are adequate for each species.

When flying any distance, Snow Geese usually form into irregular flocks of long wavy lines, seldom keeping any V formation. *Voice:* slightly higher-pitched and more nasal than the White-fronted, as *gank* or *kuahnk,* varied with more guttural notes by flocks in chorus.

Range. Breeds in ne. Siberia, n. Alaska, and arctic Canada east to nw. Greenland (the "blue" or dark phase and a larger white subspecies in eastern part of this range). Winters mainly in valleys of interior Calif. from Tehama Co. to Imperial Co. (chiefly white phase), on Gulf Coast of Tex. and La. (both phases), and Atlantic coast from N.J. to N.C. (larger subspecies); smaller numbers in B.C., w. Wash., and to c. Mexico and Japan.

Occurrence in California. Abundant from Nov.-early Mar. and Fairly Common in Oct. and Apr. (even Occ. in Sept. and to June-July) on marshy lakes and marshes plus nearby wet fields and grasslands of Central Valley south through Merced Co.; less numerous even in suitable areas inland thence south to Imperial Valley; and still fewer in coastal or bay locations. In ne. Calif., Abundant Oct.-Nov. and Feb.-Mar.; a few arrive by mid-Sept. and some stay through most winters, occasional stragglers (cripples?) through summer and have even nested.

Ross' Goose (*Chen rossii*) (figs. 15, 17)

Recognition. L. 21-25½ in. (53-65 cm.); WS. 47-54 in. (119-137 cm.); Wt. 2.3-4.2 lbs. (1.0-1.9 kg.). Almost like a smaller edition of the white-phase Snow Goose, but not enough smaller so that the size difference is evident except when the two are together and close, or the Ross' is with other geese of comparable size such as the White-front or Cackling Canada (see fig. 14). Identification of Ross' from Snow Geese in flight at any significant distance is impossible, although the size difference is apparent when they are directly overhead together at close range. On the ground or water the beak is distinctly shorter, more abrupt in profile, and lacks the black "grinning patch" of the Snow Goose (fig. 17). This last, most definite character is usually discernible only with a telescope at the distances to which they allow approach. Immature Ross' Geese, unlike the young Snow, are only slightly less white than the adults.

Habits. In feeding and flight, very similar to Snow Geese and often mixing with them. However, Ross' Geese are often also found in small flocks by themselves or associating with the same-sized Cackling subspecies of the Canada Goose. Careful study of all such separate groups of white geese will usually disclose fair numbers of Ross'. South and southwest of Merced, however, Ross' Geese sometimes outnumber the Snow Geese in midwinter, and many are shot by hunters mistaking them for the latter. Nearly the entire species population of Ross' Geese (some 12,000 to 20,000 birds) seems to be concentrated here at times, and it is therefore highly desirable to restrict hunting of white geese in this area, in order to protect this rarest of North American geese.

Range. Breeds in a highly localized area of ne. Mackenzie and on Southampton I. at n. end of Hudson Bay. Migrates southwest to winter almost entirely in the Central Valley, a few on to the Imperial Valley, and Irreg. to coastal c. and s. Calif. Casual east to Texas and west to nw. Calif.

Occurrence in California. Uncommon to Locally Abundant (see Habits, above) Nov.-Mar. in Central Valley marshy lakes and nearby wet grasslands or fields; Rare and Irreg. on c. and s. coastal area lakes, bays, and river mouths. In ne. Calif., Common to Abundant on major refuges late Oct.-Dec. and Mar.-Apr.

16. SUBFAMILY DENDROCYGNINAE (TREE OR WHISTLING DUCKS)

Fulvous Tree Duck (*Dendrocygna bicolor*) (fig. 18)

Recognition. L. 18-20 in. (46-51 cm.); WS. ca. 36 in. (92 cm.); Wt. ca. 1.7 lbs. (770 g.). A long-legged, long-necked duck with buffy-brown plumage, a light barred area on darker back, whitish slash marks or jagged line on the sides, and white around the base of the tail. The bill and feet are lead-gray. Its wings are more rounded than those of other ducks, among which the eclipse or female Pintail (figs. 18 and 19) is closest to it in shape and color. The Tree Duck's feet extend well beyond its short rounded tail in flight, and it stands high, with body more erect than other ducks, when on land.

Habits. Inappropriately named, this species rarely if ever perches in trees and in fact is seldom found in wooded areas. Tree Ducks forage mostly at night, and may be more common than the records on the Graphic Calendars (in Appendix) indicate, even though irregular from year to year in California. By day they usually retreat to dense marshes, feeding both there and on nearby open fields and shallow ponds, sometimes grubbing or poking in the mud in almost shorebird fashion. The *food* they take is almost entirely vegetable: grass and weed seeds, soft green leaves and stems, acorns, etc. *Voice*: a whistle with a wheezy quality, usually doubled, sometimes long and squealing. *Nest:* a skimpy to well-built basket of grass, usually on high ground in marshes. *Eggs:* 10-17 (or many more in "dump" nests), white or pale buff. *Downy young:* medium gray and white in a pattern similar to the Mallard (plate 5), except for a white band all across back of head; bill and feet gray.

Range. Breeds from Imperial Valley (formerly from the San Joaquin Valley) south to sc. Mexico and east to s. La. Small numbers wander somewhat farther north, Casually even to B.C. Winters mostly in Mexico, a few Irreg. in se. and sw. Calif. Also found in parts of South America, Africa, Madagascar, and India.

Occurrence in California. In Imperial Valley (and Occ. along lower Colorado R.), Uncommon to Fairly Common Apr.-mid-Sept. and in some years from late Jan. and through Oct., but Rare through winter, frequenting marshes, lakes, and streams and feeding also on wet fields; nests Apr.-Sept. in se. (also Locally inland in sw.) Calif. In the San Joaquin Valley Uncommon to Fairly, Common some years only, Apr.- early Nov., and formerly nested north to Los Banos area. Vagrants have appeared farther north, even to ne. and nw. Calif., and in c. and s. coastal locations, Feb.-Dec.

Note: Vagrants (one to five at a time) of the Black-bellied Tree Duck (*Dendrocygna autumnalis*), a species found from s. Tex. to Argentina, have occurred in the Imperial Valley June-Nov. of various years including 1972 and 1973. (See Appendix also.)

17. SUBFAMILY ANATINAE (DABBLING DUCKS)

(Picture Keys A, B, and C)

Mallard (*Anas platyrhynchos*) (fig. 19; plate 5)

Recognition. L. 19-27 in. (48-68 cm.); WS. 31-40 in. (79-102 cm.); Wt. 1.1-3.8 lbs. (500-1730 g.). A large dabbling duck with violet or blue speculum bordered by white, tail mostly white (male) or white and light brown (female), feet orange-red, and bill light greenish-yellow (male) or yellow

FIG. 18

Fulvous
Tree Duck

♀

Gadwall

♂

♀

Pintail

♂

blotched with black (female). These features identify a Mallard at all seasons, but the first three are the best marks for birds in the air. The female is coarsely mottled brown, lighter below and with a light line above the eye. After his annual molt in midsummer, the male's feathering is similar to the female's and he is said to be in "eclipse plumage," which is presumably equivalent to the dull "winter" or nonbreeding feathering of male birds of other families. Most of the body feathers are gradually replaced in Sept. and Oct., as in all our dabbling ducks, with the brighter alternate or "nuptial" plumage yielding, along with the older wing and tail quills, a breeding feathering that is then worn through the winter and into early summer. Male Mallards are unmistakable then, with bright green head, narrow white collar, reddish-brown chest, light gray sides, and black all around the base of the tail.

Habits. The Mallard is the duck best known to those who

make no study of birds, for it is widespread and the ancestor of most domestic ducks. In the Central Valley it consistently ranks among the three most abundant ducks, along with the Pintail and American Wigeon. In feeding habits it is rather typical of the subfamily, tipping-up (see Picture Key A) in shallow water, grubbing in mud on wet fields or elsewhere, and even occasionally grazing. *Food*: more than three-fourths of its food is plant materials, including seeds of certain bulrushes and other sedges, pondweeds, and grasses, and growing tips and bases of various plants; animals eaten include water insects, snails, small crustaceans, and earthworms.

Reproductive cycle: as in most dabbling ducks, this species forms into pairs during the winter. The head-bobbing and other displays on the water, and wheeling flights of courting parties in which a number of males closely follow one female, can readily be seen at park lakes as well as on the wilder marsh areas. When settled for breeding, the drake usually maintains a small territory where his mate returns to him through the egg-laying period. Soon after she is incubating steadily, however, he leaves the area and seeks protected parts of the marsh or lake for the forthcoming molt. He is flightless for several weeks then, because all the flight feathers are shed at about the same time. The female's molt is similar but later, after she has raised her brood, which she does without assistance from the male. *Voice*: of female, a loud clear *quack* uttered singly or in long series; of male, hoarser and much softer. *Nest*: a feather-lined shallow cup of grass or sedges, usually placed amid dense marsh plants or grass near water — but where such preferred cover is absent, in a great variety of situations — even at times in trees! *Eggs*: 6-15, usually about 9-10, pale greenish-buff to tan; incubated 23-29 (usually 26) days. *Downy young*: dark above with yellowish blotches and yellow line above eye, yellow below (plate 5); usually follow female until well after they can fly, but *may* be independent before full-grown.

Range. Breeds over most of the Northern Hemisphere south of the Arctic Circle, including w. U.S. and south to n. Mexico. Winters in small numbers as far north as open water remains,

FIG. 19

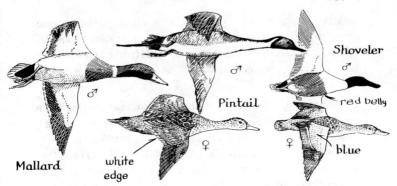

Shoveler ♂

Pintail

red belly

Mallard ♂

white edge

♀

♀

blue

commonly from w. B.C., Wash., and Great Lakes south to s. Mexico or beyond. Some migrate to Hawaii each year.

Occurrence in California. Present all year, but Very Common to Abundant chiefly Oct.-Mar., in Central Valley marshes, ponds, shallow lakes, and wet grasslands and fields; fewer in similar habitats widely over cismontane lowlands and in the Imperial Valley; Common in urban parks with lakes or ponds (where often become tame); and Fairly Common most of the year on bay marshes and tideflats, and even nearby residential areas. Nesting primarily Mar.-July, but a few may have small young as late as Oct., and early pairing is seen in Nov.-Dec. In ne. Calif., Common to Abundant Mar.-Apr. and July-Oct., plus some years through winter; also Fairly Common to Common May-Aug. as breeders there and at many mountain lakes and wet meadows widely over the state. The most widely distributed, though not always the most numerous, duck.

Gadwall (*Anas strepera*) (fig. 18)

Recognition. L. 18-23 in. (46-58 cm.); WS. 31-36 in. (79-91 cm.); Wt. 1.1-2.6 lbs. (500-1180 g.). Speculum consists of a *small square white patch* on secondaries close to the body, accentuated by black area next to it toward bend of wing; another "off-white" area on outermost secondaries is not usually noticed. Feet orange-yellow; bill gray in males, gray-brown with yellowish tinge in females. On water or land the male, even in breeding feathering, is the dullest of any of our dabbling ducks, appearing smooth gray with a brown head and a black stern around the gray tail. Fine vermiculations on the body and covert feathers are evident at close range. Females and eclipse males are like female Mallards (plate 5), except for wing pattern, less color on the bill and feet, and less white on the tail.

Habits. The Gadwall is a species that reaches maximum abundance east of the Sierra Nevada, but it has been increasing in recent years both as a wintering bird in the Central Valley and as a sparse breeder there and to the coast. The rather inconspicuous color of the males, compared to other dabblers, makes Gadwalls hard to pick out of large concentrations of ducks except in flight. However, careful scanning of marshy-bordered ponds will usually disclose many at the proper season, perhaps because they prefer more open water near their feeding grounds than some of the other species. *Voice*: of female, a rather loud *quack,* rougher and not so loud as Mallard's; of male, a soft reedy *whaack,* and less often louder double *kak* notes or shrill whistles. *Nest*: similar to Mallard's, but more often placed on an island. *Eggs*: 7-13, creamy white; incubation about 25 days. *Downy young*: very similar to Mallard (plate 5), but somewhat paler.

Range. Breeds from s. Alaska and across c. Canada and n. Mississippi Valley (a few east to Atlantic) south to c. (and Occ. sw. and se.) Calif., n. Ariz., and Tex.; also in n. and c. Eurasia. Winters from s. Alaska, s. B.C., Utah, Colo., Ill., and Va. south to s. Mexico; also to Mediterranean and se. Asia.

Occurrence in California. Common, Locally Very Common to Abundant, Nov.-Feb., and Uncommon to Fairly Common rest of year, in Central Valley ponds and lakes, marshes, and nearby wet grasslands and fields. Fewer at same seasons in c. and s. coastal lowlands and inner foothills. Nests or young in these areas Apr.-July. In Imperial and Colorado R. valleys, Fairly to Very Common Oct.-Mar., and Occ. through spring to early Aug. (has nested). In ne. Calif., usually Common Mar.-early Nov., with young out in July; in some years Locally Abundant there Sept.-Dec. Also noted Irreg. at mountain lakes, esp. in s. Calif., in fall, winter, and spring.

Pintail (*Anas acuta*) (figs. 18, 19)

Recognition. L. 20½-30 in. (52-76 cm.); WS. 25-32 in. (64-81 cm.); Wt. 1.1-3.2 lbs. (500-1450 g.). The Pintail has the longest neck of any dabbling duck (but see Fulvous Tree Duck account). Beginners should also beware of confusing them with geese. The slender neck and pointed tail are adequate marks for the expert; but the narrow white trailing border of the wing by the inconspicuous brown speculum is best for telling female or eclipse males in flight at a distance (fig. 19). The bill and feet of both sexes are light gray. The female is more finely mottled and somewhat grayer in tone than a Mallard (plate 5); the male in full breeding feathering is unmistakable, with white of underparts and foreneck extending as a crescent into the dark brown of the head. He is otherwise a light gray duck with a black stern.

Habits. A relatively few years ago, Pintails wintering in the Central Valley could literally be seen by the millions per day, and they still reach that level in fall concentrations in the Tule Lake-Lower Klamath Lake area. Not only are they often the most abundant wintering duck through most of California, they are also the earliest to arrive in numbers (see Graphic Calendars in Appendix). Along the coast and large bays, they and the American Wigeons are also the most numerous of dabblers on salt water. There they stay close to the tideflats or salt-marsh pools in which they feed, except when thoroughly disturbed. Then a surprised observer may find them rafting on deeper bays or beyond the breakers on the open ocean like so many Scoters.

In feeding, Pintails are typical dabblers (see Mallard account), but of course can reach bottom in deeper water than other species. Their *diet* is similar to the Mallard's (see account) on freshwater, but probably is largely of crustaceans and worms on the tideflats. *Voice*: of males in winter, a high, short, and mellow *quilp* oft-repeated; also other wheezy notes; females occasionally *quack* softly. *Nest*: down-lined, on dry ground in grass or marsh area. *Eggs*: 6-12, pale olive to buff; incubated 22-23 days by female. *Downy young*: white below (rather than yellow like most dabblers), brown-gray marked with white above.

Range. Breeds throughout Alaska, w. (a few to e.) Canada, and nw. U.S. south to ne. (a few to c. and s.) Calif., c. Nev., Colo., and Ill.; also in n. Eurasia. Winters from se. Alaska, w. Wash., e. Ore., s. Nev., Colo., Ohio, and s. New England and south to n. South America; in Old World south to Africa, se. Asia, and Philippines. Birds from w. U.S. migrate to Hawaii in numbers each year.

Occurrence in California. Abundant to Very Abundant Aug.-Mar. on marshes, lakes, and wet fields and grasslands of Central Valley; Very Common to Abundant at same seasons in Imperial Valley, along Colorado R., and on bayside and coastal lagoon marshes and tideflats; fewer on lowland lakes and marshes elsewhere throughout the state; and Irreg. at lakes in s. mountains. Arrivals begin in early July, well before other duck species, and numbers taper off through Mar.-Apr. as most depart. Rare to Uncommon, Occ. Fairly Common, Locally in coastal and Central Valley areas through summer, with nests or young Apr.-Aug. In ne. Calif., Abundant to Very Abundant Feb.-Apr. and Aug.-Oct., and through mild winters; Common through summer, nesting May-July. Also reported (usually in small numbers) from Sierra Nevada lakes May-Nov.

Green-winged Teal (*Anas crecca*) (fig. 20; plate 4)

Recognition. L. 12½-15½ in. (32-39 cm.); WS. 20-25 in. (51-64 cm.); Wt. 0.3-1.1 lbs. (135-500 g.). Smallest of the dabbling ducks, distinguished from other teals by lack of light blue areas on upper wing coverts (see plates 4 and 5). The speculum is shiny green close to body, black beyond this, bordered in front by buff. Together with small size, this identifies the mottled brown female or the eclipse male. Even in full breeding feathering (fall to early summer), the male often still appears dark, but has a transverse mark of white on side of chest and a buffy area under the tail. The brighter colors show up in good light, however. The Eurasian subspecies (very rare in California) lacks the white mark on the chest and has a longitudinal one above the wing instead.

FIG. 20

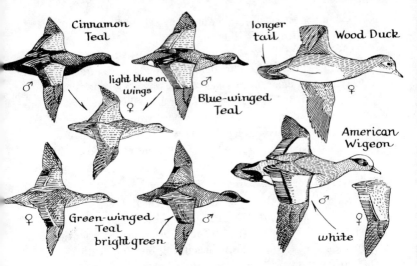

Cinnamon Teal

longer tail

Wood Duck

light blue on wings

Blue-winged Teal

Green-winged Teal bright green

American Wigeon

white

Habits. These teal often fly more erratically than the larger ducks, and because of their small size seem to be faster. In the breeding season, and to some extent in winter, teal are scattered out in marshes and on small ponds, requiring less open water than the larger ducks. The large wintering population is found mixed with the larger species in the usual waterfowl concentration areas, with highest numbers in fall in the Imperial Valley, and in March on the Modoc Co. refuges. The feeding habits of the Green-wings in these areas are similar to those of the other dabblers, although they tend to work the edges of marshy ponds more assiduously. Elsewhere they may feed a great deal on uplands near ponds or creeks, taking food up to acorns in size. *Voice*: of male, includes a high short *peep* and various lower trills; the female has a soft high-pitched *quack*. *Nest*: of grass with feather lining, in dense grass or marsh or amid willows near at least a little water. *Eggs*: 9-18, dull white to buff; incubated 21-23 days by female. *Downy young*: similar to Mallard (plate 5), but with two dusky streaks on the side of the head rather than one.

Range. Breeds from c. Alaska, n. Mackenzie, and se. Canada south regularly to Wash. and across n. U.S. to New England, Occ. to ne., c., and sw. Calif., e. Ariz., and N.M. Winters from

s. B.C. and most of n. U.S. (where any open water remains) south to s. Mexico and Caribbean.

Occurrence in California. Fairly Common Aug.-Sept. and Apr., and Common to Abundant Oct.-Mar., in Central and Imperial Valley marshes or lakes and nearby grasslands or wet fields; also in ne. Calif., except fewer in midwinter. Widely found in lower numbers on smaller ponds, ditches, and slow streams in lowlands throughout the state and at lakes of s. mountains. They nest sparsely May-July near the coast, in inland valleys, and in ne. Calif.

Blue-winged Teal (*Anas discors*) (fig. 20; plate 5)

Recognition. L. 14½-16½ in. (37-42 cm.); WS. 23-26 in. (58-66 cm.); Wt. 0.5-1.2 lbs. (227-540 g.). Females are indistinguishable in the field (and barely so in hand) from female Cinnamon Teal (see fig. 20). Males in full breeding feathering (fall to early summer) are easily told by the large crescent-shaped white mark between eye and bill (but see differently shaped mark in the female scaup, fig. 22, and male goldeneye, fig. 23). Otherwise the drake is dark, with the same pattern of light blue on the wing coverts as the Cinnamon Teal (fig. 20 and plate 4) and Shoveler (fig. 19 and plate 5), and a dark green speculum.

Habits. This teal is a regular nester east of the Sierra Nevada and a few occur all summer in the Central Valley, where it has been known to nest. In addition to the records there (see Graphic Calendar in Appendix), a study at Lake Earl, Del Norte Co., suggests that this species may have raised up to forty broods there in 1963-65, judging from total Blue-winged and/or Cinnamon Teal broods found and the number of males in postbreeding groups. Most of the Blue-winged Teal seen in California, however, are individuals or pairs on small marshy ponds, or picked out from amid the wintering ducks massed on the big valley refuges. In voice and other habits, including nesting, this species is very similar to its near relative, the Cinnamon Teal.

Range. Breeds from s. B.C., n. Sask., and se. Canada across n. U.S., and sparingly south to ne. and extreme ce. Calif. (Mono-Inyo cos.), a few Irreg. to nw. and c. Calif., s. N.M., and the Gulf of Mexico. Winters from c. (a few in nw.) Calif., c. Tex., and s.e. U.S. south to Brazil and Chile.

Occurrence in California. Rare to Uncommon, Sept.-Apr. on marshes, lakes, smaller ponds, and at times fields widely over lowlands, most frequent in the ne. Calif. area (except winter) and in Imperial and Colorado R. valleys, where Irreg. Fairly Common. Also Uncommon but Regular Mar.-Aug. on ne. plateau, and sometimes a few summer in Central Valley or on nw. coast; eggs or young have been found in June-July. Also reported from mountain lakes June-Nov.

Cinnamon Teal (*Anas cyanoptera*) (fig. 20; plate 4)

Recognition. L. 14½-17 in. (37-43 cm.); WS. 23-26 in. (58-66 cm.); Wt. 0.6-1.1 lbs. (270-500 g.). Males in breeding feathering, which they begin to assume earlier in fall than most ducks, are cinnamon red over the whole head and body and could be confused with only the male Ruddy Duck (plate 8), but are much slenderer and lack its white cheeks. The Cinnamon Teal's bill is dark gray, its feet yellowish in both sexes. The female and eclipse male have mottled brown feathering like the Blue-winged Teal (plate 5). At all seasons they show a large light blue area on the upper wing coverts, from the base to beyond the bend of wing (like the Blue-winged Teal and Shoveler; see fig. 19), and a dark green speculum. The wing linings are white, as in those other species also.

Habits. Formerly considered a summer resident only, this beautiful duck of the west is now found throughout the year in suitable habitat in most of California. Its breeding population continually suffers from drainage and filling of marshes, but it is still probably second only to the Mallard as a nesting duck here. Its preferred feeding habitat is along the edge of tall marsh growth bordering open water of a pond or slow creek, where it takes mostly seeds and new shoots of plants of many sorts. On migration, this predilection often results in their segregation from other ducks; but some Cinnamons are found amid the larger concentrations on bigger waters also. *Voice*: of male, a low chatter and occasionally a louder peeping whistle; of female, a soft, high-pitched *quack*. *Nest*: a variably skimpy or complete bowl of grasses or other plants, lined with feathers, usually placed amid dense cattails or sedges but sometimes in grass up to 100 feet from water. *Eggs*: 6-14, usually about 10, dull white or pinkish buff; incubation by female, probably for 21-23 days. *Downy young*: upperparts less extensively dark, and markings of lighter hue, than in the Mallard (plate 5). Male parent is said to accompany the female with her brood, at least at times, this being quite unusual among dabbling ducks.

Range. Breeds from se. B.C., e. Mont., and w. Nebr. south through states from Rocky Mts. westward (but not west of Cascades in Ore. and Wash.) and south to c. Mexico. Winters from c. (rarely n.) Calif., s. Ariz., N.M., and s. Tex. to n. South America. A separate population occurs in c. and s. South America.

Occurrence in California. Common to Locally Abundant in Oct.-Dec. and Feb.-Mar. migration periods, Fairly Common balance of year, on marshes, lakes, and ponds, and fewer on nearby grasslands or fields, of cismontane lowlands generally; Uncommon at s. mountain lakes (a few winter records there and in nw. Calif.). Nests widely in lowlands Apr.-July. Common to Abundant in similar habitat of se. Calif. Aug.-Mar., a few to early May; a few have nested. On ne. marshes, Fairly Common to Common Mar.-Aug. (and then Abundant at times in fall); eggs or young there in May-July. Also noted in summer at mountain lakes.

Northern Shoveler (*Anas clypeata*) (fig. 19; plate 5)

Recognition. L. 17-20½ in. (43-52 cm.); WS. 27-33 in. (68-84 cm.); Wt. 1-1.8 lbs. (450-820 g.). Essentially the same pattern of light blue upper wing coverts (usually showing only in flight) and dark green speculum as in the Cinnamon and Blue-winged teals (see plates 4 and 5). At any season the Shoveler can be told from these smaller species by its bright orange feet and enlarged gray bill (because of which hunters often call them "Spoonbills"). The female and eclipse male are otherwise similar to, but smaller than, a female Mallard (plate 5), while the male in breeding feathering (late fall to early summer) is conspicuously patterned white, reddish-brown, white, and black from fore to aft below, and has a whitish tail and green head, the latter darker than the Mallard's.

Habits. Because of their unique bill, Shovelers are well adapted for water feeding and may often be seen swimming steadily along with the beak held in the water in front as a combined scoop and strainer. They seldom come out of the water to feed, and do not join with other ducks on the long flights to feeding grounds in fields. They seem to have been derived in evolution from the teals, to which their color pattern and marsh-associated habits both relate them. They have simply become specialists at the art of plankton-straining, a little of which various other dabbling ducks also do. The migrant and winter populations thus build up to high levels on particular shallow lakes, salt evaporators, oxidation ponds of sewage plants, etc., where conditions are at least temporarily "just right" for the growth of the tiny suspended plants and animals they can filter out. Very few *nest* in California, making a skimpy lining of grass and down in a hollow in or near a marsh. *Eggs*: 6-14, pale buff to greenish gray; incubation by female for 21-23 days. *Voice*: of male, a low-pitched, soft *kukuk* or *konk*; of female, a feeble *quack*.

Range. Breeds from c. Alaska, n. Yukon, Sask., and Manitoba and Pa. south to Ore., ne. Calif., N.M., Iowa, and N.C., rarely to c. and sw. Calif.; also across much of temperate Eurasia. Winters from sw. B.C., Wash., Nev., and c. and se. U.S. south to n. South America; also to tropics in Old World, and Hawaii.

Occurrence in California. Fairly Common to Common on marshy lakes and ponds of Central Valley, coast slope valleys, and s. mtn. lakes (fewer in foothills), late Aug.-early May, and Rare through summer, Occ. nesting Mar.-July. Very Common to Abundant on some salt evaporating ponds, brackish pools, and parts of Salton Sea Sept.-early May, on ne. marshes, lakes, and pools Mar.-Apr. and Sept.-Nov.; also Common there through mild winters; Fairly Common through summer, and breeds. Recorded on Sierra Nevada lakes July-early Nov.

European Wigeon (*Anas penelope*) (fig. 21)

Recognition and Habits. L. 16½-21 in. (42-53 cm.); WS. ca. 32 in. (81 cm.); Wt. 1-2 lbs. (450-900 g.). Males of this primarily Old World species can be told from the abundant American Wigeon (fig. 20 and plate 5) by their mostly *gray body* and *reddish-brown head* with buffy crown. Even females show a browner head than females of the American, in which the head is grayer than the back. The upper wing pattern, brownish chest, and white belly are similar in both species, but the European has darker wing linings. Diligent search of flocks of wintering wigeons will often disclose one or a few Europeans among them. Most California records are from the large concentrations in the Central Valley refuges or from park lakes in or near cities — probably because of greater observer effort there. Individual birds often stay for long periods in such areas, where both aquatic forage and grass for grazing are available.

Range. Breeds all across n. Asia and Europe, including Iceland and Kamchatka. Winters through much of Europe and s. Asia to c. Africa and Philippines; also migrates regularly but in small numbers over most of North America, wintering usually with flocks of American Wigeons.

Occurrence in California. Regular but Uncommon on lakes, marshes, and nearby grasslands of cismontane lowlands Oct.-Apr.; Rare in se. Calif. Nov.-Mar. and in ne. Calif. Oct.-Mar.

American Wigeon (*Anas americana*) (figs. 20, 21; plate 5)

Recognition. L. 17-23 in. (43-58 cm.); WS. 30-36 in. (76-91 cm.); Wt. 0.8-2.5 lbs. (360-1130 g.). Wigeons in flight (fig. 20) can be identified far off by the white patch on the upper wing surface forward of the green speculum (white area smaller in adult females than in males, and faint in immature females); or, if seen from below, by the pointed tail, white belly abruptly separated from dark chest, and lack of square white area in speculum (see Gadwall account and fig. 18). The wing linings are partly white in this species, dark in the European Wigeon. On the water, the pinkish-brown body and gray head with

FIG. 21

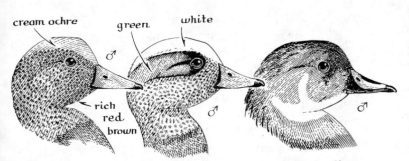

cream ochre   green   white

♂

rich red brown

♂

♂

European Wigeon   American Wigeon   Wood Duck - eclipse

green stripe distinguish the American from the European (fig. 21). Because of the white crown of the male, the American Wigeon has also been known as the Baldpate. The bill and feet are gray.

Habits. The American Wigeon is among the top three, and locally the most abundant, of dabbling ducks wintering in California. In the Central Valley refuges and shooting grounds, where their numbers seem to be holding up better than those of the Pintail, American Wigeons often feed in shallow water or wet cultivated fields like the Mallard; but they also do much more grazing than the other dabblers, even on lawns near city park lakes such as in San Francisco. They also join with Pintails on the tideflats, and are noted for pilfering food from the beaks of just-surfaced diving ducks. This adaptability to differing diets may be one of the significant factors in their sustained population levels. The "courting parties" (see Mallard account) are much in evidence in midwinter among American Wigeons, as many as eight to ten males following one female about for considerable periods. *Voice*: of male, high-pitched wheezy whistles, usually in threes, as *whew-whi-whi,* in chorus on water or in the air; of female, low-pitched, raspy quacks. *Nest*: a hollow on dry ground, often some distance from water, lined with grass or plant stems, plus down. *Eggs*: 6-12, usually about 10, creamy white; incubation 24-25 days, by female.

Range. Breeds from w. Alaska and far nw. Canada south through e. Wash. and Ore. to ne. Calif. and east to Manitoba,

Minn., and Nebr. Winters near the coast from s. Alaska and interiorly from n. Calif. and sw. Utah, Ill., and se. N.Y. south to Central America.

Occurrence in California. Common in Sept. and Apr. to Very Common to Very Abundant Oct.-Mar. in Central and Imperial Valleys, and somewhat fewer elsewhere throughout cismontane lowlands, on lakes, marshes, and nearby grasslands or fields. Fairly Common to Uncommon at same seasons on salt ponds, bays, and Occ. to ocean near shore. Occasional stragglers through summer. Recorded Sept.-May on s. mountain lakes, and Oct.-Nov. and early May on Sierran lakes. In ne. Calif., Very Common to Abundant late Aug.-Nov. and Mar.-early Apr. and through mild winters; Rare to Uncommon as breeder May-July.

Wood Duck (*Aix sponsa*) (figs. 20, 21; plate 8)

Recognition. L. 15½-21 in. (39-53 cm.); WS. 27-30 in. (68-76 cm.); Wt. 0.8-1.9 lbs. (360-860 g.). The male in brightly colored, crested, white-marked breeding feathering is unmistakable (plate 8). In eclipse males, the white of the throat with extensions onto the cheeks is still evident (fig. 21). Otherwise, females or eclipse males can be told by the brown-gray body, somewhat streaked whitish below, bluish speculum, longish *square-ended tail,* and large gray head, usually slightly crested. In flight the bill is usually held at a more downward angle than in other ducks (fig. 20). The female also has conspicuous white "spectacles."

Habits. This beautiful inhabitant of waters with wooded shores was formerly a widespread breeder in California, but is now found in summer only locally except in the inner Coast Ranges near and north of San Francisco Bay. Constant disturbance or outright destruction of riparian trees throughout cultivated and urban areas, as well as human and domestic animal predation on the young, seem to be the factors involved. When not breeding, Wood Ducks are more widespread in small groups on ponds and streams or in coves of larger lakes. The largest fall concentrations are found on lakes with oak woodland near the shores.

Food: nearly all vegetable matter, including duckweed, pondweeds, seeds of sedges and various plants in and out of water and, especially in fall, acorns. *Voice*: loud, strident *hooeek*, and various squeals and chuckling notes (male); alarmed females utter a sharp *cr-r-rek*. *Nest*: in natural cavities of trees or in boxes provided by man, less often in woodpecker holes (Flicker or Pileated), at anywhere from 2 to 50 feet above ground or water (and up to half a mile from water). *Eggs*: 10-15, white to buff; incubated for 28-31 days by female. *Downy young*: dusky above with only small light spots on rump, light yellow below and on side of head; they jump from the nest hole, as mother calls on ground or water below, within first day of life.

Range. Breeds from s. B.C. and nw. Mont. south through w. Ore. and Regular to c. (Occ. s.) cismontane Calif.; also over most of s. Canada and e. U.S. Winters from w. Wash. to s. Calif. and in se. U.S., Cuba, and Tex. south to c. Mexico.

Occurrence in California. Fairly Common to Locally Common Sept.-early Apr. on ponds and lakes with wooded borders in middle (between w. and e.) Coast Ranges south through Alameda and Santa Clara cos.; fewer at same seasons elsewhere on still and slow tree-bordered waters of n. and c. cismontane lowlands, and still fewer in s. Calif.; sometimes to lakes of urban parks. Rare to Locally Uncommon Apr.-Aug. as breeder in woodlands near ponds, lakes, or larger slow streams (recently reestablished in c. San Joaquin Valley). Recorded on Sierran lakes and streams (formerly nested in Yosemite Valley) and on ne. plateau and Mt. Shasta area at varying times throughout the year.

18. SUBFAMILY AYTHYINAE (DIVING DUCKS)

(Picture Keys A, B, and C)

Redhead (*Aythya americana*) (fig. 22; plate 6)

Recognition. L. 18-22 in. (46-56 cm.); WS. 29-35 in. (74-89 cm.); Wt. 1½-3 lbs. (0.7-1.4 kg.). A diving duck of about average proportions, with black-tipped blue-gray bill and an abrupt rounded forehead. Adult males have a reddish head, black chest and under-tail area, with light gray sides and back (compare with Canvasback, plate 6). Females are smooth brownish with vaguely demarcated whitish forepart of face. They are very similar to female Ring-necks, but usually more buffy-brown and with a less distinct eye-ring and paler bill. In flight, Redheads show a paler gray stripe along the secondaries and inner primaries of the wing.

Habits. More of an inland duck than most of the others in its subfamily, wintering Redheads are most often found on large reservoirs with shallow ends or coves. Here, or in the still shallower waters of ponds and lakes in the Central Valley refuges, they dive to grub in the bottom mud. They eat more of the leaves and stems of aquatic plants and less seeds and insects than other ducks. The whole species population of the Redhead has been seriously reduced in numbers because of drought and draining in its main breeding grounds, but it is still common east and north of the Sierra Nevada. A few breed occasionally south to the lower Colorado R. and (formerly at least) in undisturbed marshy ponds near San Francisco Bay and south to San Diego Co. *Voice*: of male, a deep, cat-like *meow*. *Nest*: of marsh plants, placed amid sedges or cattails close to or over shallow water. *Eggs*: 10-15, cream-colored to olive-buff; incubation 24-28 days, by female. *Downy young*: lighter brown above than dabbling ducks (plate 5), but with similar pattern and also basically yellow.

Range. Breeds from c. B.C., s. Mackenzie, and nw. Minn. south, mostly east of the Cascades and Sierra Nevada to s. Calif., Ariz., N.M. and east to s. Wisc. Winters from s. B.C., Nev., Utah, and across c. and s. U.S. south to c. Mexico and Caribbean.

Occurrence in California. Fairly to Locally Common Oct.-Mar. on lakes of sw. and se. Calif. (including Salton Sea) and inner c. Coast Ranges, with fewer in Central Valley floor and outer coast areas; Uncommon on bays. Irreg. Rare to Uncommon at marshy lakes of same areas through summer; eggs or young noted Apr.-Aug. On ne. plateau, Fairly Common to Common Mar. and Aug. (to Oct. ?); also Irreg. Uncommon to

FIG. 22

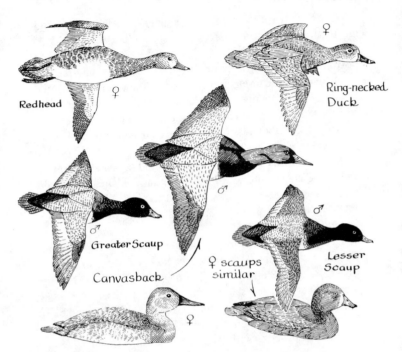

Redhead ♀

Ring-necked Duck ♀

Greater Scaup ♂

Canvasback

♀ scaups similar

Lesser Scaup ♂

♀

Common in winter and Fairly to Locally Common through summer, with eggs or young May-July. Recorded on Sierran lakes May-Aug. and Nov.

Ring-necked Duck (*Aythya collaris*) (fig. 22; plate 6)

Recognition. L. 15-18 in. (38-46 cm.); WS. 24-30 in. (61-76 cm.); Wt. 1.2-2.0 lbs. (540-910 g.). Similar to Lesser Scaup in size and shape, but crown decidedly puffier, even raised as a rounded low crest at times. The dark brown ring around the neck is rarely visible in the field, but at close range a broad white band is conspicuous on the blue-gray bill, just short of the black tip. Males are black-backed, with pale gray sides that become white in a crescent in front of each wing, conspicuous even far off. The dark brown female (fig. 22) is best told by the gradually paler facial area, white eye-ring, and bill color as in the male. In flight, Ring-necks show a light gray wing stripe like the Redhead.

Habits. This species is also preeminently a lake and pond inhabitant — as restricted to such habitats as the Redhead, if not more so. On good-sized lakes and reservoirs with mud bottoms, small companies of Ring-necks can be found regularly all winter, and even ponds of only a few acres are visited briefly. *Food*: mainly various aquatic and marsh plants, plus some insects, water snails, etc. *Voice*: similar to Scaups. *Nest*: of sedges or other plants and debris, in wet places close to open water. *Eggs*: 8-12, greenish-buff.

Range. Breeds from s. Mackenzie (and s. Alaska) across c. and se. Canada and south in mountains to Ore. (rarely ne. Calif.) and Ariz., and across nc. and ne. U.S. Winters from s. B.C., N.M., Ark., s. Ill., and Mass. south through Central America and Caribbean.

Occurrence in California. Fairly to Locally Very Common Oct.-early Apr., and Uncommon in May and Sept., on lakes of Coast Ranges, Sierra foothills, and s. mountain and desert areas; fewer at same seasons on outer coast itself and in sw. Calif., but Occ. to both c. and s. Calif. bays. On ne. and Sierra Nevada lakes, Common to Abundant Nov.-Dec., Rare Mar.-June and Sept.-Oct., with two known nestings in June.

Note: The **Tufted Duck** (*Aythya fuligula*), a northern Eurasian species similar to the Ring-necked but with a decided tuft of feathers on the mid- to hind-crown of the male, and a vestige of the same in the female, has been found at least six times on lakes in c. and nw. Calif. late Dec.-early Mar. (plus one male that stayed at Arcata Apr.-early. Nov. 1968. One was also present in e. Ventura Co. for three winters as of 1975. While some of these are possibly escaped captives, others are probably truly wild birds, as the species has been found in Alaska, B.C., and Ore., and in ne. U.S.

Canvasback (*Aythya valisineria*) (fig. 22; plate 6)

Recognition. L. 19-22 in. (48-56 cm.); WS. 29-36 in. (74-91 cm.); Wt. 1.9-3.6 lbs. (0.86-1.6 kg.). A large diving duck with distinctive head profile, sloping smoothly into long beak. The male is nearly white above, between two black "ends," and has a reddish head and neck, while the female shows less contrast, with light gray and brown in the same pattern. The very light tone of the black and wings is prominent even in flight, as are the long neck and bill.

Habits. The favorite winter feeding areas of "Cans" are the shallow waters over and near the intertidal mudflats of large bays. The scarcity of such areas elsewhere brings over half the Canvasbacks of the Pacific Flyway to winter on San Francisco Bay (and most of the others stop there on migration). On the Bay these flocks often spend a major part of their time sleeping, but feed actively in the very shallow water as it first floods the tideflats. What *food* they take there is little known, but it must be mostly invertebrate animals such as small clams, crustaceans, and worms. In fresh water, however, this species is primarily vegetarian, and indeed is named from the aquatic "wild celery" (*Valisneria*) it prefers in the northeast. Maintenance of the populations of this favorite western game bird through the winter, however, depends in large part on the continued availability of extensive shallow bay foraging areas. The new San Francisco Bay National Wildlife Refuge is a giant step toward this goal. Hundreds of Canvasbacks also come to brackish Lake Merritt in downtown Oakland, and large numbers also are found, but more irregularly, on certain inland lakes of central and southern California, particularly Big Bear and Baldwin lakes, even when these are partly frozen.

When resting, Canvasbacks gather in close rafts on deep water. In flight, they are among the fastest of ducks (50 mph or more). They bunch irregularly or, if going far, form into irregular V's and rise to considerable heights. Before they leave in spring, courtship displays can often be seen among the flocks, a male stretching his neck far forward close to the water and now and then tossing his head up and over the back while uttering a weak *ik-ik-coo* call. *Nest:* a large mound of sedges, etc., just above water level in a marsh near open water. *Eggs:* 7-9 or more, grayish-green, moderately dark; incubated 24-28 days by female. *Downy young:* yellow and dusky in a pattern similar to Mallard (plate 5), but side of head all yellowish and beak profile nearly as in adult; feet large and gray.

Range. Breeds from c. Alaska, n. Mackenzie, and se. Manitoba south, east of the Cascades, to ne. Calif., n. Nev., Utah, Nebr., and Minn. Winters from s. B.C. and nw. Mont. across c. and ne. U.S. and south to n. Baja Calif. and c. Mexico, but predominantly in Calif. and middle Atlantic states.

Occurrence in California. Abundant late Nov.-Mar., Uncommon to Locally Common Sept.-Oct. and Apr.-May, on bays of c. and n. Calif.; somewhat fewer at similar seasons on Salton Sea and Central Valley and s. mountain lakes; fewer yet elsewhere in lowlands on lakes and marshes, and s. Calif. bays and coastal lagoons; stragglers frequent most of summer, mostly along coast. In ne. Calif., Abundant Mar.-early Apr. and Oct.-Nov., even Dec.-Jan. in mild winters, and a few remain to nest (young reported June-Aug.). Also noted on Sierran and Cascade lakes in May-June and Oct.-Nov.

Greater Scaup (*Aythya marila*) (fig. 22)

Recognition. L. 15½-20 in. (39-51 cm.); WS. 28-35 in. (71-89 cm.); Wt. 1.3-3.0 lbs. (0.6-1.4 kg.). Much like the Lesser Scaup (see account and fig. 22 and plate 6), but with the white wing stripe extending across base of most of the primaries as well as the secondaries. This is about the only way a female Greater Scaup can be told from a female Lesser in the field, as the slight difference in size is scarcely apparent even if they are together (fig. 22). Adult male Greater Scaups in good light show green gloss all over the head, not just in the ear region, and the head is also less puffy in appearance than the Lesser's. The sides and back of the Greater are also whiter.

Habits. Both species of scaups are common on San Francisco Bay and northward in winter, but because of the difficulty in telling them apart under most field conditions in the mixed flocks, their relative status on the open bays is not well known. The high winter numbers shown on the Graphic Calendar (in Appendix) are therefore merely best estimates. The Greater is, in general, a more northern bird, less prone to visiting shallow coves and tide channels and decidedly less common on fresh water in winter than the Lesser. Otherwise its habits are similar to those of its slightly smaller relative (see account).

Range. Breeds in arctic and subarctic areas around the world, in America south to nw. B.C., N.D., Mich., and Que. Winters along the Pacific coast from Aleutians to s. Calif., and rarely inland to nw. Mexico; also from Que. to Cuba, and in Old World south to s. Europe, n. India, and Japan.

Occurrence in California. Fairly to Locally Very Common on larger bays of c. and n. Calif. Nov.-Mar. (a few in Oct., Apr.-May, and stragglers to July); decidedly fewer on inner lagoons, salt ponds, and the ocean itself, as well as bays to s. Calif. Noted Occ. Oct.-May on inland lakes east through Central Valley and in ne. and se. Calif.

Lesser Scaup *(Aythya affinis)* (fig. 22; plate 6)

Recognition. L. 14½-18 in. (37-46 cm.); WS. 24-33 in. (61-84 cm.); Wt. 1.0-2.4 lbs. (450-1090 g.). Male scaups on the water can be told from far off by their pattern of black on the rear and front ends (including the head) with white between, including the back, which at closer range is seen to be finely barred with dark. Either sex of this species can be distinguished from the slightly larger Greater Scaup by the shorter white part of the wing stripe, which becomes light brown at the innermost primary. Male Lessers show purplish reflections over most of the head (though often a little green in the ear region also), and the head is higher crowned, and the sides of the body somewhat more barred, than the Greater's. Female scaups are dark brown ducks with an abruptly demarcated broad white patch above the base of the bill.

Habits. The "Little Bluebill," as this species is usually known to hunters, is one of the most widespread of the diving ducks in North America. They visit small ponds and larger streams in winter, and lakes of some size have a more or less continual population then. At urban park lakes where the ducks are fed regularly (e.g., Oakland, San Francisco, Los Angeles), Lesser Scaups often become so tame they will feed almost from the hand. But the flocks of thousands are to be found on open waters of shallow bays, feeding with Canvasbacks and Ruddy Ducks in water over the tideflats, and also with scoters where the water is up to twenty feet or so in depth. All these diving ducks go to the bottom for their *food*, which in the Lesser Scaup includes many small mollusks and crustaceans, and probably worms sieved from the mud. When in fresh water they also take seeds and leaves and soft stems of a variety of water plants. *Voice*: low *purrr* notes, often given in flight; when startled, a loud, rough *scaup*; others during courtship, beginnings of which are seen before they depart in spring.

Range. Breeds from c. Alaska, n. Mackenzie, and n. Manitoba south to c. B.C., ne. Colo., and ne. Iowa. Winters from s. B.C. and ne. Colo. and across most of U.S. south to n. South America.

Occurrence in California. Very Common to Very Abundant, Oct.-Apr., Fairly Common Sept. and May, with stragglers Rare to Uncommon all summer, on bays along entire coast and on the Salton Sea; fewer at same seasons on coastal and some inland lakes throughout lowlands; Occ. on rivers. On ne. lakes, Common to Abundant in at least Nov.-Dec. and Mar., sometimes midwinter also; a few persist into July and have been known to nest (perhaps cripples). Recorded on Tahoe and other Sierran lakes May-early June and Oct.-Nov.

Common Goldeneye (*Bucephala clangula*) (fig. 23)

Recognition. L. 16-20 in. (41-51 cm.); WS. 25-32 in. (63-81 cm.); Wt. 1.2-3.2 lbs. (0.5-1.5 kg.). At any season goldeneyes can be recognized by their shape — chunky body, short thick neck, and large puffy head with abrupt forehead above the tapered bill. Only the smaller Bufflehead (fig. 23 and plate 6) has similar contours. Adult males at a distance show conspicuous white sides, chests, and necks (as does the differently shaped Common Merganser, fig. 25 and plate 7), the Common Goldeneye having also a green-glossed head with a circular white spot between eye and bill. Females and first-winter males are gray with a white belly, white ring all around the neck below the rich brown head, and large white speculum plus adjacent white on middle secondary coverts. This extensive white basally in the wing is even larger in the male, and distinguishes goldeneyes in flight from other ducks except mergansers (fig. 25).

FIG. 23

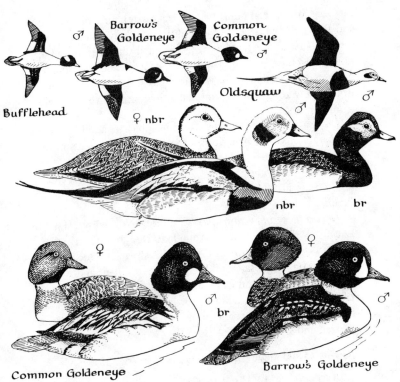

Common Goldeneye

Barrow's Goldeneye

Habits. Often called "whistlers" by hunters, from the distinctive shrill sound made by their wings, goldeneyes are rapid and agile flyers for a diving duck. In a good wind they can, if necessary, rise from the water with very little takeoff run; and they even *nest* in cavities in trees, often high above the ground, so they must alight on limbs. In California they are, like the Lesser Scaup, widespread on both salt and fresh water in winter, but seem to shift from one feeding area to another quite commonly. Goldeneyes are less prone than scaups and scoters to form compact rafts on open water, but they are found scattered in and near the large concentrations of these other divers, as well as by themselves, even far up tidal channels. Their *food* when on salt water has not been studied, but even on fresh water is largely of animal matter obtained from the bottom or edges of submerged plants. From Feb. onward they are often seen courting, the male stretching his neck up and forward, the head feathers puffed out, or with head held forward just above the water. Occasionally he does a head-up, tail-up display, uttering a quick *zee-at* as his head is snapped back over his shoulders, beak pointed skyward. *Voice*: at other times a low rolling note from males, low-pitched rough *quacks* from females.

Range. Breeds around the world in subarctic and high north temperate areas, in America south to s. B.C. and across n. U.S. to New England and Nova Scotia. Winters, at north, from wherever open water is available, south to n. Baja Calif. and se. U.S. coast; in Old World to Mediterranean, n. India, Burma, Japan.

Occurrence in California. Common Nov.-early Mar. to Locally Very Common or Abundant Dec.-Jan., Fairly Common to Uncommon through Apr., with stragglers Occ. through summer, on bays, salt ponds, and lagoons of n. and c. Calif. and on Salton Sea and Colorado R. and nearby lakes; fewer on lakes elsewhere throughout lowlands; Occ. on rivers and marshy ponds. Fairly Common to Rare in ne. Calif. and on Sierran and s. mountain lakes, mostly Nov.-Apr.

Barrow's Goldeneye (*Bucephala islandica*) (fig. 23)

Recognition. L. 16-20 in. (41-51 cm.); WS. ca. 29 in. (74 cm.); Wt. ca. 1.1-2.9 lbs. (0.5-1.3 kg.). This species, scarce in California, can be easily distinguished from the Common Goldeneye at close range if in adult male feathering, which includes a purple-glossed head with a *crescent-shaped* white mark between eye and bill. Females and first-winter males are practically identical to the Common Goldeneye, but at very close range the more abrupt forehead (when feathers are relaxed), slightly shorter beak, and larger nail at its tip are clues to the Barrow's. It also has less white in the middle upper wing coverts than the Common — in all comparable plumages.

Habits. Most California records of Barrow's Goldeneye are from lagoons near or coves of San Francisco Bay and the coast of Marin and Sonoma cos. Their behavior here seems quite similar to that of the Common Goldeneye, with which they usually associate. As a breeding bird in the state, however, they occur (or did prior to 1940 at least) only at tree-bordered lakes and large streams of the northern to central Sierra Nevada and the Lassen Peak area. The *nest* is placed in a cavity in trees. *Eggs*: 6-15, pale green; incubation about 20 days, by female. *Downy young*: dark dusky and white in Mallard pattern (plate 5), except whole top of head to below eyes black; no yellow except dull yellowish green feet.

Range. Breeds from s. Alaska and n. Mackenzie south to sw. Ore. and mountains of ne. and ce. Calif. (formerly at least), and in Rocky Mts. to Colo.; also in e. Canada to Greenland. Winters along or near coasts from s. Alaska to c. (rarely s.) Calif. and from e. Quebec to N.Y.

Occurrence in California. Uncommon, or Occ. Locally Fairly Common, late Oct.-Mar. on lagoons, brackish lakes, and bays of c. and n. Calif.; usually fewer and Irreg. on lakes and rivers inland. Casual in s. Calif. Also Rare (formerly more Regular) late Jan.-Aug. and Nov.-Dec. on mountain lakes and larger streams of Mt. Lassen to Yosemite area, where small young have been found June-Aug.

Bufflehead (*Bucephala albeola*) (fig. 23; plate 6)

Recognition. L. 13-15½ in. (33-39 cm.); WS. 21-24 in. (53-61 cm.); Wt. 0.6-1.4 lbs. (270-635 g.). A duck of goldeneye shape (see Common Goldeneye account and fig. 23), but decidedly smaller, about the size of a Green-winged Teal, and thus one of the smallest ducks. Adult male Buffleheads are easily distinguished by the white sides and chest, plus a large white patch across the hind crown. Females and first-year males are all dark gray above, with a distinctive smaller white mark just below and behind the eye. Both sexes have a white belly and small white speculum, the adult male showing white additionally up to the bend of the wing and along the junction of wing and body.

Habits. In its feeding, the Bufflehead is much like its larger relative, the Common Goldeneye, with which it is often associated on San Francisco Bay and tributary sloughs. The Bufflehead reaches peak numbers in that area, however, on certain large salt evaporator ponds with concentrated brine shrimp (and water-boatmen and brine-fly larvae), upon which they apparently feed avidly. On the open bay they tend more to scatter like goldeneyes. Although all diving ducks fly rapidly, Buffleheads appear to be going faster than any others because of their small size. The weak *voice,* rarely heard in winter, consists of gutteral and squeaky notes from the male and hoarse, low *quacks* from the female. A few *nest* at northeastern California lakes, using cavities as much as 50 feet up in trees (as the goldeneyes do). Being small, Buffleheads can utilize the holes made by the Flicker, a common woodpecker. *Eggs:* 6-14, yellowish to pale buff, incubation period 28-33 days. *Downy young:* very like Barrow's Goldeneye (see account), but feet plain gray.

Range. Breeds from s. Alaska, n. Mackenzie, and nw. Ont. south to s. B.C., n. Mont. and c. Manitoba, and sparingly in mountains of Ore. and ne. Calif. Winters from Aleutians, B.C., Mont., the Great Lakes, and Maine south to Baja Calif., c. Mexico, and se. U.S. coast.

Occurrence in California. Common, Locally Very Common, late Oct.-Apr., on salt evaporating ponds, lagoons, and bays throughout the state, though fewer in s. Calif. and on Salton Sea and in Colorado R. valley; Uncommon to Locally Fairly Common on lakes (Occ. rivers) in cismontane lowlands at same seasons. Stragglers Rare, chiefly along coast Occ. all summer. On lakes of ne. plateau, Abundant Mar.-early Apr. and Nov., Irreg. Fairly Common to Common through winter, when open water is present, to Apr.; stragglers Occ. through June. Also recorded from Sierran and s. mountain lakes Nov.-Dec. and Mar.-July, a few with nests or young May-June in Lassen Co.

Oldsquaw (*Clangula hyemalis*) (fig. 23)

Recognition. L. 15-23 in. (38-58 cm.); WS. 26-31 in. (56-79 cm.); Wt. 1.6-2.3 lbs. (725-1040 g.). A northern diving duck that has distinct breeding (summer) and nonbreeding (winter) feathering in both sexes. Males have greatly elongated central tail feathers and are largely white in winter, with dark brown chest and back and a brown patch on the side of the head. Females lack the long tail feathers, and the brown and white areas are less distinct. Both sexes are dark-necked in summer and have unpatterned dark wings at all seasons.

Habits. Oldsquaws favor rough water by the rocky coasts, or fairly deep but calmer bays or coves. They can dive to 200 feet to reach food when necessary. Most of those seen in California are single individuals, but small groups have been found in Tomales and San Francisco bays and near Pt. Reyes and Monterey. The species is notably noisy in flocks.

Range. Breeds on arctic coasts around the world, in America south to Aleutians, sw. Hudson Bay, and Labrador. Winters on both coasts from limit of ice south to Wash. and S.C., on the Great Lakes, and a few across n. and c. U.S. to s. Calif. and Fla.; also in Old World to c. Europe, sw. and c. Asia, and Japan.

Occurrence in California. Rare to Uncommon Oct.-Mar. (Occ. to May or even July) on bays, large lagoons, and ocean, usually near rocky shores, and on coastal and se. lakes and Salton Sea. Recorded on lakes of Central Valley and nearby foothills Nov.-Feb.; on ne. plateau May and Oct.-Dec.; on large rivers of nw. Calif.Nov.-Dec.; and at L. Tahoe mid-May.

Harlequin Duck (*Histrionicus histrionicus*) (plate 11)

Recognition. L. 14½-21 in. (37-53 cm.); WS. 23-28 in. (58-71 cm.); Wt. 1.0-1.6 lbs. (450-720 g.). The adult male in breeding feathering (winter-spring) is unmistakable — a bluish-slate duck with orange-brown sides and clownish arrangement of white markings. The female is all dark brown except for paler belly and *three* whitish patches on the side of the head, differing thereby from the somewhat similar but larger female scoters (fig. 24).

Habits. A bird preferring turbulent waters, the Harlequin Duck is found foraging in small companies amid the surf or in choppy areas in coves of our rockiest headlands. Tomales Pt. in Pt. Reyes National Seashore has the most regular population, and relatively few ever straggle far into even adjacent bays. Their *food* is mostly taken from the edges of underwater rocks or pilings, and includes a great variety of marine crustaceans, snails, limpets, and chitons. Where undisturbed, they rest on rocks above the splashing waves. A few Harlequins nested in former years along larger streams in the Sierra Nevada, from the Stanislaus to the San Joaquin headwaters, and may still do so in a few places. Their typical nesting habitat farther north is along similar streams, the young being led onto rough water even while still small and unable to fly. *Nest*: sparsely arranged cup of grass and down in hollow of tree, cliff, or on ground. *Eggs*: 5-10, buffy white; incubated about 25 days by female. *Downy young*: dusky and white like Barrow's Goldeneye (see account), but black on top of head not onto cheeks; bill yellow except basally.

Range. Breeds from e. Siberia, w. Alaska, Baffin I., Greenland, and Iceland south to s. B.C., ec. Calif. (formerly, at least), in the Rockies to Colo., and to Quebec and Labrador. Winters from Aleutians south to s. Japan, s. Calif., and from Labrador to Mass.

Occurrence in California. Chiefly Rare to Uncommon, Oct.-early Apr. on ocean on and near rocky coasts, with stragglers all summer; Rare at same season on c. and n. bays and anywhere on coast south of San Luis Obispo Co. Casual as transients on lakes or rivers of inland c. Calif., and formerly nested May-Aug. on larger rivers of c. Sierra Nevada.

King Eider (*Somateria spectabilis*) (fig. 24)

Recognition. L. 18½-25 in. (46-53 cm.); WS. 34-40 in. (86-102 cm.); Wt. 2.8-4.4 lbs. (1.3-2.0 kg.). A large, heavy-bodied sea duck with distinctive feathering in the adult male: white upper back and neck, nearly white chest and most of head, the body and wings otherwise black with white patch on flanks and on inner wing coverts. There is a large roundish orange-yellow shield extending onto the forehead from a same-colored bill. The female is rich buffy brown, with small dusky bars on the body feathers, fine streaks on the head and neck, and a tiny white line along the edge of the speculum. She has only a small round-ended frontal shield extending about halfway to the eye from the nostril, and shows a slight forehead above it. First-winter males are intermediate between these patterns, showing a whitish chest by early spring.

Habits. All the California records through 1975 are of

FIG. 24

♂ br ♀ King Eider

♂ 1st yr

♀ Surf Scoter ♀

White-winged ♂ Scoter

Black Scoter

females or immature males near harbors such as Bodega and Monterey and on rough-water areas along the same stretch of mostly rocky outer coast. Several individuals have remained in one area for months. Eiders are deep divers (to 180 feet or more) with a feeding preference for mollusks, crustaceans, and even sea urchins, which they take from underwater rocks or pilings.

Range and California Occurrence. Breeds nearly around world in Arctic. Winters from limit of sea ice south to Europe, Kuriles, s. Alaska, and ne. U.S. coast; Rare and Irreg. south on coasts to N.J. and c. (once s.) Calif., where recorded 1-2 at a time Dec.-Apr., (mostly since 1955), with stragglers Occ. persisting through summer.

White-winged Scoter (*Melanitta deglandi*) (fig. 24)

Recognition. L. 19-24 in. (48-61 cm.); WS. 34-40 in. (86-102 cm.); Wt. 2.1-3.9 lbs. (0.95-1.8 kg.). Scoters in general are large, heavy-bodied diving ducks with large bills. Adult male feathering is largely or all black; females and first-year males are mostly dark brown. Scoters are distinguished from cormorants (fig. 8 and plate 2) readily by shape and behavioral characteristics (see Picture Keys A, B, and C). The large white speculum identifies the White-winged from the other scoters at any season, but it is often hidden when the bird is on the water. Adult males show a small white patch around the eye and curved up behind it, also a blackish knob over the base of the yellowish bill. Immature birds and some adult females have two whitish patches on each side of the head, variably distinct or obscure, as in the Surf Scoter (fig. 24). First-year birds are lighter brown, sometimes even whitish on the belly also. The feather line at the base of the bill is convex toward the bill in all plumages.

Habits. In long, irregular flocks or lines, the White-winged can be seen migrating along the outer coast with the more abundant Surf Scoter. The birds usually fly fairly close to the water, many flocks curving into the larger bays while others seem to go from headland to headland. Most fly close to the water, but on occasion some groups are up to several hundred feet above it. When flying overland to or from their northern breeding areas (or from the Gulf of California?), they rise much higher, but small numbers nevertheless stop at various waters enroute – or are at times grounded by weather. When exceptionally calm conditions prevail, the whistling noise made by the wings of scoters is audible far across the water. In taking off, they require a long run over the surface under such conditions and always seem to get under way with difficulty, but their flight is rapid and direct when finally up to speed. In its feeding, the White-winged is similar to the Surf Scoter (see account). Occasional short, low *croaks* are usually the only vocalizations heard from them in winter.

Range. Breeds from nw. Alaska, far nw. Canada, Manitoba and n. Ont. south to ne. Wash. and N.D.; nonbreeding birds, mostly subadults, are widespread well south in summer. Winters along seacoasts from e. Aleutians to Baja Calif. and from Quebec to S.C., also on the Great Lakes and casually elsewhere inland.

Occurrence in California. Common to Locally Very Common Oct.-mid-Apr., Fairly Common Sept. and late Apr.-May, on ocean and large bays, fewer into river mouths and around offshore islands, with nonbreeders Locally but Regularly Uncommon to Fairly Common through summer, chiefly at the mouths of rivers or tidal sloughs. Transients also noted on ne., c., and sw. Calif. lakes Oct.-Jan., and on Salton Sea and se. lakes Oct.-Feb., plus stragglers in May-Aug.

Surf Scoter (*Melanitta perspicillata*) (fig. 24; plate 6)

Recognition. L. 17-21 in. (43-53 cm.); WS. 30-34 in. (76-86 cm.); Wt. 1½-2½ lbs. (0.7-1.1 kg.). Adult male Surf Scoters are very distinctively marked with two white head patches, but first-year males are sometimes solidly black toward spring and thus confused with the Black Scoter by beginners. Females and the more immature males are identical with the White-winged (see account) except for the lack of white in the wings, a smoother "Canvasback-like" head profile (see fig. 22), and less extensive feathering over the base of the bill. In the Surf Scoter the bill extends quite close to the eye, so the feather line is concave toward the bill (closely around a black circle on the bill in the adult male).

Habits. Scoters are powerful divers, going to the bottom in up to 40 feet or more of water to obtain their favorite *foods*: mussels, small or soft-shelled clams, snails, limpets, crabs, and other crustacea, and a few fish and minor amounts of plant parts. When diving deeply, the birds sometimes use the primary portion of the wing to aid their progress, keeping the rest closed; but in shallower water only the feet are used. The Surf Scoter is so named from its ability to dive under the large breaking waves along the outer beaches, or to ride over them, as momentary conditions warrant. They feed more commonly, however, just beyond these breakers and in the gentler waters of large bays where the other scoters, scaups, and other diving ducks are also found. When not feeding, scoters usually assemble in loose to compact "rafts" on deeper water, many of the birds in such a flock sleeping while maintaining approximate location by swimming in small circles. Only in areas free from disturbance by man or large land animals do these and other diving ducks come ashore. *Voice*: usually none in winter, sometimes low croaking notes. *Breeding* in this and other large diving ducks does not take place (in many individuals anyway) until the second year following hatching, thus accounting for the relatively large number of summer "stragglers" regularly found at such places as Seal Beach, Moss Landing, and Morro, Bolinas, and Bodega bays, and at coastal river mouths.

Range. Breeds from w. Alaska and nw. Canada coast south to n. B.C., s. Northwest Territories, and Locally east to Labrador; nonbreeders widespread in summer across Canada and along n. U.S. coasts. Winters on seacoasts from Aleutians to Gulf of California, Mexico, and from Nova Scotia to Fla., and Occ. inland.

Occurrence in California. Abundant on ocean near shore and on large bays, Oct.-Apr., Common by late Sept. and to mid-May; fewer on smaller estuaries, with nonbreeders Uncommon to Locally Common all summer at river mouths and on outer coast. Massed flocks in migration along coast in Oct.-Dec. and Mar.-May, at which times also noted Occ. on inland lakes. Regular transient at Salton Sea (Irreg. to se. lakes and Colorado R.) Oct.-Nov. and late Mar.-May, but recorded there also through summer and Occ. in winter.

Black Scoter (*Melanitta nigra*) (fig. 24)

Recognition and Habits. L. 17-21 in. (43-53 cm.); WS. 30-34 in. (76-86 cm.); Wt. 1.9-2.8 lbs. (0.9-1.3 kg.). Formerly called the "Common Scoter" (appropriate in more northern areas), this is decidedly the least common of the scoters in California, although in a few recent years it has nearly equaled the White-winged at some locations. Black Scoters are found mostly in company with the other species on larger bays and near outer coast headlands. Adult males are solid black except for a large bright yellow knob over the base of the bill (but see note on immature Surf Scoter in that account). Females have a dark crown, but the whole side of the head and upper neck is pale brown, not broken into separate patches as in the other scoters. There is no white in the wing.

Range. Breeds in arctic and subarctic regions around the world, in America south to s. Alaska, with nonbreeders to James Bay and Newfoundland. Winters along coasts from Aleutians to c. (Occ. s.) Calif. and from Newfoundland to S.C., on the Great Lakes, and elsewhere inland; also in Europe and sw. and e. Asia.

Occurrence in California. Uncommon to Occ. Common Oct.-Apr. on bays and ocean near shore; stragglers Occ. through summer; recorded on inland lakes in Nov. and May.

19. SUBFAMILY OXYURINAE (STIFF-TAILED DUCKS)

(Picture Key C)

Ruddy Duck (*Oxyura jamaicensis*) (fig. 25; plate 8)

Recognition. L. 14½-16 in. (37-41 cm.); WS. 21-24 in. (53-61 cm.); Wt. 0.6-1.4 lbs. (270-635 g.). The only member of its subfamily in most of the United States, the Ruddy is easily told by the rounded tail when it is held stiffly erect, as it commonly is, especially by males. Otherwise the small size, chunky build, and dark feathering with white throat and cheeks identify it. The female has a dark line from the bill across the cheek. Males show the bright reddish-brown plumage and bright blue bill from sometime in Mar. or Apr. to Aug. or early Sept. They are dull brown like the females at other seasons, except for the pure white cheek.

Habits. The wings of a Ruddy Duck are relatively small, and it requires a long run over the surface to become airborne, flying with furiously fast wingbeats. It is not surprising that they seldom fly from an approaching person or boat unless there is a good wind. Ruddies are superb divers, however, often feeding with Canvasbacks and scaups on the mudflats covered by tide waters, and in deeper waters also. Flocks of several thousand winter regularly where such feeding grounds are extensive and protection from strong onshore winds is available. The *foods* taken by Ruddies on salt water are little known, presumably mostly annelid worms, crustaceans, and small mollusks. On fresh water some three-fourths of their diet is of pond or marsh plants.

The elaborate *courtship* display of the bright-plumaged male is seen only when he is settled on a pond bordered by tall tules or cattails. Swimming toward or around a female (or sometimes merely as a seeming declaration of territory), he raises his spread tail so sharply it is over the back, and stretches his head up and then back so that the turquoise bill is directly above his chest (see plate 8). Then with a quick series of down-and-up jerks of the head he slaps his bill against his chest or into the water, uttering a muffled *chut-chut-chut-chut-chut-che-ch'b'bzt.* Another display includes a forward rush while standing half erect, somewhat in the fashion of a grebe. *Nest*: a basket woven from and attached to tall marsh plants, just above water. *Eggs*: 6-10 (Occ. up to 20!), dull white or cream-colored, and exceptionally large (one-fifth or more the weight of the mother). *Downy young*: gray and white, with a dark stripe across the cheek; they dive well immediately after hatching. Unlike the males of most ducks, the drake Ruddy remains with the female and assists in raising the brood.

Range. Breeds from B.C. and c. (rarely e.) Canada south through w. U.S. to sc. Mexico, c. Tex., Iowa, Ill., and w. Pa. Winters from s. B.C. south along w. coast and across s. and e. U.S. to Central America and the Bahamas. Also resident in Guatemala and in Andes of South America.

Occurrence in California. Common to Very Abundant late Sept.-Apr. on bays and some salt ponds (few Occ. to ocean), and a few on such salt water through May-Sept. On cismontane lakes and ponds Fairly Common to Common all year, breeding Apr.-Oct. (mostly May-Aug.) in tall marshy borders. Also Abundant or Very Abundant Sept.-Feb., and Common to May, on Salton Sea; Fairly to Very Common then on se. lakes and ponds, but few remain to nest (young in Apr.-May). In ne. Calif., Common to Locally Very Common June-Aug. (nesting);

Abundant to Very Abundant late Feb.-Mar. and Oct., and Irreg. Fairly Common to Abundant through winter.

20. SUBFAMILY MERGINAE (MERGANSERS)

(Picture Keys A and C)

Hooded Merganser (*Lophodytes cucullatus*) (fig. 25)

Recognition. L. 16-19 in. (41-48 cm.); WS. 22-27 in. (56-68 cm.), Wt. 1-1.9 lbs. (450-860 g.). A small merganser with fanlike crest on the head that may be spread or depressed so that it is apparent only on the nape. The male is black above except for a white patch in the thin crest (differently arranged than in the Bufflehead, fig. 23 and plate 6) and partly white speculum and upper wing coverts, orange-brown on the sides and white elsewhere below the two black marks extending from the shoulders toward the chest. Compared to the females of other mergansers (see plate 7), the Hooded female is grayer on the foreneck and bill and has a smoother, duller (light reddish-brown) crest.

Habits. The Hooded Merganser is very partial to ponds and small reservoirs, rarely going to salt water. Large flocks are not encountered in California, but small groups come to the same location year after year, even though frequently frightened away from the pond by too much human activity. *Food*: small fish, crayfish, tadpoles and frogs, and aquatic insects, pursued and captured while the bird swims beneath the surface. *Nest*: usually in a cavity of a tree near water. *Eggs*: 6-18, usually 10-12, white; incubation by female, estimated at 31 days.

Range. Breeds from Alaska to sw. Ore. and across s. Canada and n. and c. U.S. to Tenn. Winters from s. B.C., Utah, Nebr., and e. U.S. south to Mexico.

Occurrence in California. Fairly Common Nov.-early Jan., Uncommon to Apr., Rare and Irreg. other months, on ponds, some lakes, and rivers in interior foothills and the San Francisco Bay area; somewhat fewer elsewhere in n. and c. Calif. lowlands then, and Rare in s. Calif. and on bays anywhere at same seasons. Recorded on Sierran and s. mountain lakes or ponds chiefly Nov.-Mar.

20. MERGANSERS

Common Merganser (*Mergus merganser*) (fig. 25; plate 7)

Recognition. L. 21½-27 in. (54-68 cm.); WS. 33-38 in. (84-96 cm.), Wt. 2.0-4.5 lbs. (0.9-2.0 kg.). The largest merganser. Males are easily identified by their pure white underparts, sides, chests, and lower neck, only the much shorter-bodied goldeneyes (fig. 23) showing as much white from a distance. The male also lacks a crest, but the female has a small one that may be two-parted. She is best told from the female Red-breasted by the sharply demarcated edge of the reddish-brown color on the neck and around the white chin. The female Common is also usually a purer gray tone above, and richer red-brown on the head. In their first fall, immatures are like the female, but may lack the reddish-brown of the fore-neck and thus resemble the Red-breasted (fig. 25 and plate 7) even more.

Habits. Predominantly a freshwater species at all seasons, the Common Merganser forages in any waters deep enough for it to pursue its fish prey beneath the surface. Because they sometimes catch even fast-swimming fish such as trout if these are numerous, many fishermen feel that these birds should be destroyed. In many streams and lakes it has been shown that the fish available for the sportsman's harvest are larger if there is a regular cropping of the too numerous smaller sizes. Under such circumstances, the predation by mergansers and other eaters of small fish is actually beneficial ultimately to a quality fishing experience by man. Most fish eaten by this species are under eight inches in length. In hatchery or rearing ponds it is necessary, of course, to keep such predation to a low level; but this can be done by screening these areas or merely by erecting single parallel wires at proper intervals over the water. Mildly gregarious, flocks of up to 20 or 30 mergansers are encountered in good habitat. *Voice:* hoarse, low-pitched *gak* or *grrk* notes; otherwise usually silent in winter. *Nest:* usually in cavity in dead tree or log, less often in hole in bank or among rocks, the amount of material variable. *Eggs:* 6-17, usually 9-12, pale buff; incubated by female for about 33 days. *Downy young* have dark stripes and a reddish area on side of head; they jump from tree nests as do those of other tree-nesting ducks.

FIG. 25

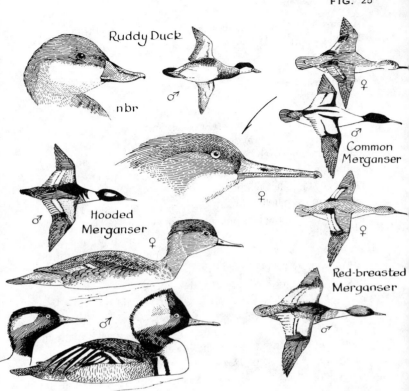

Ruddy Duck

nbr

♂

♀

♂

Common Merganser

♀

Hooded Merganser

♂

♀

♀

Red-breasted Merganser

♂

♂

Range. Breeds in subarctic and cold temperate zones across Eurasia and in America south to s. Canada and ne. U.S. and in w. mountains to c. Calif. and through Rockies to n. Mexico. Winters from s. B.C., Mont., Mo., the Great Lakes, and e. Canada south to c. Mexico, Fla., and Bermuda; also in Old World to n. Africa and se. Asia.

Occurrence in California. Very Common (Occ. Abundant) on Salton Sea and Colorado R. late Dec.-early Mar., a few as early as Nov. and through Apr., and Casual May-Oct. Fairly Common to Common Nov.-Mar. on lakes, ponds, and larger streams widely through cismontane lowlands; Rare in same areas to early May and late July-Oct., and Irreg. as breeder, with young noted May-Aug., on rivers or lakes of Coast Ranges south to San Luis Obispo Co. In the Sierra Nevada and ne. plateau, Uncommon to Locally Fairly Common on lakes and streams Mar.-Nov., breeding late Mar.-Aug.; Locally Fairly to Very Common Dec.-Jan. (concentrated by frozen water elsewhere?).

Red-breasted Merganser (*Mergus serrator*) (fig. 25; plate 7)

Recognition. L. 16-19 in. (41-48 cm.); WS. 28-34 in. (71-86 cm.), Wt. 1.3-1.9 lbs. (590-860 g.). Adult males are distinctively marked with reddish-brown chest (also streaked with dark), finely barred gray sides, white lower breast and belly, white ring around neck, and a racy two-parted crest on the black head. The feathering of the female and first-winter immature is much more often seen in California. These are best told from females or young of the Common Merganser by the shaded-out junctions of reddish-brown, gray, and white about the neck and head, and the generally dingier tone of the grayish upperparts. In addition, this is the only merganser *regularly* encountered on salt water.

Habits. When settled for the winter in California, the Red-breasted Merganser shows strong preference for salt water. However, there is a major migration of this species, with the Common Mergansers, along the lower Colorado R., particularly in spring (see second Graphic Calendar, in Appendix). Since few of these reach the Salton Sea or other points well inland, it seems their main route takes them northward east of the Sierra. This species is usually found singly or in small groups that shift from place to place for foraging. In exceptional cases, as many as 100 or more may congregate where fishing is good. The species of fish taken on bays and in the ocean differ, of course, from those taken on freshwater, and more crustaceans are eaten by this species. Like the diving ducks, the few mergansers remaining in the state in summer are mostly nonbreeding yearlings, chiefly at river mouths.

Range. Breeds from n. Alaska, n. Canada, and Labrador south to nw. B.C., c. Alberta, Minn., the Great Lakes area, and Maine; also in Greenland and across n. Eurasia. Winters from se. Alaska and Quebec south chiefly on seacoasts (a few on Great Lakes and elsewhere well inland) to nw. Mexico, across s. U.S., and on coasts of Europe and sw. and e. Asia.

Occurrence in California. Fairly to Occ. Common Nov.-Apr., Irreg. Rare to Uncommon rest of year, on bays, coastal lagoons, and ocean near shore, and fewer near islands well offshore and on cismontane lakes. In se. Calif., Common to Very Common on Colorado R. in Apr. and Oct.; few Irreg. elsewhere on lakes, streams, and Salton Sea, Oct.-June.

21. FAMILY GRUIDAE (CRANES)

(Picture Keys B and C)

Sandhill Crane (*Grus canadensis*) (fig. 26)

Recognition. L. 34-48 in. (86-122 cm.); WS. 6-7 ft. (2.0-2.3 m.). Although poorly informed persons often call herons "cranes," these true cranes can be readily distinguished from herons (see Picture Keys B and C) by their flight with long necks extended, by the long curved feathers of the lower back and the bare red skin area on the head of adults, and by their voice and behavior. The very long legs and neck and long bill, coupled with their size, permit confusion with no other birds in the state. Two subspecies of Sandhills occur here, recognizably different in size when they are together at close range. Except for a whitish face, the feathering of adults is entirely medium gray, or variably stained with rusty. Immatures are mostly brownish, including the featherless forecrown area.

Habits. These great birds are one of the truly thrilling wildlife attractions still to be found in central California, yet are unknown to most people because of their limited areas of occurrence. Broad stretches of short grass or corn stubble on land that is wet most of the winter are their favorite feeding grounds, although they do also invade the interspersed open marshlands and some alfalfa and other croplands. When feeding, as well as in flight, wintering cranes are strongly gregarious and walk along in an irregular group in which most birds have their necks low as they move, while a few with necks up serve as lookouts for the moment. Waste corn and other seeds and young green grassblades constitute most of their *food*; but bulbs, sedge stems or roots, and some large insects, snails, and amphibians are taken when they feed in marshes. Very wary by nature, cranes often will not allow a person to walk within good binocular range of them without flying, and they prefer to remain in relatively treeless country where danger can be watched for. In the air they are excellent fliers, usually quite vocal, and may even soar in circles when rising thermals are strong. Their typical wingbeat includes a quick upward flip of the wing-tip, followed by a slower downstroke.

During their main migration periods (Feb.-Mar., Oct.-Dec.) such flocks can be seen at many points along the Central Valley and northward to Modoc Co. When settled in midwinter, the chief populations are near the Sutter Buttes, in the Delta area near Thornton (the large subspecies is common here), on Merced Co. grasslands, and in most years on the Carrizo Plain. By Feb. *courtship* activities begin in the flocks, the participating birds deeply lowering and raising their heads, flapping wings loosely, and frequently hopping high into the air with legs dangling (fig. 26). Two to ten or more may join in such antics for several minutes at a time. *Nest*: a mound of marsh plants 4 to 5 feet across, placed in remote parts of extensive marshlands. *Eggs*: 1-2 (rarely 3), light gray-brown with sparse darker spots; the parents incubate alternately for a total of 28-30 days. *Downy young* leave nest within first few days, walk about marsh and vicinity with parents until able to fly at 9-10 weeks.

Range. Breeds from ne. Siberia and across arctic America to Baffin I. south to the Aleutians and nc. Canada (small subspecies); also from B.C. to Manitoba and south in scattered areas to w. Minn., S.D., Colo., and ne. Calif. (large subspecies); also resident in c. Fla. Winters in Calif. regularly from Tehama Co. south, chiefly interiorly, to e. San Luis Obispo Co. and Irreg. to Imperial Co.; also from s. Tex. to c. Mexico.

Occurrence in California. Locally Abundant, Oct.-early Mar., Fairly Common to early Apr. and from mid Sept., on moist grassland and some croplands and nearby marshes of Central Valley and on Carrizo Plain, San Luis Obispo Co. Irreg. Uncommon (formerly Common) at same seasons in Imperial Valley, and small numbers recorded Sept.-Apr. on coastal (and

FIG. 26

Sandhill Crane

a few on inner foothill) grasslands, especially of sw. Calif., and twice in June in Merced Co. On the ne. plateau Fairly Common to Common, Occ. Abundant, in migrations late Feb.-Mar. and Oct.-early Nov., the large subspecies remaining all summer as Uncommon breeder Apr.-July.

22. FAMILY RALLIDAE (RAILS, GALLINULES, COOTS)

(Picture Keys A, B, and C)

Clapper Rail (*Rallus longirostris*) (plate 3)

Recognition. L. 14-16½ in. (35-42 cm.); WS. ca. 20 in. (51 cm.). A bird somewhat like a small chicken, but slenderer, with a bill over 2 inches long and slightly downcurved, thicker than the bill of large shorebirds. The general coloration is medium brown with a buffy tone on the foreparts, light and dark vaguely formed streaks on the back, and prominent *white bars* on the *gray-brown sides*. The stout, long-toed feet are dull yellowish-gray, and the bill is usually partly yellow.

Habits. In California the Clapper Rail is now listed as endangered, as it is among the most highly restricted of all our birds in its choice of habitat. Those of the coastal populations are rarely found beyond easy running distance from the dense pickleweed or cordgrass salt marshes. They forage in this marsh, particularly along small channels in it, and at times hesitantly onto nearby open tideflats. The populations of southeast California have been found associated with dense riparian brush as well as marshes. Like all rails, the Clapper is adept at skulking amid dense vegetation, but its habit of jerking its short tail quickly upward sometimes reveals its location. These birds swim well for short distances. If surprised or cornered, they run into the marsh or fly a short way over it with legs dangling (see Picture Key C) and drop quickly out of sight.

Food: crabs, amphipod and isopod crustacea, polychaete worms, small snails, and occasional fish, and parts of plants; the mud snail now so abundant on San Francisco Bay tideflats is, however, apparently too thick-shelled. *Nest*: of grass or other marsh plants, placed on slightly higher points in the salt marsh, often flooded by the excessively high tides of early summer; earlier nestings (in April) are usually more successful, though some birds renest when the tides are less high again (two full E periods of Graphic Calendar; see Appendix). *Eggs*: 5-14, pale yellowish-brown, spotted with darker; incubation 21-23 days, by both parents. *Downy young*: solid black; they are sometimes mistakenly called "Black Rails" by novices. *Voice*: a somewhat raspy *jek,* or *chek-a-kerCHEKa-kaCHEK-kaCHAK-er,* etc., in irregular cadence, the calling of one bird (or a human imitation or any sudden noise) often triggering the calls of others.

Range. Resident, mostly in salt marshes, along the Pacific coast from Marin Co., Calif., south to n. South America, and on the Atlantic coast from Conn. south.

Occurrence in California. In coastal Calif., now Regular only in salt marshes of Tomales, San Pablo, and San Francisco bays, near Seal Beach, and on Upper Newport and San Diego bays; small or intermittent populations also at Elkhorn Slough (n. Monterey Co.), Morro Bay, and possibly in other marsh remnants of San Diego Co.; also (another subspecies) near s. end of Salton Sea and lower Colorado R., at least near Parker Dam and below.

Appearing Uncommon, they are seen to be Fairly Common

in coastal marshes when extra-high tides, or special techniques such as dragging a rope across the marsh, disclose them. Nests with eggs late Mar.-July, young recorded Apr.-Aug., probably two broods. Uncommon to Rare in marshes and dense shrubbery near s. end of Salton Sea and along se. Calif. streams, esp. Colorado R.; eggs noted in late May, young in July. Dispersing individuals Occ. at unusual, even freshwater, locations.

Virginia Rail (*Rallus limicola*) (plate 3)

Recognition. L. 8½-10½ in. (21½-27 cm.); WS. ca. 14 in. (36 cm.). Similar to the larger Clapper Rail in shape, but more brightly colored — light reddish-brown on foreparts, blacker streaks on back, and especially darker (nearly black) between the white bars on the flanks. The bill and legs are pale reddish-brown. Juveniles into early fall are much darker, even nearly black except for a paler throat.

Habits. Although probably not as uncommon in marshes with plants two feet or more tall as the Graphic Calendar (in Appendix) would indicate, this rail is difficult to see "on order" in such a habitat, because it hears the searcher coming. Good views are sometimes obtained quietly by chance, as the bird forages on occasions in a muddy opening or along a marsh-pond border. In fall or winter they may also be found on floating debris in salt marshes at high tides, like the Clapper Rail. *Food*: insects and their aquatic young stages, small crustaceans, worms, occasional small fish and, in fall, seeds of marsh plants. *Voice*: *wrrag-wragh-wrag-wrak-wrak-wrak* in medium but descending pitch is the usual response to a sudden noise (as to a stone tossed into the marsh, or clapping of hands), but is also given at other times; also a sharper *kid-ik* or higher *tick-et*, uttered irregularly or in long series, and various squeals and lesser notes. *Nest*: a shallow cup of sedges, stems, etc., woven along or attached to dense marsh plants, usually over shallow water. *Eggs*: 5-12, buff with scattered red-brown spots; incubation 19 days. *Downy young*: all black except partly yellow bill (plate 3).

Range. Breeds from sc. B.C. across s. Canada and south through Calif. and most of the U.S. to c. Mexico and various parts of Central and South America. Winters in s. B.C., w. Wash., Ore., Calif., and in se. U.S. south through Latin America.

Occurrence in California. Uncommon all year in freshwater marshes throughout cismontane lowlands, and Locally Fairly Common in these and in salt marshes Sept.-Apr. Nests in freshwater marshes Apr.-early June, with growing young noted until early Aug. Recorded in se. Calif. marshes nearly every month, but probably does not nest. Noted in higher mountains May-Dec., with a few nest records; and on ne. plateau Apr.-early Oct., nesting June-July.

Sora (*Porzana carolina*) (fig. 27; plate 3)

Recognition. L. 8-10 in. (20-25½ cm.); WS. ca. 13-14 in. (33-36 cm.). A chunkier rail than the Virginia, with bill heavier and shorter than the head, as in all rails of its group (called Crakes in the Old World). Adults are slate gray on the neck and chest, brownish and mottled on the back, barred gray and white on the sides, and have a *yellow bill* and *black face and middle of the throat.* The legs are light green. Immatures are browner, less distinctly barred on the sides, and lack the black on face and throat. They thus resemble the rare Yellow Rail but are not so yellowish, are less distinctly striped above, and have no white in the wing.

Habits. Very similar in its habitat and foraging to the Virginia Rail (see account). Either of these species is much more often heard than seen. Some individuals can sometimes be alarmed into flight by crashing through a marsh, although this is difficult except in the shorter types of marsh where they do not usually stay long. As in all rails, the flight appears weak. These two species are both known to migrate considerable distances nevertheless, flying mostly at night, as do the small songbirds. When grounded on such a flight, rails are sometimes found in such unexpected places as city dooryards or at an open pool in a park or field. *Voice*: a sharp, high-pitched *keek* is the usual response to a sudden noise near its hidden location in the marsh; also a short "whinnying" call of about a dozen high-pitched but descending notes and a

FIG. 27

1st yr Sora

Yellow Rail

ad

Black Rail

pleasant clearly whistled *ter-weeeee,* with the second note higher. *Nest:* a sturdy cup or mound of grass, sometimes arched over, anchored to emergent plants or sometimes on top of a mound, in marsh. *Eggs:* 6-15, buffy, with many brown spots; incubation 19 days. *Downy young:* all black except yellowish chin.

Range. Breeds from c. B.C., s. Mackenzie, and across s. Canada south over all but se. part of U.S., and to n. Baja Calif. Winters from nc. (rarely nw.) Calif., Ariz., and through Gulf states and S.C. south to West Indies and n. South America.

Occurrence in California. Probably Uncommon all year, Occ. Fairly Common Dec.-Mar., in freshwater marshes of cismontane lowlands; eggs recorded Apr.-June, and growing young as late as Oct. In Sept.-Apr. also in coastal salt marshes, and noted in se. marshes nearly all year. Noted late Mar.-Sept. at marshes of ne. plateau and in mountains, nesting at L. Tahoe and in s. mountains in late May-June.

Yellow Rail (*Coturnicops noveboracensis*) (fig. 27)

Recognition and Habits. L. 7 in. (18 cm.); WS. 12-13 in. (30½-33 cm.). This very secretive inhabitant of wet grassy meadows or thin sedge marshes is probably a regular migrant and occasional or local breeder east and north of the Sierra Nevada. They are nocturnal on their North Dakota breeding grounds, where a study of their behavior in an enclosure in a marsh was under way in 1972 — the only other individuals ever seen being those captured by a trained dog. Yellow Rails were collected in California occasionally prior to 1918 and there are a few verified sight records since 1960 (see Graphic Calendar in Appendix). Normally this species calls only at night. A Yellow Rail is similar to the immature Sora (fig. 27) but smaller, yellower on the foreparts, and has a more sharply striped back and a small white patch in the secondaries visible in flight. Its *voice* includes *tic* or *kuk* notes of clicking quality, usually given in long series if at all.

Range. Breeds from c. and e. Canada south to Dakotas and ne. U.S., and (formerly at least) in ce. Calif. and possibly Colo. Winters chiefly in La. to Fla., rarely in Ore. and Calif.

Occurrence in California. Very Rare, Oct.-Feb. in cismontane lowlands, preferring wet grassy meadows but recorded in true marsh, even in salt marsh. Recorded twice in Apr. near Quincy and twice on plateaus east of Sierra, with eggs in early June and "nesting" in summer.

Black Rail (*Laterallus jamaicensis*) (fig. 27)

Recognition. L. 5-6 in. (12½-15 cm.); WS. ca. 11 in. (28 cm.). The sparrow-sized body, typical short-necked, short-tailed shape of a rail, and tapered *black* beak are distinctive, except from some *downy* young of larger rails, all of which are black (see plate 3). The dark slate general color, with faint white bars on the sides and chestnut nape, and the prominently white-*spotted* back are the absolute proof of identity of this very secretive marsh denizen.

Habits. Few California bird students had ever seen a Black Rail until recent years, although they are supposedly regular in pickleweed (*Salicornia*) salt marshes about San Francisco, Tomales, and San Diego bays. They have long been known to nest in this type of habitat on San Diego Bay, and young were seen with adults at Inverness on Tomales Bay in 1966. However, nests have been found primarily by use of trained dogs that search out the birds as though they were mice. Otherwise, the best time to search for this species in salt marshes is during an exceptionally high tide, when the pickleweed zone is completely flooded and they are driven to the uppermost edges of the marsh or onto floating debris. The scattered records inland in California prior to 1970 included a few from Imperial Valley and the Colorado River. In 1970 and 1971, however, so many were heard calling on various dates ("c" records of Graphic Calendar in Appendix) at West Pond, near Bard on the lower Colorado R., that it appears they must nest there commonly. The characteristic *voice* of the species is a quickly uttered *kik-kik-groo*, the last note abruptly lower in pitch. The female is said to give a short series of hollow *coo* notes. *Nest*: a well-woven cup of marsh plants, well concealed within dense marsh growth. *Eggs*: 6-13, pale buff or pinkish finely dotted with brown.

Range. Breeds Locally from Kansas, Ohio, and Mass. south to Fla., and from c. Calif. to near San Diego and n. Baja, Calif., and along lower Colorado R. Winters in coastal marshes of c. and s. Calif. and in se. U.S. Also found in West Indies, Peru, and Chile.

Occurrence in California. Probably Rare (Occ. Locally Fairly Common) all year in pickleweed salt marshes of c. to sw. Calif. and in se. marshes; most records Apr.-Nov. Nests or young noted years ago near San Diego late Mar.-May; young with parents at Tomales Bay late July 1966; and calling birds were probably nesting June-July on lower Colorado R. Migrants or vagrants have appeared inland from late July to early May, mostly in nw. Calif.

Common Gallinule (*Gallinula chloropus*) (plate 8)

Recognition. L. 12-14½ in. (30½-37 cm.); WS. ca. 21 in. (53 cm.). A dark gray and brown bird that combines some of the characteristics of rails and some of coots (see Picture Keys A, B, and C) appearing when on the water more like coots, but readily distinguished by the dull brown tone of back and wings and the white marks along sides. The frontal shield and bill of the adult are bright red, the tip of the bill yellow; in immatures the bill and smaller shield are dull yellowish-gray or orangish. The large, long-toed greenish feet are not webbed or lobed.

Habits. When feeding in dense marshy growth or along

nearby muddy shores, the gallinule behaves much like the related rails. However, it also swims a great deal, tipping-up or even diving if necessary to get at the aquatic plants it prefers as *food*. Its swimming is usually jerkier even than a coot's, and it seldom ventures very far from a marshy retreat, although the tules or cattails of an irrigation ditch are sufficient. This species has been gradually spreading its range northward in the coastal area, but it is still primarily a bird of the Central Valley refuge and hunting areas and others with "managed" water channels and marsh borders. Gallinules are decidedly less pliable in habitats utilized than the coot, but their numbers are probably somewhat higher than the Graphic Calendar (in Appendix) indicates, except for the Dec.-Jan. records, which have been of concentrations at particularly good spots. *Voice*: a low *kup,* and a great variety of *kaks, k-k-k-krunks* in series, and other explosive notes. *Nest*: a large shallow cup of sedges or cattails, anchored at or slightly above water surface by at least a narrow channel leading to open water. *Eggs*: 6-15, buff, with irregular brown markings; incubation about 21 days. *Downy young*: blackish, with dark reddish bill and center of crown.

Range. Breeds from Marin (and Sonoma?) Co. and n. Sacramento Valley of Calif., c. Ariz., Okla., c. Minn., and se. Canada, south to Argentina. Withdraws from c. and ne. U.S. in winter; resident elsewhere. Also found in Hawaii and over most of Old World, including various island groups.

Occurrence in California. Uncommon to Locally Fairly Common in freshwater marshes and marsh-bordered ditches and ponds of Central Valley, where Occ. Locally Common Dec.-Feb., but more widespread rest of year; fewer in cw., sw., and se. marshes. Nests with eggs noted late Apr.-July, growing young as late as Aug. Casual in nw. and ne. Calif.

American Coot (*Fulica americana*) (fig. 2; plate 8)

Recognition. L. 14-16 in. (35-41 cm.); WS. ca 26 in. (66 cm.). Although coots behave on the water quite like many ducks, and are often confused with them or with grebes by beginners, the very different bill lacking strainers and terminal nail, the frontal shield, and the large toes with flat lobes along them, rather than webs, distinguish them readily (compare with ducks and grebes in Picture Keys A and C and fig. 2). Adult coots are dark slate-gray with a blackish head and neck, white bill, and light green to yellow-green legs. The white outer undertail coverts form an inverted V when displayed, and there is a narrow white trailing edge of the secondaries. Juveniles (fig. 2) are lighter gray above, merging to whitish below and, when small, over the neck also. Their bill is pale gray and has only a small extension onto the forehead.

Habits. Known also as "mud hen" to many people, the American Coot is so abundant and widespread in California that it is one of the few water birds known to almost everyone. In swimming, coots usually bob the head back and forth, but not necessarily in unison with their foot movements. Their progress thus often appears jerky when they are looking for food. Favorite items of *diet* are the foliage of pondweeds, the larger algae, and growing tips of grasses that are often obtained by grazing quite far from water. Small aquatic animals are also eaten if readily available, as on bay shores. Although they swim high, coots can dive to considerable depths when necessary. They are often considered unworthy of the sportsman's efforts, but are said to be quite suitable for the table if skinned carefully and then prepared (Calif. Fish and Game Dept. distributes "whitebill" recipes). *Voice*: of male, a short *puhk* or, if alarmed, *puhlk*; comparable calls of female, *punk* or *poonk* or *punt-unt,* the members of a pair being distinguished by voice; also a variety of harsh cackling and grunting notes and an explosive descending call of similar quality.

Highly territorial in the breeding season and monogamous on their territory, coots have an elaborate set of *displays* despite their plain plumage. A head lowered toward an invader, rapid swimming, running noisily over the surface, and actual grappling and fighting represent four increasing degrees of aggressiveness. When the opponents are evenly matched, however, they often merely raise their wings a bit and lift and spread the tail while pivoting about each other. *Courting and mating* are restricted to a specially built platform of marsh plants just at the water surface.* *Nest*: a mound of cattails or tules on a massive foundation in shallow water, lined with finer plant materials (eggs laid in only one of several built). *Eggs*: 8-13 (rarely to 20+), pinkish-buff with many fine dark brown spots; incubation 23-26 days. *Downy young*: blackish with orange-red head and part of neck (plate 8); they are brooded on special brood nests when young, chiefly by the male, and become independent at 20-30 days of age, when far from full-grown.

Range. Breeds from c. B.C., s. Mackenzie, Sask., and se. Canada south throughout the U.S. (local in ne.) and Mexico to West Indies and Guatemala. Winters from B.C. south in Pacific states and from Ariz., Tex., s. Ill., and Md. south. Also found in South America and Hawaii.

Occurrence in California. Abundant to Very Abundant on Central Valley lakes, marshes, and wet grasslands and fields Sept.-Apr., and Common to Very Common rest of year, nesting in marshes; eggs Mar.-early Aug. and growing young Apr.-Aug. At same seasons usually somewhat fewer in same habitats in other cismontane areas and at s. mountain lakes (esp. shallow weedy ones). Uncommon to Locally Common on bay tideflats and salt marshes late Aug.-Apr. Uncommon on rivers and Rare on ocean and to offshore islands Sept.-Apr. In se. Calif., Very Common to Abundant on Salton Sea Sept.-Apr., and fewer on lakes and major streams into at least July; young noted Apr.-May. In ne. Calif., Very Common to Abundant on marshy lakes and nearby grasslands Mar.-Oct., and Irreg. Fairly to Very Common where lakes are open through winter; eggs noted June-July and dependent young June-Aug.

23. FAMILY HAEMATOPODIDAE (OYSTERCATCHERS)

(Picture Key B)

Black Oystercatcher (*Haematopus bachmani*) (plate 11)

Recognition. L. 17-19 in. (43-48 cm.); WS. ca. 33 in. (84 cm.). A very large shorebird (size of small duck) with all black

*Most of the above from various research papers by Gordon W. Gullion.

feathering, a long, bright red, chisel-shaped beak, and pink legs; unmistakable if these features are seen. In juveniles the beak is dull pink and gray.

Habits. Highly restricted to the surf-swept rocky shore of the outer coast, this species is Rare to absent even along much of our rocky shore. Using its stout but chisel-like beak to pry off or break open the protective coverings of its prey, the oystercatcher *feeds* on the mussels, limpets, and barnacles that compose the great beds of the most exposed ledges and off-shore rocks. It also takes a variety of other small animals found among the mussels. When the tide covers its feeding areas, or when resting at other times, these birds select a niche of a rocky pinnacle or offshore islet to rest not far above the splashing waves. The darkness of many of the rocks, especially when wet, makes these dark birds quite inconspicuous when not in flight. During the spring courtship period, however, they are quite noisy and fly about conspicuously in small groups. *Voice*: shrill piping *kyip* notes in short to long series, varying little in pitch and easily audible at a distance despite noise of surf; also lower, mellow calls. *Nest*: a slight hollow in a rock ledge close to the high tide or splash zone, usually lined with rock or shell chips. *Eggs*: 2-3, buff or light olive, irregularly marked with brown and black; incubated 25-26 days. *Downy young*: dark gray with blackish and buff markings.

Range. Breeds from w. Aleutians east and south along Pacific coast to c. Baja Calif. Mostly resident, but some of n. populations move south in winter, and small numbers appear in nonbreeding areas in fall and winter.

Occurrence in California. Uncommon to Locally Fairly Common all year on outer rocky coast of c. and n. Calif. and s. islands (Rare on mainland of s. Calif.), Regular at Fairly Common level Oct.-Feb. in best places, when probable migrants from north are present; but seeing ten or more in a day is not likely except in a few favored localities such as near Crescent City, Trinidad, the Farallon Is., Tomales Pt., and the Pt. Lobos and Montana de Oro state parks. Eggs usually May-June, young from June-early Sept.

Note: The **American Oystercatcher** (*Haematopus palliatus*) is similar to the Black except for its abruptly white underparts from chest to tail, white upper tail coverts, and broad white strip along the wing. It breeds from n. Baja Calif. southward and from N.J. south on the Atlantic coast, being frequent on sandy beaches as well as rocks in both areas. Stragglers on the s. Calif. Channel Is., and rarely the mainland coast, were found a number of times between 1862 and 1910, then not for over fifty years. However, since 1963 there have been at least twenty-eight sightings from Avila Beach southward, of a dozen or more individual birds, on dates ranging throughout the year (see Appendix).

24. FAMILY RECURVIROSTRIDAE (AVOCETS, STILTS)

(Picture Keys B and C)

American Avocet (*Recurvirostra americana*) (fig. 28; plate 9)

Recognition. L. 16-20 in. (41-51 cm.); Bill 3½-4 in. (9-10 cm.); Legs ca. 6 in. (15 cm.); WS. 30-35 in. (76-89 cm.). The only long-legged, long-necked shorebird with back *and* wings showing large black and white areas above (compare with stilt, fig. 28, and Willet, fig. 31). In addition, the slender *upturned* black bill with needle-like tip is distinctive. The cinnamon color of head and neck of the breeding feathering is replaced by white to dull light tan at other seasons. At a distance, avocets appear almost as conspicuously white as gulls.

Habits. Avocets are shorebirds of adaptable behavior, yet with several unique features. They often obtain food from mudflats by probing much like a large sandpiper, their long legs of course enabling them to do so in deeper water than most other species — so a line of avocets is often the first to follow out the receding tide. Sometimes they feed by tipping-up like ducks, or by dabbing at the water with short pecks, like a phalarope, while swimming (their toes are connected by sizable partial webs). But most impressive is their method of feeding in soupy mud: walking along and sweeping the bill vigorously sidewise below the surface with the mandibles almost closed. The thin beak is smooth-edged, yet their frequent

FIG. 28

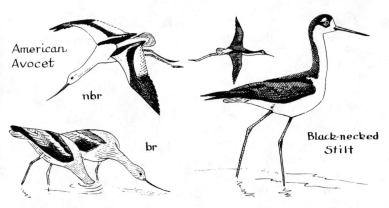

American Avocet

nbr

br

Black-necked Stilt

swallowing shows how successful this technique is (for getting worms from the mud?).

The actual *food* by these various techniques and from different habitats varies, of course: swimming or surface-dwelling insects and their larvae, small crustacea, tiny snails, worms, and even seeds of aquatic or marsh plants in amounts unusual for a shorebird. The birds feed so much in concert that there appears to be cooperative advance across an area. When not feeding, they mass closely in or near quiet pools or salt ponds. Only when spread out on their nesting territories are avocets aggressive toward each other; and even then, at the first approach of a predator or person, they congregate to swoop at and scold the intruder. Their diversionary display if the nest is approached is very spectacular, with wings and gangly legs moving awry as though half broken. *Voice*: loud, high-pitched *kleek* or *peep* notes, long-repeated when alarmed. *Nest*: a hollow on open flat or dike or in edge of low marsh, scantily lined with nearby plant bits or debris or built up to a mound if water rises; usually in loose colonies. *Eggs*: 3-5, buff with many irregular markings of brown or blackish; incubation 23 or more days, apparently by female. *Downy young*: buff and gray above, with a few darker markings (plate 9); run and swim soon after hatching.

Range. Breeds from prairie areas of cs. Canada south through Wash. and Ore. east of Cascades to Calif. lowlands, except for coast belt north of San Francisco Bay, and east to

Nebr. and Tex. Winters from coastal n. Calif. and s. Tex. south along w. coast of Mexico to Guatemala; Casual migrant to e. U.S.

Occurrence in California. Common May-mid-July, increasing to Abundant Sept.-Apr., in salt evaporating ponds, muddy tideflats, and pools, including those in marshes; nests with eggs Apr.-June, dependent young mostly May-early Aug. Very Abundant in winter, Uncommon or Fairly Common in summer, at Salton Sea, and somewhat fewer in Colorado R. valley and other se. Calif. waters. In the Central Valley, Fairly Common through winter to Very Common or Locally Abundant Apr.-Sept.; somewhat fewer at comparable seasons inland in wc. and sw. Calif.; eggs primarily Apr.-June, dependent young May-July. In ne. Calif. Fairly Common to Locally Abundant Apr.-mid-Oct. (Ore. and Nev. dates); eggs here primarily late May-June; dependent young June-July. Migrants recorded at lakes in the Sierra and s. mountains.

Black-necked Stilt (*Himantopus mexicanus*) (fig. 28)

Recognition. L. 13-17 in. (33-43 cm.); Bill 2-3 in. (5-7½ cm.); Legs 7½ in. (19 cm.); WS. ca. 28 in. (71 cm.). An even longer-legged bird than the avocet, though smaller of body. The upperparts, including the whole wing surface, are solid black (or dark brown in immatures) except for a white rump, gray tail, and white forehead and patch behind the eye. This plus the pure white underparts, straight needle-like black bill, and *red* or *pinkish* legs and red eye are like no other bird.

Habits. This relative of the avocet is like it in many respects, but less adaptable in its feeding and nesting. With a straight bill, it does not engage in the side-sweep method of *feeding,* but probes in mud and picks many insects from amid marsh plants, or insects and brine shrimp from open water. Drainage for agriculture of many areas of seasonal low marsh growth and the large beds of shallow Buena Vista and Tulare lakes destroyed much of the Central Valley nesting habitat for stilts; but this is partly offset by the spread of rice-growing, for these birds find many parts of the green, flooded fields within their leg-length, and have taken to nesting on the small dikes that retain the water. Even more evident as a recent change (since about 1960) has been the great increase in the use of salt-evaporation ponds throughout the year, and this now seems to be their prime wintering habitat in California.

The flight of stilts is direct, with steady flaps, the bill pointing straight forward and the long legs trailing behind (sometimes somewhat drooping). *Nest*: similar to avocet's, but more often in marshy situations and thus with more material; colonies also usually denser, though often small. *Voice* of adults when such an area is invaded is sharper and more incisive than the avocet's similar call; stilts engage in similar distraction displays. *Eggs*: 3-5 (occasionally 6-7), yellowish or buff, irregularly spotted and blotched with brownish black; incubation about 25 days, shared by both parents. *Downy young*: light buffy-gray above, mottled with dusky plus a few larger black blotches, buffy-white below.

Range. Breeds from s. Ore., Idaho, and s. Sask. south through interior n. and c. Calif., and near the coast south from San Francisco Bay, east to Colo., s. Tex. and Locally se. U.S., and south to n. South America. Winters from c. Calif. to n. Mexico and coasts of Tex. and La. south to n. South America.

Occurrence in California. Common, Locally Very Common, Mar.-Sept., and Fairly to Locally Common rest of year, in salt ponds and various types of coastal marshes and lakes with shallow water, from San Francisco Bay area south. In the Central Valley and se. Calif., Common to Locally Abundant Apr.-Sept. and Irreg. Uncommon to Fairly Common rest of year; fewer at similar seasons inland in sw. Calif. and at some s. mountain lakes. Most eggs (both coastally and inland) reported May-early July, dependent young June-July. Recorded on ne. plateau June-Aug. (probably Apr.-May also), and a few in nw. Calif. Apr.-June and in Sierra Nevada late Aug.

25. FAMILY CHARADRIIDAE (PLOVERS)

(Picture Keys B and C)

Semipalmated Plover (*Charadrius semipalmatus*) (fig. 29; plate 10)

Recognition. L. 6½-8 in (16½-20 cm.); WS. ca. 15 in. (38 cm.). A small plover with medium brown upperparts, white collar, white forehead and "eyebrow" stripe, and a *single dark chest band* that is black in breeding feathering to brown and sometimes incomplete centrally in fall and winter. The orangish feet and base of the bill are also duller then, or the bill is all black. In flight the wings show a lengthwise white stripe and the tail is fully bordered by white in a unique pattern.

Habits. Primarily a mud or sand flat species when in California, these little plovers feed in the "run, stop, look" manner characteristic of their family. Almost any small animal they can thus obtain from the surface is utilized as *food,* except those with hard shells. "Semipalmateds" (the name refers to the small webs between bases of the toes) are aggressive toward each other when feeding, and each bird thus helps to space out the whole group for more effective scanning of a large area. Such a behavioral feature is advantageous for surface feeders, although probably not for those that probe the substrate randomly. In April there is often much bickering among these plovers, even when assembled on their roosts at high-tide periods. *Voice*: a whistled *chee-wee,* given quickly, the second note higher or slightly lower than the first; also, in spring, a stuttered series of mellow notes.

Range. Breeds from w. Alaska and nw. B.C. across most of Canada to Nova Scotia. Winters chiefly coastally from c. (Occ. nw.) Calif., Tex. and S.C. south to Patagonia.

Occurrence in California. Common, Occ. Very Common, Apr.-May and late July-Oct., Fairly to Locally Common Nov.-Mar., and Irreg. Rare through summer on mudflats of bays and lagoons. Somewhat fewer at same seasons on sandy beaches, and Rare to Uncommon on lake and stream shores or wet grasslands and fields of coastal area and Central Valley. In se. Calif., Fairly Common to Common Apr.-early May and Aug.-Oct. (Occ. in July and to Dec.) on Salton Sea and lake and river shores. A few migrants noted inland in sw. Calif. Apr.-May and Dec., and at lakes of Sierra Nevada or ne. Calif. Apr.-May and late July-Aug.

Snowy Plover (*Charadrius alexandrinus*) (plate 9)

Recognition. L. 6-7 in. (15-18 cm.); WS. 13-14 in. (33-36 cm.). About the size of the Semipalmated and of similar

FIG. 29

orange

Killdeer

ad

Semipalmated Plover

nbr

general pattern of feathering (see fig. 29 and plate 10), but the Snowy's hue is decidedly paler above (the color of dry beach sand), the chest band is always represented only by dark marks at the side (black in breeding feathering only), and the bill and feet are always black. In flight, the Snowy shows a white line in the wing and white sides of the short tail.

Habits. In their favorite habitat on the dry sand of the upper levels of open beaches, hundreds of these plovers can be present and yet go undetected by the casual beach visitor, so well do they match the substrate. Even their movements are not conspicuous in the bright sun, for they adroitly keep their white underparts turned away from view and often huddle low until a person passes. The *food* of Snowy Plovers on the high beaches includes beach hopper amphipods, ground and rove beetles, and other insects attracted to the seaweeds and driftwood cast ashore. Occasionally they do forage in the wet sand, especially on young sand crabs. Most of their original nesting habitat in low dunes above the outer beaches is now so subject to human activity that the populations have dwindled. However, this is partly compensated by their spread to the man-made salt-pond habitats of the inner parts of San Francisco, Newport, and San Diego bays. Here they feed in part on the brine flies, but still use the run, stop, look, and grab method, although the flies may be in a cloud around them on the ground. *Voice*: a three-noted mellow whistle; and a low raspy *gwit-gwit* or a long grating trill when their nest area is invaded. *Nest*: a hollow in sand or bare earth, usually skimpily lined with bits of shell, gravel, or grass. *Eggs*: 2-3, sand-colored — i.e., pale buff with tiny blackish markings; incubation 24 days. *Downy young*: mixed light buff and gray-white above, with fine black spots (also matching most beach sand), white below; they follow parents to feeding areas within one day.

Range. Breeds along coasts from s. Wash. and from La. and Fla. south to n. Mexico and Cuba, and Locally inland in various western states including Central Valley of Calif. and Salton Sea; also in parts of Central and South America and across much of Eurasia. In winter retires from most inland and northern areas to coastal sectors of breeding range, in w. U.S. north to n. Ore.

Occurrence in California. Common Sept.-Mar. on sandy beaches and bayshore sandflats; fewer June-Aug., though still nests Locally in this habitat. Uncommon to Fairly Common all year (fewer in winter) on salt pond dikes, where major nesting is now located. Eggs Apr.-early Aug.; dependent young, May-Aug. On shores of Salton Sea apparently Uncommon all year, but Irreg. and sometimes Fairly Common late Apr.-early Aug.; nesting reported. Migrants also recorded at various times of year on barren shores of inland lakes or rivers, mostly in sw. Calif., on offshore islands, and in ne. Calif. (Apr.-Aug. only). Noted at L. Tahoe in Aug. and Nov., and has nested in nearby w. Nev.

Note: The **Piping Plover** (*Charadrius melodus*) and the **Wilson's Plover** (*C. wilsonia*), both of cs. to e. U.S., have each been found on s. Calif. beaches. (See Appendix for details.)

Killdeer (*Charadrius vociferus*) (fig. 29)

Recognition. L. 9-11 in. (23-28 cm.); WS. ca. 20 in. (51 cm.). A medium-sized plover easily recognized by the *two black chest bands*, black and white facial areas, and white collar. Otherwise a Killdeer appears brown above and white below when at rest; but in flight it shows a white line lengthwise on the wing, and a large rusty area is conspicuous above the base of the rounded tail, itself dark and then edged with white. The bill is black, the legs and feet pale brown.

Habits. Most widespread and adaptable of all California shorebirds, the Killdeer is a familiar sight and sound on farmland and about fresh waters all across the continent. Its foraging method is typical of plovers in general (see notes for Picture Key B). Its *diet* consists mostly of insects, including many of the worst crop pests. Although up to a few hundred individuals may gather to feed in a particularly good field, as when it is being irrigated, the Killdeer does not move about in the compact flocks so typical of its relatives. It is, however, usually the first of any species in a field or along a shore to give vocal notice of its alarm at the approach of an intruder, the persistence of its calls having given rise to its species name. *Voice*: the extreme alarm note is a ringing *kil-DEE, kil-DEER,* etc., but often only a milder *dee* is given; these are also combined in long series, and a long musical trill is heard in the breeding season. *Nest*: a slight hollow with little or no lining, scraped in the ground, usually on gravel (sometimes even gravel roofs). *Eggs*: 3-5, usually 4, ivory to buff irregularly marked with black and brown; incubation 24-29 days, much of it by the male. *Downy young*: gray-brown, rather longtailed, with black markings in some of the adult locations and elsewhere; parental care continues for 5-6 weeks.

Range. Breeds from nw. B.C. and c. and se. Canada south throughout U.S. to n. Mexico and West Indies, and in Peru and Chile. Winters from B.C. and Ore. across c. U.S. to N.Y. and south into South America.

Occurrence in California. Common all year, Locally Fairly Common in summer and Very Common in winter, throughout cismontane lowlands and foothills on grassland and fields with low vegetation near any sort of water, along streams, ponds, lakes, by stock tanks, even Locally in towns, and Uncommonly to bayshores. Eggs reported late Feb.-July; dependent young Apr.-July. Also present all year in se. Calif., but few through summer; and in s. mountains, but few through winter.

On lake and stream shores and moist grass or fields of ne. Calif. and in Sierra Nevada, Fairly Common late Mar.-Sept., fewer through winter (only mild ones?), with nests or young reported June-July.

Mountain Plover (*Charadrius montanus*) (fig. 30)

Recognition. L. 8½-10 in. (21½-25 cm.); WS. ca. 18 in. (46 cm.). A medium-sized plover, smooth brown above like the Killdeer but somewhat paler and with only a lighter brown wash on foreneck and chest. It is white or near white on the breast, belly, forehead, and variable eyebrow stripe, and there is a thin light margin on the rounded tail and a thin white wing stripe. However, all the other distinctive marks of the Killdeer and smaller plovers (see fig. 29 and plates 9 and 10) are missing (no chest bands, no collar). The pale legs of the Mountain Plover are longer than those of other plovers.

Habits. Preferring the most featureless plains or gently sloping mesas possible, with little or no vegetation and certainly no "mountains" to hinder their distant vision and rapid running, this species is most inappropriately named. They are also spotted in broad cultivated fields (e.g., Imperial Valley), but can be found most regularly in areas such as the Carrizo Plain and along old California Route 33 in nw. Kern Co. From some of the gentle rises of the highway there, a scanning with binoculars of the most desolate terrain in winter may disclose a hundred or more spread out to feed. They feed in the surface pick-up method of all plovers, but run farther between stops and are usually more widely spaced — in keeping with their open habitat. When alarmed, they group into a straggling flock as they fly off low to the ground. *Voice*: in winter, chiefly a low *kwerr*.

Range. Breeds from n. Mont. and N.D. south in the Great Plains area to se. N.M. and Tex. Winters from c. Calif., s. Ariz., and c. Tex. south to nc. Mexico.

Occurrence in California. Fairly Common or Common Oct.-Mar., stragglers from July, on very short grassland or mostly barren fields of Central Valley (chiefly San Joaquin

PLATE 2

Brown Pelican
br
⑦

Pelagic
Cormorant
br
⑨

Brandt's
Cormorant
br
⑨

Double-crested
Cormorant
⑨

FIG. 30

Mountain Plover

white stripe
nbr

black axillar
white

nbr

no stripe
no black

Black-bellied
Plover
br

nbr

American
Golden
Plover
br

nbr

Valley), Salinas Valley, Carrizo Plain, and in Imperial Valley.
Stragglers also recorded on coastal grasslands of c. and s.
(Casually nw.) Calif. Aug.-Feb.

American Golden Plover (*Pluvialis dominica*) (fig. 30)

Recognition and Habits. L. 9½-11 in. (24-28 cm.); WS. ca.
22 in. (56 cm.). Formerly considered a very rare vagrant in
California, this moderately large plover with the famous over-
seas migration routes has been found with increasing frequency here in recent years, particularly on pastures near the
outer coast. In general color pattern it is similar to the abundant Black-bellied Plover; but in breeding feathering the light
spots of the upperparts are *golden yellow,* and in nonbreeding

183

PLATE 1

25. PLOVERS

feathering they are buffy and there is a buffy wash on the chest. Some immature Black-bellies are brownish, so reliable identification of the Golden in fall and winter requires noting the *absence* of black under the wing and of white in the wing and tail. In its normal wintering grounds, the Golden is found widely over open fields, golf courses, and on mud to sand shores as well. The few that reach California are of course only wanderers from this amazing annual movement of a species to thousands of pinpoint destinations across thousands of miles of inhospitable ocean. *Voice:* most commonly a somewhat rough *que* or *queedle* in flight; also a more whistled *QUE-que-que,* but not plaintive like the Black-bellied's call.

Range. Breeds in arctic tundra from Siberia and Alaska east to Baffin I. Winters mainly in s. South America, se. Asia to Australia, and on c. and s. Pacific islands. Migrates regularly in spring through Mexico and c. U.S. and Canada; in fall their passage is mostly well out to sea both east and west of U.S., as is the e. Pacific population's at both seasons. Stragglers occur fairly regularly in Calif. and elsewhere near the coast.

Occurrence in California. Regularly Uncommon to Fairly Common Sept.-Oct., and Irreg. Uncommon July-Aug. and Nov.-June (Occ. Locally Fairly Common Nov.-Feb.), on moist grasslands near coast; fewer to actual tideflats or sandy beaches, including those of Salton Sea. Recorded also in Central Valley and inland in sw. Calif., mainly Sept.-Oct.

Black-bellied Plover (*Pluvialis squatarola*) (fig. 30)

Recognition. L. 11-13½ in. (28-34 cm.); WS. ca. 23 in. (58 cm.). The largest of our plovers, with gray to brown upperparts finely *spotted with white* and in nonbreeding feathering a variable wash or obscure streaking on the whitish underparts. These aspects of the standing bird will distinguish it from all other plovers except the Golden, from which it is told by the white wing stripe, whitish base of the barred tail, and the black axillar feathers under the wing. Adults of both species in or nearing breeding condition have black on the face (below a

PLATE 3

Emperor Goose
(15)   *adult*

(15)   White-fronted
Goose

adult

(15)   Snow Goose

adult

(15)   "Blue" Goose

adult

Clapper Rail   (22)

Virginia Rail   (22)

Sora

(22)

dny

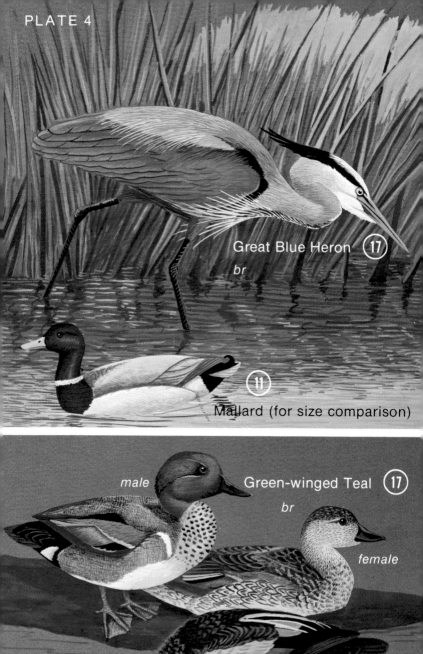

PLATE 4

Great Blue Heron ⑰
br

Mallard (for size comparison) ⑪

male   Green-winged Teal ⑰
br
female

male
Cinnamon Teal ⑰
br

PLATE 5

Mallard (17)

female
br

male

dny

female

male

Northern Shoveler (17)
br

(17)
American Wigeon   *br*   *male*

Blue-winged Teal
br
(17)

female

male

PLATE 6

Canvasback (18) *female*

male

Redhead (18) *male*

Surf Scoter (18) *male*

Ring-necked Duck (18) *male*

Lesser Scaup (18) *male*

Bufflehead (18) *male* *female*

PLATE 7

Common Merganser (20)

female

male

male

female

Red-breasted Merganser (20)

Common Murre (35)

br

nbr

(35)

br

Rhinoceros Auklet

br

nbr

Marbled Murrelet (35)

PLATE 8

Green Heron (11)
adult

Common Gallinule (22)
adult

dny

dny

American Coot   adult (22)

female

Wood Duck (17)

br

male

male

Ruddy Duck   br

female

(19)

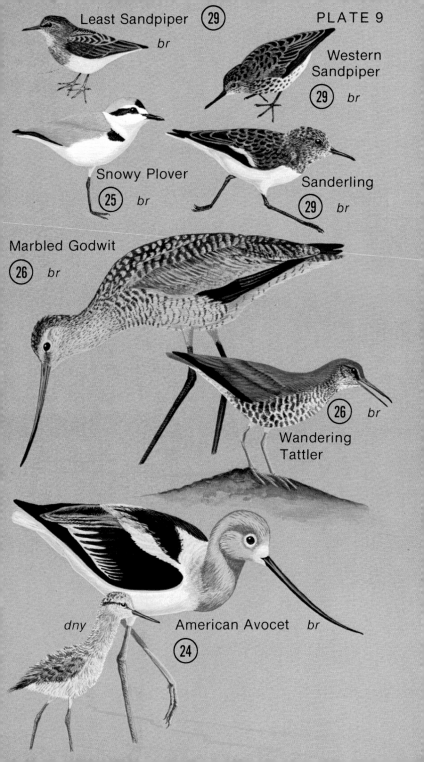

Least Sandpiper ㉙ *br*

PLATE 9

Western Sandpiper ㉙ *br*

Snowy Plover ㉕ *br*

Sanderling ㉙ *br*

Marbled Godwit ㉖ *br*

Wandering Tattler ㉖ *br*

American Avocet *br* ㉔

dny

PLATE 10

Dunlin (29) br

Semipalmated Plover (25) br

Red Knot (29) br

Willet (26) br

Greater Yellowlegs br (26)

Short-billed Dowitcher (27) br

Long-billed Dowitcher br (27)

PLATE 11

Pigeon Guillemot *br* (35)

Tufted Puffin *br* (35)

Ruddy Turnstone *br* (28)

Black Oystercatcher (23)

Harlequin Duck (28)

female

male

PLATE 12

nbr

adults

br

California Gull

32

foot

1st winter

PLUMAGES
OF THE
WESTERN
GULL

3rd winter

2nd winter

adult

32

1st winter

adult

Heermann's
Gull

white zone) and most of the underparts, but the summer stragglers and early migrants may be in the nonbreeding or an intermediate, partially black, pattern. The bill is black, the legs gray.

Habits. Largely a tideflat forager when in California, these big "Grey Plovers" (as the British name them) *eat* many small crabs, amphipods and other crustacea, snails, and polychaete worms from the open flats, as well as a variety of insects from salt-marsh borders or on upland areas. The large eyes of this plover indicate good vision even in dim light, and they may often be found searching for food under a bright moon or where city lights reflect from bay waters. As with other plovers, they spread out on the flats, and in at least some cases particular individuals establish territories there to which they return at each ebbtide. During high-tide periods, however, this species is quite sociable, and massed flocks can be found on protected roost areas. These are either on open dry flats or short-grass areas (if these are normally not subject to disturbance by man or predators), or in shallow pools or on islands with other shorebirds. *Voice*: a distinctive high-pitched and plaintive whistle, as *tee-a-wee-e,* often used by birds about to alight or take flight.

Range. Breeds on arctic tundra from n. Russia east to Alaska and n. Canada islands (barely onto mainland). Winters along seacoasts from sw. B.C. and N.J., and in some interior valleys from c. Calif. and s. Tex. south through s. Brazil; also in Old World on coasts of s. Eurasia to s. Africa and Australia.

Occurrence in California. Very Common late July-Apr., Locally Abundant Aug.-Oct. and Occ. to Apr., on tideflats of bays and nearby barren flats, short grasslands, marshes, and salt-pond dikes; at same seasons also Fairly Common to Common on outer coast sandy, and Occ. rocky, shores and offshore islands (formerly Abundant on grasslands there). Numbers decline May-early July. In se. Calif., Salton Sea to the Colorado R., recorded nearly all year, Very Common to Abundant spring and fall. In ne. Calif. recorded Apr.-May and late July-Dec.

26.-29. FAMILY SCOLOPACIDAE
(PROBING SHOREBIRDS)*

26. SUBFAMILY TRINGINAE (TRINGINE SANDPIPERS, CURLEWS, GODWITS, AND ALLIES)

(Picture Keys B and C)

Note: The **Upland Sandpiper** (*Bartramia longicauda*) is a bird of curlew head and body shape and general color pattern (fig. 31), but smaller (L. 12 in. = 30 cm.; WS. 22 in. = 55 cm.), with a slender, straight bill shorter than the head. Its normal range is from Alaska to Maine and through c. U.S., with a few known to have bred in s. Ore. It has been found in Calif. at least eight times at scattered locations (see Appendix).

Whimbrel (*Numenius phaeopus*) (fig. 31)

Recognition. L. 15-19 in. (38-48 cm.); Bill 3-4 in. (7½-10 cm.) or as short as 1½ in. (3.8 cm.) in some immatures; WS. ca. 32 in. (81 cm.). Distinguishable as a curlew by its slender, downcurved beak, and from the Long-billed by its general dusky-brown coloration, *dark and light striped crown*, smaller size (about that of the Willet), and usually shorter bill. Both curlews lack white in the wings and have gray legs.

Habits. This species, formerly known in America as the Hudsonian Curlew, is similar in much of its foraging to the larger Long-billed Curlew, but is more varied in that it also pokes among seaweeds and in rocky crevices of the outer coast or on bayside levees. Compact flocks are often seen in the interior valleys in spring. The fall passage, however, is predominantly along the coast (see Graphic Calendars in Appendix) and seems to be composed of birds in less gregarious "mood," for there are usually less than ten together. When foraging along the tideflats and beaches, the Whimbrel often associates with Willets and Marbled Godwits. Its high-tide roosts are also usually shared with these species, or sometimes it joins with other shorebirds. *Voice:* three to eight clear staccato *kwik-kwik-kwik* notes in rapid series, all on one pitch.

*See note to Picture Key B.

FIG. 31

Willet

Whimbrel

Long-billed Curlew

Marbled Godwit

Range. Breeds almost around the world in arctic regions (except Greenland), south in North America to w. Alaska and s. Hudson Bay. Winters along coast from c. Calif. and Caribbean (rarely se. U.S.) south to Chile and Brazil; also from s. Eurasia to s. Africa, Australia, and w. Pacific and Atlantic islands.

Occurrence in California. Coastally, Fairly Common Mar.-May, Fairly Common to Common July-Sept., and Uncommon to Locally Fairly Common Sept.-Mar. on bay tideflats and outer beaches, fewer to rocky shores and fields, lakeshores, and offshore islands; Rare in June. Inland, esp. Central and Imperial Valleys, Common to Abundant mid-Mar.-May during northward migration on wet fields, grasslands, and at times to lakes and river margins; Irreg. Uncommon to Fairly Common inland at other seasons.

Long-billed Curlew (*Numenius americanus*) (fig. 31)

Recognition. L. 20-26 in. (51-66 cm.); Bill 3-9 in. (7.6-23 cm.) (usually 5-8 in., juveniles shorter); WS. ca. 38 in. (96 cm.). The longest of our shorebirds, with a body about the size of a medium-small duck or an oystercatcher, but with slenderer neck and very long, downcurved bill. Such a bill and the general brown-mottled feathering rules out all birds of other families (e.g., ibises; see Picture Keys B and C and fig. 12). Among large shorebirds, the buffy-brown tone, even the cinnamon in the wing, is similar to the Marbled Godwit (fig. 31 and plate 9). The curlew, however, is larger, shows darker primaries, and has the dark markings arranged more as short streaks than bars, both on the back and sides. When the bill is hidden, such features help distinguish these two; but the cinnamon tone is sufficient to tell either from the gray to gray-brown of a Willet (plate 10) or the dusky brown of a Whimbrel, the only other curlew in California, from which the Long-billed is further distinguished by the lack of prominent dark and light stripes on its head (see fig. 31). Individual birds with extremely long bills can be told by that feature alone.

Habits. By virtue of its longest of shorebird bills, this species is able to reach food items no other can. On tidal mudflats, these curlews often wade belly-deep while probing with head below the surface of the murky water (fig. 31). Polychaete worms, burrowing crustaceans, and small snails and bivalve mollusks are their chief *foods* in such places. It is interesting that in at least some mudflats such animals are distinctly scarcer below the approximate depth to which this bird can probe, the lack of oxygen at deeper levels requiring efficient water circulation through a burrow. The largest flocks of Long-billed Curlews in California, however, are found in the Central Valley and Imperial Valley, mostly when the fields there are quite wet. When not feeding, Long-bills repair to safe resting grounds in shallow water or on an open flat, sometimes near but usually keeping somewhat apart from other large shorebirds. Their flight is steady and duck-like. *Voice*: a ringing *cur-LEE*, rising in pitch; also a musical *k-k-kr-r-ring* when taking flight, and a rapid *klee-kle-le-le*. *Nest*: a shallow 8-inch cup in tall grass or on dry spot in marsh. *Eggs*: 4 (or 5), pale buff to greenish, spotted with darker browns to grays; incubation shared by pair, about 30 days. *Downy young*: buff with dark brown to nearly black markings on upperparts.

Range. Breeds from s. B.C. east to Manitoba and south to ne. Calif., n. Utah, N.M., and c. Tex. Winters from n. coast and n. Sacramento Valley of Calif. and s. Tex., La., and S.C. south to Guatemala.

Occurrence in California. Common to Locally Very Common July-early Apr., Uncommon by mid-May and Rare in June, in Central and Imperial Valley fields and grasslands, usually wet ones; fewer on lakeshores and in similar habitats toward coast and on bay tideflats; fewer yet on outer coast beaches and offshore islands. In ne. Calif., Fairly Common to Common Apr.-Sept. nesting Locally in wet meadows Apr.-May.

Marbled Godwit (*Limosa fedoa*) (fig. 31; plate 9)

Recognition. L. 16-20 in. (41-51 cm.); Bill 4-4½ in. (10-11½ cm.); WS. ca. 32 in. (81 cm.). A large shorebird with long gray legs and a long, slightly upturned (sometimes almost straight), pinkish beak with a black terminal half. The feathers are buffy- to cinnamon-brown with small darker markings everywhere except on the paler midbelly and wing linings, while the outer primaries are dusky. At a distance, or when the bill is not visible, this godwit is thus much like the still larger Long-billed Curlew, but the curlew's wing-tips are even darker brown. At freshwater localities, most large, cinnamon-brown shorebirds are Long-billed Curlews; at saltwater localities, flocks of any size are more likely to be Marbled Godwits. However, the two species sometimes mix, especially on and near tideflats.

Habits. Along with the Willets, with which they are usually associated, these godwits make up the bulk of the large probers on the tideflats and sandy beaches. They do not display the preference for surface-dwelling crabs that is true of the Willets, but usually stick diligently to probing for small snails and clams, sand crabs, amphipods, and various worms. Where such prey is found at bill-length depth, these birds often feed with head below the surface of the water, or even into the mud a bit. Only when nearby grasslands are soggy from rains do these coastal godwit flocks gather on them to probe (for earthworms?), but those at more inland localities sometimes feed in drier sites, even taking grasshoppers. Well over 90 percent of the California population of godwits is associated with tidal areas, however, and they often mass closely on their customary high-tide roosts at the upper tide limit in low marshes, or on undisturbed islands or dikes or in shallow ponds. They are more consistently gregarious than some of the other species that frequently join them, such as Willets, curlews, or dowitchers. All these species feed during low-tide periods and roost during high-tide periods, regardless of the day-night cycle. The flight of the godwits is direct and duck-like, except for occasional zigzag tumbling over water by some individuals. *Voice*: loud, abrupt *quk, kerWUK* or *KORkoit* calls; also an easily imitated *koWADica-WADica-WADica-WADica* with rather mellow quality.

Range. Breeds from c. Alberta, s. Sask., and Manitoba south to c. Mont., N.D., and wc. Minn. Winters along coast from c. Calif. and sparsely inland from Sacramento Valley and w. Nev., and from se. Tex. and S.C. south to Central America, rarely to Chile.

Occurrence in California. Common by mid-July, Abundant Aug.-Apr., and Locally Common through May (few in June) on bay or lagoon tideflats, with roosts in nearby marshes, salt ponds, etc.; fewer at same seasons at Salton Sea and on outer coast sand beaches. Far fewer (except Irreg. in spring and July-Sept.) at lakes and on wet fields or grasslands of Central Valley and sw. and se. Calif. Recorded in ne. Calif. Apr.-early May and late July-Sept., at Sierran or s. mountain lakes July-Sept., and Occ. to offshore islands.

Solitary Sandpiper (*Tringa solitaria*) (fig. 32)

Recognition and Habits. L. 7½-9 in. (19-23 cm.); WS. ca. 16 in. (41 cm.). Appropriately named, this sandpiper is never a flocking species even where common. In the far west it is quite rare, but is apparently more regular east of the Sierra Nevada and in southern California than toward our north coast. It should be looked for on small ponds and lakes or even mere puddles, especially those with sparse, low marsh borders, at any altitude. It forages much like the yellowlegs (see next two accounts) and resembles them in general coloration (plate 10), but is smaller and darker olive-brown on the back; all but the central pair of tail feathers are barred black and white like a Spotted Sandpiper, but there is no white line in the wing. The

FIG. 32

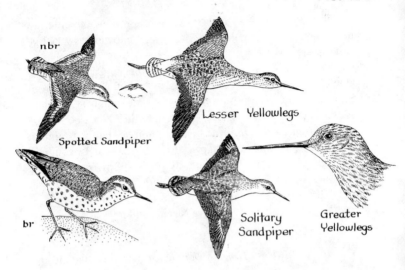

nbr

Lesser Yellowlegs

Spotted Sandpiper

br

Solitary
Sandpiper

Greater
Yellowlegs

legs of a Solitary Sandpiper are dull light green to gray, the bill black. Its flight is often quite erratic, and its characteristic *voice* is a sharp *peep, weet-weet,* two- or three-noted and higher pitched than the call of the Spotted, with a less whistled quality than that of a yellowlegs.

Range. Breeds from c. Alaska and n. Mackenzie south to nw. B.C. and across c. Canada to Labrador. Winters from Tex. and Fla. (Casual s. Calif.) south through Mexico to Argentina.

Occurrence in California. A Rare to Uncommon transient late Mar.-May and mid-July-Oct., more Regular in s. and e. than in c. or n. Calif., on marshy or weedy ponds, lake shallows, and streams; also recorded a few times Nov.-Feb.

Lesser Yellowlegs (*Tringa flavipes*) (fig. 32)

Recognition and Habits. L. 9½-11 in. (24-28 cm.); Bill 1½ in. (3.8 cm.) or less; WS. ca. 20 in. (51 cm.). Very similar to the more common Greater Yellowlegs (see account and plate 10), except slightly smaller with a proportionately shorter, slenderer bill about equal to head length or slightly longer. Difference in size alone is difficult to judge unless the two species are together (as they often are); but the *voice* of the Lesser is distinctly higher in pitch, of thinner quality, and usually consists of only one or two notes on the same pitch. In its foraging and flight habits, the Lesser Yellowlegs is very like its larger relative.

Range. Breeds from nc. Alaska across c. Canada to wc. Quebec and south to ec. B.C. and s. Manitoba. Winters from S.C., La., and Tex. south through e. Mexico to Chile and Argentina, fewer near w. coast from n. Calif. to Mexico.

Occurrence in California. Uncommon to Fairly Common Aug.-mid-Oct. and Mar.-Apr., Irreg. Rare to Uncommon through winter and May-July, in pools or marsh and lake shallows widely through lowlands; perhaps most frequent in se. Calif. Irreg. Rare at any time on tideflats. Recorded Apr. and July-Sept. at lakes in Sierra Nevada and ne. Calif.

Greater Yellowlegs (*Tringa melanoleuca*) (fig. 32; plate 10)

Recognition. L. 12½-15 in. (32-38 cm.); Bill 2-2 1/3 in. (5-6 cm.); WS. ca. 25 in. (63 cm.). A fairly large prober of more slender build than the Willet (fig. 31 and plate 10), mottled gray above with a white eye-ring and belly, straight black bill, and *bright yellow legs*. In flight the tail shows whitish (narrowly barred with dark) and the upper tail converts pure white. The wings are all dark. All of these features are also true of the slightly smaller Lesser Yellowlegs, from which this commoner species can be told by its longer bill (about one and a half times head length) and different voice.

Habits. The two species of yellowlegs are by preference freshwater foragers, this one being the most widespread of the larger shorebirds in fall and winter in the Central Valley. They are very active when seeking *food*, wading in shallow water or isolated pools and catching swimming insects, crustacea, or even small fish by quick lunges. On occasion they also probe in mud substrates, as do related species. When alarmed or migrating, yellowlegs may bunch together in flight, but otherwise they are much less gregarious than most probing shorebirds, a feature possibly associated with their special food-getting method. *Voice*: in flight, a series of three or four clearly whistled notes, as *whee-whee-wheu*, the last note dropping slightly in pitch; single notes may also be given when on the ground.

Range. Breeds from sc. Alaska and c. B.C. east in the spruce muskeg belt across Canada to Newfoundland. Winters from w. Ore. and n. Calif. (coast and Sacramento Valley), across s. U.S. to S.C. and south through Middle and South America.

Occurrence in California. Fairly or Locally Common Aug.-Apr., Uncommon in May and July, and Rare in June, in shallow waters and even to grassland pools of Central and Imperial valleys; fewer at same seasons in other cismontane lowlands, and Irreg. on bay and coastal lagoon margins. Also recorded at lakes in Sierra Nevada Apr.-May and July-Aug. and in s. Calif. mountains Sept.-Dec., and on ne. plateau Aug.-early Sept. (though no doubt also there in spring).

Willet (*Catoptrophorus semipalmatus*) (fig. 31; plate 10)

Recognition. L. 14-17 in. (36-43 cm.); Bill 2-2½ in. (5-6 cm.); WS. 25-28 in. (63-71 cm.). In flight the easiest of all large shorebirds to identify, because of the *broad white band along the blackish wings* and the white upper tail coverts (see fig. 31). In nonbreeding feathering the upperparts are otherwise rather smooth gray including most of the secondary coverts, which usually conceal the wing pattern when the bird is at rest. There is a narrow white ring around the eye and a broad white stripe from above the eye to the bill. The underparts are whitish with a variable gray wash on the chest and throat, the legs are *blue-gray*, and the straight bill is black or with gray basally. In breeding feathering (Apr.-July) there are numerous small dusky bars on the chest and sides, and fine dusky streaks on the upperparts.

Habits. Most widespread of the common large shorebirds in the state, and thus the usual standard to which others can be compared, the Willet is also one of the most southerly nesters in its family. They breed south to just east of the Sierra Nevada, and farther south on the Atlantic coast. Late spring birds often take up territory on some marshes about San Francisco Bay near grassland or other partly dry flats; they might well begin to breed if undisturbed through a number of

193

seasons, as may come about through the new San Francisco Bay National Wildlife Refuge. Most of the year, the large flocks in and near tidal areas show feeding and roosting routines similar to those of the Marbled Godwit, but Willets in smaller numbers are also scattered about salt marshes, ponds, sloughs, etc. In these tidal areas the Willet's *food* shows a high percentage of crabs, but it also probes deep into the mud for worms and small mollusks when necessary. Inland, aquatic or mud-burrowing insects and crustaceans are more important foods.

Voice: a loud, harsh *krrak*, or extended to *ker-r-r-RAK-er-RAK-er-k-k-kerRAKit* in long series when alarmed; also shorter *kak* and *keleek* notes, and when on territory a musically whistled *ter-WILL-WILL-it* repeated in long series. *Nest*: a hollow 6-7 inches in diameter, lined with dry grasses, in grassland near or on mound in low parts of marsh growth. *Eggs*: 4, buff to light greenish, irregularly marked with variably sized dark to medium browns and grays. *Downy young*: underparts buffy-white, upperparts pale to medium buff or light brown with dark brownish markings, including on head.

Range. Breeds from s. Ore., Idaho, and prairie areas of s. Canada south to ne. Calif. w. Nev., n. Utah, Colo., and S.D.; also along e. coast from Nova Scotia to Fla. and Tex. and in Bahamas and West Indies. Winters from nw. and interior c. Calif. south on or near Pacific Coast to Chile, and from Va. and Tex. south to n. South America.

Occurrence in California. Abundant July-Apr., and Locally Common to Uncommon May-June, on tideflats of bays and coastal lagoons and in nearby salt marshes and salt ponds (especially for roosting); somewhat fewer at same seasons on shores of Salton Sea and on sandy outer coast beaches; still fewer on rocky shores and coastal grasslands and freshwater. Irreg. Rare to Locally Common in Central Valley and inland in wc. and sw. Calif. about lakes and marshes and on wet grassland and fields. In ne. Calif., Uncommon to Locally Fairly Common Apr.-Aug. (probably occurs to Sept. or Oct.) nesting May-June.

FIG. 33

nbr

Stilt Sandpiper

Wandering Tattler

nbr

br

br

Surfbird

nbr

Sanderling

Wandering Tattler (*Heteroscelus incanus*) (fig. 33; plate 9)

Recognition. L. ca. 11 in. (28 cm.); WS. unknown. A medium-large shorebird of rocky areas, with straight black beak about as long as the head, unmarked slate-gray upperparts including wings and tail, white line above the eye, and dull yellowish to pale green legs. In fall and winter, the underparts are whitish with a gray wash across the chest. In breeding feathering, the whole underparts are barred gray and white, and the side of the neck is streaked with white.

Habits. On migration and in winter, the Tattler is mostly found on the outer rocky coast, but a few (esp. Aug.-Oct.) regularly enter large bays such as Humboldt, Tomales, and San Francisco bays and are then to be found along breakwaters and edges of landfills, or they may forage at times on smooth sand or mud flats. On the Pacific islands, this species occurs on almost all types of shores; but in its preferred habitat on the California coast, with much dark rock and seaweed, it gains the same protection by its dark plumage as do the Oystercatcher, Surfbird, and Black Turnstone. Tattlers occur singly or in small groups in this habitat, and may be difficult to find as they move slowly about just above the surf, pecking amid the algae, mussels, etc., for any small animals as food. Their flight is strong and rather direct. *Voice*: a high-pitched, rapid whistle *li-li-li*, more liquid than the calls of other species on the rocky shore.

Range. Breeds above timberline in mountains of sc. Alaska to nw. B.C. Winters on seacoast from s. Calif. (Occ. to Ore. and Wash.) south to nw. South America and on islands of c. and cs. Pacific Ocean.

Occurrence in California. Primarily an Uncommon transient Apr.-May and July-Oct., Fairly Common late Apr.-May and Aug., on rocky shores of outer coast and offshore islands; Rare to Uncommon at same seasons on rocky bayshores (Occ. even to tideflats). Irreg. Rare through winter and in June in same habitats.

Spotted Sandpiper (*Actitis macularia*) (fig. 32)

Recognition. L. 7-8 in. (18-20 cm.); WS. 13-14 in. (33-36 cm.). A medium-sized sandpiper that "teeters" its hind parts or its whole body as though balancing on a wire. Many shorebirds "bob" the head and foreparts suddenly up and down, but this is the only shorebird in the U.S. that teeters (though members of the songbird family of pipits and wagtails have a somewhat similar mannerism). In color the Spotted Sandpiper is smooth olive-brown above and white below, with *round* dark brown spots over most of the underparts in breeding feathering, unspotted in the immature and nonbreeding adult. A white line over the eye, a white line along the wing fading out before the tip, and black and white barred sides of the tail are further marks. The legs are dull yellow-gray, and the bill is black with some pinkish at the base.

Habits. The "Spotty" is not only unique in its teetering habit, but is also distinctly nongregarious and prefers shores with a jumble of wood, rocks, or debris—all quite unlike most sandpipers. In winter, however, it may be found foraging between the tide levels of a rocky coast frequented by Surf-birds and turnstones, but even there behaves independently of these others. Its *food*, obtained from the surface or by prob-ing, is any of the small or soft-bodied invertebrates that lodge or hide amid the debris or on nearby sandy areas. In summer, a Spotted Sandpiper sometimes extends its foraging a consider-able distance from the water's edge, when attracted by insects in abundance.

Local flights of a Spotted are usually low over water, with quick series of a few wingbeats each interspersed with brief glides on downcurved wings in a unique pattern (fig. 32). When traveling a distance or going over uplands, however, it proceeds with vigorous flaps on the slightly erratic course typical of most shorebirds. *Voice*: a clear, sharp *peet-weet*, the second note often lower in pitch; also this "stuttered" and various softer notes, some extended into long series. *Nest*: a hollow in the ground, sparsely lined with grass or drift mate-rials, in the open or amid logs, snags, rocks, or clumps of vegetation. *Eggs*: usually 4, pale buff, spotted and blotched with dark brown; incubation mostly by the male, about 21 days. *Downy young*: mottled gray and brown; swim and even dive if pursued into water; teeter from the very first day.

Range. Breeds from nw. Alaska and across Canada (except ne.) south to Monterey Co. and s. mountains of Calif., c. Ariz., Tex., and N.C. Winters along coast from B.C. and inland from n. Calif. and across s. U.S. south to sc. South America.

Occurrence in California. Uncommon to Locally Fairly Common Aug.-May on sw. rocky outer coast, lake shores, and river mouths, especially debris-littered ones; somewhat fewer at same season along streams inland and in similar habitats of se. Calif. and through winter in c. and nw. lowlands. In the Sierra Nevada and n. Coast Ranges, Uncommon Apr.-Sept., Locally Fairly Common in Aug., along streams and lakeshores; fewer on wooded lake and stream margins in s. mts. and wc. Calif. Statewide records of eggs, May-July; of dependent young, June-Aug. Recorded also in ne. Calif. Apr.-Aug.

27. SUBFAMILY SCOLOPACINAE (SNIPE, DOWITCHERS)

(Picture Key B)

Common Snipe (*Capella gallinago*) (fig. 34)

Recognition. L. 10½-11½ in. (27-29 cm.); WS. ca. 15 in. (38 cm.). A medium-sized prober with a very long, straight bill about twice as long as the head. The light buff line through the center of the dark crown and the buff streaks on the dark brown back distinguish the snipe from all shorebirds with similar bill and body shape. Bill pinkish-brown with a blackish tip; legs dull green. In flight, the wings show all dark, but the tail is rusty with two narrow black bars.

Habits. Until an intruder in its soggy habitat comes quite close, a snipe remains motionless amid the grass and usually escapes detection even by persons looking for it. When it flies up suddenly, however, it almost always utters a raspy *scaipe* note or two as it flies off in an erratic course, only to pitch as suddenly down and swerve to alight at a new hiding place. Even when many are flushed from one good feeding area, snipe usually do not fly in a compact flock. Because of the difficult target that snipe thus make, they have been kept on the game list with an open hunting season. This is not true of most shorebirds, which have flocking and feeding habits that make them especially vulnerable to extirpation through hunting, when coupled with their low reproductive potential. In their food-getting, snipe walk slowly along and probe quickly here and there in the mud, even full bill-length. *Food*: earthworms, burrowing larvae of beetles, moths, cicadas, etc., some surface insects and even occasional seeds of sedges and grasses.

On the breeding grounds, male snipe perform an aerial display in which "winnowing" sounds are produced by the stiff lateral tail feathers as the wings beat forcefully past them about fifteen times, while the bird performs a shallow dive. They also call from the ground or atop a fencepost, *teek-a teek-a teek-a* . . . in a long see-saw rhythm. *Nest*: a grassy cup in a tussock of vegetation in or at edge of marsh. *Eggs*: 4 (rarely 5), light brown blotched with dark brown; incubation

FIG. 34

Short-billed Dowitcher

nbr

Common Snipe

Red Knot

18-20 days. *Downy young*: marked heavily with varied shades of brown plus some white spots.

Range. Breeds in subarctic areas around the world, in America south to ce. (rarely s.) Calif., Ariz., Colo., and nc. and ne. U.S. Winters from s. B.C., Ore., Utah, Colo., and ce. U.S. south to Brazil, and from Europe and s. and e. Asia to Africa and Philippines.

Occurrence in California. Uncommon to Locally Common Oct.-Apr., Irreg. and Uncommon Sept. and May, on wet grasslands and short marshes widely through lowlands and cismontane foothills; Uncommon to Rare same season in upper part of salt marshes. Locally Rare through summer (breeding Apr.-May) in Central Valley short marshes and rice fields. Recorded July-Jan. on wet meadows of s. mountains. In ne. Calif. chiefly Mar.-Dec., Fairly Common in summer, nesting chiefly June-early Aug., though small young noted as early as mid-May; fewer on Sierra Nevada wet meadows at similar seasons.

Short-billed Dowitcher (*Limnodromus griseus*) (fig. 34; plate 10)

Recognition. L. 10½-12 in. (27-30½ cm.); Bill ca. 2-2½ in. (5-6½ cm.); WS. ca. 19 in. (48 cm.). Dowitchers are chunky, medium-sized probers with long, straight bills, light greenish legs, and dark and light barred tails with whiter upper tail coverts merging to a pure white tapered patch extending well onto the back. Although usually visible only from above as the bird flies, this white patch and the nearly all-dark wings are characteristic. In nonbreeding feathering, both species of dowitchers show this pattern, are brownish-gray fading to a whitish belly, and have a white line over the eye. Since only a minority of individuals can be distinguished by their extremely long or short bills, dowitchers in this feathering can usually not be identified to species in the field except by voice. In breeding feathering, evident in most by mid-April, both species show cinnamon-brown on the underparts. This varies greatly in extent in the Short-billed, which has only faint to distinct black *spots* (not bars) on the *sides of the breast* (plate 10), though these group into bars on the flanks.

Habits. This species is the largest of the several species of oft-associated rapid probers of tidal mudflats (see also an average sandpiper in Picture Key B), and often mixes with the slower but larger waders such as the Willet, Marbled Godwit, and avocet. Until a detailed study of their morphology and habitat differences was made by Frank A. Pitelka in 1950,[*] the two species of dowitchers were often confused even in museums. Each is known to occur in the habitat preferred by the other, especially during migrations (see Graphic Calendar in Appendix), but when settled for the winter the Short-billed is almost strictly a saltwater bird on the California coast. It is very gregarious, and flocks mass closely in flight and pack even more densely when waiting out the high tide on their roosts in marshy or open pools or on salt-pond dikes near the bay or lagoon. *Voice*: a mellow *tu-tu-tu* or *kew-kew-kewp*, usually uttered in series of up to six or so notes when taking flight, irregularly or singly at other times, and distinguishable from calls of the Long-billed by their lower pitch and more resonant quality. Both species can be heard and compared directly in spring and fall where tideflats and freshwater pools are close together.

[*]Univ. Calif. Publ. Zool. 50:1-108.

Range. Breeds near the coast in s. Alaska and from s. Mackenzie across c. Canada probably to n. Quebec and south to ec. B.C. and probably c. Alberta. Winters along coasts from c. Calif., S.C. and Tex. south to Peru and Brazil, and in small numbers inland in sw. U.S.

Occurrence in California. Abundant mid-July-Oct. and late Mar.-Apr., Very Common or Common through winter (except Rare in n. Calif.) to mid-May, and few stragglers in June, feeding on tidal mudflats and roosting on nearby salt marshes or salt-pond dikes, pools, etc.; far fewer on sandy beaches and at lakes and ponds near coast. In se. Calif., Regular in spring and July-Oct. on shores of Salton Sea, fewer on freshwater and irrigated fields. Migrants Irreg. Rare to Uncommon in Central Valley and in Sierra Nevada. Inland, numbers relative to Long-billed often not established.

Long-billed Dowitcher (*Limnodromus scolopaceus*) (plate 10)

Recognition. L. 11-12½ in. (28-32 cm.); Bill ca. 2½-3 in. (5.7-7.6 cm.); WS. ca. 20 in. (51 cm.). For distinctive features of dowitchers as a group, see Picture Key B and Short-billed account above. Only the extremely long-billed individuals of this species, in which the bill is distinctly more than twice as long as the head in normal alert posture, can be told from the Short-billed (fig. 34 and plate 10) by this feature alone. Spring birds (and some in late summer) in breeding feathering can be distinguished by their entirely cinnamon underparts with distinct *bars* of black on the *sides of the breast* as well as the flanks. When once learned well, the voice is the best field mark and is reliable for birds at any season.

Habits. From Dec. through Feb. any dowitchers seen at inland points in California (except possibly Salton Sea) are almost certainly Long-bills. They are also then in coastal lowlands, and some even occasionally forage in salt water along with the Short-bills, as they do fairly commonly in spring and fall when on migration. The *food* taken by Long-bills on their more typical freshwater habitat consists largely of midge and fly larvae, aquatic or wet ground "earth"-worms, small burrowing crustacea, etc. — all obtained by probing in muddy places. *Voice*: a sharp *kip* or *keep* more often uttered singly than is the lower-pitched call of the Short-billed, but also given in series frequently when alarmed; both species give soft, rather liquid calls when feeding.

Range. Breeds in ne. Siberia, nw. Alaska, and nw. Mackenzie. Winters from nw. and nc. Calif., w. Nev., s. Ariz., and N.M., to Fla., and south to Guatemala.

Occurrence in California. Very Common to Occ. Abundant Aug.-May, fewer from mid-July, in shallows of Central Valley and se. Calif. lakes, short marshes, and wet fields; somewhat fewer in same habitats of cismontane lowlands elsewhere, and fewer yet to bay tideflats and Occ. to river margins. In ne. Calif., Common or Very Common Apr.-May and Very Common to Abundant Aug.-Oct. Recorded at s. mountain lakes mid-Apr. and late July-Dec.

28. SUBFAMILY ARENARIINAE (TURNSTONES)

(Picture Keys B and C)

Ruddy Turnstone (*Arenaria interpres*) (fig. 35; plate 11)

Recognition. L. 8-9½ in. (20-24 cm.); WS. ca. 18 in. (46 cm.). Both Ruddy and Black Turnstones show a distinctive set of white areas when seen from above in flight — basal half of tail, line along each wing like that of many shorebirds but wider, triangular patches on the inner forepart of wing, and an oval area on rump and lower back. Turnstones also have a stout but bluntly pointed and slightly upturned beak. In fall and winter, the Ruddy on the ground appears medium brown on the head and upperparts (all the white patches being covered) and white below except for brown to dusky patches on each side of the chest. Its legs are dull orangish brown. In breeding feathering, the calico pattern and bright orange legs are unmistakable.

Habits. The unique turnstone foraging method, for which its beak is specially adapted, is to search the nooks and crevices on a pebble-strewn or seaweed-covered shore, flipping over these objects to expose the small animals that often abound there. At other times, Ruddy Turnstones may be seen foraging from the surface of a smooth beach or tideflat much

FIG. 35

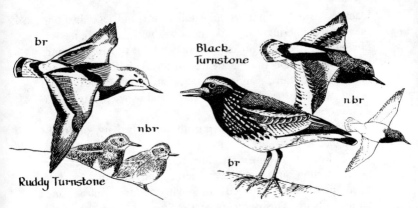

as plovers do, but with occasional probings also, like a sand-piper. They sometimes even excavate pits in sand or stiff mud in pursuit of burrowing forms. A great variety of invertebrates is thus taken as *food*, but most are the amphipods, isopods, young crabs, and insects that gather about the drift line or hide in seaweed masses. *Voice:* a strident, several-noted call that is well approximated by the Hawaiian name of the species *akekeke*, uttered very rapidly.

Range. Breeds around the world in the arctic tundra, in America south to w. Alaska and n. Hudson Bay. Winters along seacoasts from c. (Occ. n.) Calif. and S.C. South to Chile and Brazil, and in Old World from w. and s. Eurasia to South Africa, Australia, and many Pacific islands. A few vagrants inland.

Occurrence in California. Fairly Common Aug.-Sept. and Apr.-early May, Uncommon rest of year, on rocky shores and sand beaches, including bays and offshore islands; fewer on tidal mudflats; stragglers rare through summer. On San Diego Bay Occ. Common Dec.-Feb., but Rare to Uncommon in c. Calif. then. In se. Calif. reported Apr.-May and late July-Sept., chiefly at Salton Sea. Also noted in Central Valley in May and late July-early Oct.

Black Turnstone (*Arenaria melanocephala*) (fig. 35)

Recognition. L. ca. 9 in. (23 cm.); WS. ca. 18 in. (46 cm.). Distinctive white areas in flight as in the Ruddy Turnstone (see account), from which it can be told by its dark slate to blackish color, including the whole chest and legs. In non-breeding feathering the dark areas are without markings or obvious light edgings of feathers; only a faint whitish chin and short mark behind the eye are exceptions. In breeding feathering (Apr.-May into summer), there are conspicuous white feather tips on the chest, side of neck, and wing coverts, and a white area in front of the eye.

Habits. Although it performs the same variety of food-getting actions as the Ruddy Turnstone (see account), the Black shows a stronger attachment to rocky coastlines and hence is more apt to be seen peering and thrusting its way amid the attached seaweeds, or poking about in mussel and barnacle beds. The darkness of its plumage, with the white belly turned away, matches the color of many such areas well and makes it difficult to pick out even the usual small flock of slowly moving birds, as is true also of the Surfbirds and Wandering Tattlers often associated with them. When forced to fly, by the imminent surge of a larger wave or by intruders on its rocky domain, the flock suddenly becomes conspicuously *black and white* as they converge to seek safer ground. Often they call as they take wing, their typical *voice* being similar to that of the Ruddy, but somewhat higher pitched.

Range. Breeds along the coast of w. and s. Alaska south to vicinity of Sitka. Winters on the coast from se. Alaska to nw. Mexico. Casual inland in migration.

Occurrence in California. Fairly Common from mid-July and Common to Locally Very Common mid-Aug.-Oct. and Apr.-mid May, Locally Common through winter, on rocky to muddy or sandy shores of bays. A few migrants recorded inland, mainly on Salton Sea.

29. SUBFAMILY CALIDRIDINAE (CALIDRINE SANDPIPERS AND ALLIES)

(Picture Keys B and C)

Surfbird (*Aphriza virgata*) (fig. 33)

Recognition. L. 10 in. (25 cm.); WS. Unknown. A chunky, medium-sized shorebird (large sandpiper) of plover proportions, except for the somewhat longer beak without a distinctly swollen middle. In fall and winter, the head and body are smooth gray except for an elongate white spot above the lores and gray streaks on the white belly and whitish throat. The tail is blackish with an abruptly white basal half, this and a white line along the wing showing conspicuously in flight. However, there are no other large white areas as the frequently associated turnstones have (see fig. 35 and plate 11). The beak of the Surfbird is dusky to somewhat yellowish in the basal half, the legs dull yellow-green. Spring migrants may be in breeding feathering, with white-bordered wedge-like black spots in ill-defined streaks on the head and body.

Habits. This aptly named species is almost as highly restricted to the surfswept outer shore rocks as is the Black Oystercatcher, but is much more generally distributed. When *feeding* on the seaweed-covered rocks or in the mussel and barnacle beds, Surfbirds pick from the surface like plovers or poke into crevices for almost any of the abundant small invertebrate animals so numerous there. Where a rocky headland or jetty provides enough forage area along an otherwise sandy shore, flocks of Surfbirds and Black Turnstones (usually mixed) often stay on such an "island" of their habitat all winter. If disturbed from one part, they wheel about over the nearby rough water and realight in another part of the rocky area. When the tide covers their feeding grounds, they usually assemble for rest and sleep on some protected rock a little above the splash zone. *Voice:* a plaintive two- or three-noted *ke-week* or *kee-weeh-a*; less often vocal than the turnstones in winter.

Range. Breeds in the *alpine* tundra of sc. Alaska mountains. Winters along the w. coast from se. Alaska to s. tip of South America.

Occurrence in California. Fairly Common late July-Apr., sometimes Locally Common Aug.-Dec. and Apr., Uncommon in May, on rocky shores of outer coast and offshore islands. Rare on sandy beaches, and recorded Oct.-Apr. in large bays.

Red Knot (*Calidris canutus*) (fig. 34; plate 10)

Recognition. L. 10-11 in. (25-28 cm.); Bill ca. 1½ in. (3.8 cm.); WS. 20-21 in. (51-54 cm.). A large sandpiper of quite stocky build, the neck and legs of only moderate length and the straight bill about equal in length to the rather large head. In fall and winter, they are grayer above and paler on the head than the similar-sized dowitchers (fig. 34 and plate 10), with light feather edges in immatures. In spring birds approaching breeding condition, and some of the returning fall birds, there is extensive light reddish-brown over the underparts, face, and sometimes the back. The bill is gray and the legs dull to bright pale greenish. In flight, Red Knots can be told from dowitchers by the lack of a conspicuous white area on the lower back, although the tail base shows whitish and there is a narrow white line along the base of the secondaries.

Habits. Formerly considered quite an uncommon migrant in California, especially in spring, the Red Knot has apparently increased its use of the Pacific coastal migration route over the past fifty years and now reaches concentrations of 500 to over 1000, especially on San Diego Bay. They are prominently gregarious on their preferred foraging habitat of broad sandy-mud tideflats, probing as do others of their family (see Dunlin account). They also usually mix with dowitchers or the smaller sandpipers on safe high-tide roosts when the tideflats are covered with water. In flight, however, they frequently separate from the others, traveling faster in tight flocks in which their long wings beat amazingly close to their neighbors. *Voice:* low-pitched, of nasal to slightly raspy quality, either single notes or somewhat rolled.

Range. Breeds in n. arctic regions around the world, south in America only to n. Canada mainland. Winters on coasts from Mass. to Fla. and Tex. and from c. Calif. (Occ. Ore.) south to s. Argentina. Much commoner as migrant on Atlantic coast than in Calif. In Old World, winters from s. Eurasia to s. Africa and New Zealand.

Occurrence in California. Fairly Common to Common all year but only Locally through winter, Irreg. Very Common Aug.-Jan., and Irreg. Uncommon late May-early July, on sandy-mud tideflats of bays and coastal lagoons of c. and s. Calif.; somewhat fewer on n. coast, and on soft mudflats and outer coast beaches anywhere and at Salton Sea (but no record there Nov.-Feb.). Migrants also recorded in Central Valley and se. lakes.

Sanderling (*Calidris alba*) (figs. 33, 36; plate 9)

Recognition. L. 7-8½ in. (18-22 cm.); Bill ca. 1 in. (2.5 cm.); WS. ca. 15 in. (38 cm.). A medium-sized sandpiper (large "peep") with black legs and a straight black bill shorter than the head and somewhat stouter toward the tip than in most sandpipers. In nonbreeding feathering, Sanderlings are the whitest of sandpipers — pure white below, white forehead and broad line above the eye, and the upperparts otherwise pale gray. The flight feathers are dusky with a prominent white line along their bases made still more contrasty by the dark primary coverts, the last sometimes showing when the wing is folded (fig. 36). In breeding feathering (shown by some birds in Apr.-May) there is a bright buff suffusion with small black-ish spots over the head, neck, and chest, and on the back and upper wing coverts.

Habits. This is the one kind of sandpiper that is usually noticed and admired by the casual visitor to sandy beaches, for Sanderlings are the only ones that regularly feed in the back-wash of the big "comber" waves. A whole flock moves in to the ebbing water on twinkling feet, probes avidly for the brief time available, and then retreats up the beach again to escape the next crashing wave. Their *food* in such places consists predominantly of the younger stages of the abundant burrow-ing sand crab (*Emerita analoga*), itself a plankton strainer dependent on the surf for both food and oxygen. On calmer shores, and even occasionally on the open outer beaches, Sanderlings also feed on amphipods and other small prey by more leisurely search and probing as they walk along in usual sandpiper fashion. Even in the larger bays and wherever they occur inland, however, they show strong preference for sandy substrates. Sanderlings are also persistently gregarious and will mix with any other flocking species present, but other probers usually fail to keep up with them in a waveswept area. When the tide is high or the wind too strong, Sanderlings mass together on the high beach or gather in small groups in the lee of the wispy plants of nearby dunes. *Voice:* sharp *twik* or *quit* notes as a flock flies, and rather liquid twitterings when feeding.

Range. Breeds around the world in the high Arctic, even to n. Greenland, south in America only to n. part of Hudson Bay. Winters from s. B.C. and Mass. south along the coasts to s. Chile and Argentina; also on shores of w., s., and e. Eurasia, Africa, and through most Pacific Islands to Australia — the most cosmopolitan of shorebirds.

Occurrence in California. Very Common to Locally Abundant late July-Apr., Common through May, and few Occ. to mid-July, on sandy beaches of outer coast; somewhat fewer on bayside sandflats and beaches; fewer yet reach fully rocky or muddy shores. Uncommon, Occ. Fairly Common, chiefly as migrant, at coastal lakes and at Salton Sea and lakes of se. Calif. Recorded in Central Valley, ne. plateau, mountain lakes, inland in sw. Calif., and on offshore islands.

Western Sandpiper (*Calidris mauri*) (fig. 36; plate 9)

Recognition. L. 6-7 in. (15-18 cm.); Bill 1-1½ in. (2.5-3.8 cm.), or only 2.0 cm. in immatures; WS. ca. 13 in. (33 cm.). This species, the most abundant of the small sandpipers or "peeps" on the mudflats of our bayshores, is between the Dunlin and Least in size. It is distinguished by black legs and a black beak about equal to the head or a *little* more in length, rather thick at the base and slightly downcurved toward the tip (less than in the Dunlin). In nonbreeding feathering it is grayer and somewhat paler above than either the Dunlin or Least, but darker than the Sanderling. There is only a faint wash or fine streaking on the upper chest. As in all peeps, there is a broad whitish line over the eye. Spring birds approaching breeding condition (Apr.-May) and some of those returning in late summer after presumably interrupted nestings are bright rusty above, especially on the crown and upper back, and are more streaked on the chest. In flight, the typical "peep" pattern of white sides on the upper tail coverts and a narrow white line lengthwise in the wing are evident.

Habits. Among the most gregarious of all our water birds,

FIG. 36

SANDPIPERS (calidridine) in non-breeding feathering

Western Sandpiper
small, 6-7", bill as long
as head, curves
down slightly

Least Sandpiper
smallest, 5-6½"
bill short, legs dull
pale green

Sanderling
medium, 7-8½", pale
bill short straight, legs black
sandy beaches

Baird's Sandpiper
medium, 7-8", bill
almost as long as head
"scaly" back, legs black

Dunlin

medium, 8-9"
bill down
curved, longer
than head
legs black

nbr

Rock Sandpiper
large, 9", bill yellowish
legs gray-green
rocky coast

Pectoral Sandpiper
large, 8-9½" bill &
legs yellowish

these little travelers from the Arctic can still be found in flocks of 50,000 or so in the most favorable parts of the San Francisco Bay region. In late September, this many may be packed shoulder to shoulder on the open ground of a diked-off flat, in shallow pools of a low marsh, or on an isolated sandspit or island as they gather to wait out the period of high tides which cover their chief feeding grounds. Their eating habits and *diet* are similar to those of the Dunlin (see account), often associated with them. It is when these massed flocks are disturbed before their feeding grounds are available that one can thrill to the spectacular sight of their wheeling and turning in unison, showing white underparts and gray backs in quick alternation as they course about low over the area. Continued harassment on such roosts is probably an important factor in limiting their period of stay in that locality – if not, indeed, the actual number that ultimately survive. If the extensive tideflats of our larger bayshores are to continue to support these birds in any abundance, not only those tidal areas but also suitable roosts free from disturbance must be preserved. Where protected places are not available nearer to their feeding areas, these little birds may fly up to 5 miles or more to reach them and return to the roost twice each day according to the tide schedules. Enroute, these and other sandpipers fly in loosely bunched or long sinuous flocks on a rather direct course, except for deviations so as to fly over water as much of the way as possible.

Toward spring, Western Sandpipers become quite aggressive toward each other and many can be seen carrying the tail high even as they feed, then running frequently with lowered head and upraised tail to drive off a neighbor. Recent studies by banders have shown that the start of the fall migration consists entirely of adults, with the first young-of-the-year appearing in mid-Aug. *Voice:* high-pitched, slightly rattled *kreet* notes, not so thin and drawn-out as the call of the Least Sandpiper.

Range. Breeds on the coast of w. and n. Alaska, and winters chiefly on coasts from n. Calif., N.C., and Gulf of Mexico south to n. South America, also in coast slope valleys of Calif. and southward.

Occurrence in California. Abundant to Very Abundant July-early May, declining to Uncommon in mid-June and then rising again, on muddy tideflats of bays and coastal lagoons, roosting (or some feeding) in nearby salt marshes, salt ponds, or other open land with pools. Also Abundant at same seasons on shores of Salton Sea; fewer on sandy shores both in bays and outer coast; and fewer yet on rocky shores. On lakeshores and muddy pools in wet fields in cismontane lowlands (chiefly Central Valley and wc. and sw. Calif.) apparently Common July-May (though few Oct. and no Nov. records), Occ. Abundant Apr. and July-Sept.; Irreg. to shoals and margins of streams. Recorded also in same seasons at mountain lakes and on ne. plateau, where Irreg. Abundant May and Sept.

Note: The **Semipalmated Sandpiper** (*Calidris pusilla*) has been reliably reported in small numbers (mostly single individuals) some thirty times in California, Apr.-May and July-Sept. (See Graphic Calendar in Appendix.) Except for the lack of reddish-brown above, this species is very similar in size and color to the Western (fig. 36 and plate 9), but has a beak slightly shorter than the head. Fall and winter adults of the Western usually have small traces of the rusty above, but immatures do not, and some of them are almost as short-beaked as the Semipalmated — making field identification then very difficult.

Least Sandpiper (*Calidris minutilla*) (fig. 36; plate 9)

Recognition. L. 5-6½ in. (13-16½ cm.); Bill 2/3-3/4 in. (1.6-1.9 cm.); WS. 11-12 in. (28-31 cm.). Smallest, by a bit, of the small sandpipers or "peeps," but not safely distinguishable by size alone except when compared directly at similar distances. Both the Least and Western sandpipers change from rather gray-brown plumage in fall and winter to a richer brown with more extensive wash on the chest in spring, beginning in late March in some individuals. The Least is darker above than the Western in the dull feathering and less rusty in spring, and always has more of a brown wash on the chest than the Western in comparable feathering. The surest identification, however, is by the thinner, shorter bill (somewhat less than head length) and the dull yellow-green legs of the Least, the latter being visible only at close range and in good light. In flight, the Least shows the pattern of most peeps — white on sides of the upper tail coverts and a narrow white line in the wing.

Habits. In keeping with its small size, this species displays a preference for *feeding* along narrow channels and in mudholes or at the edge of a salt marsh. The birds are usually scattered singly or in small flocks while they probe for small fly larvae, amphipods, and worms in the soft mud or pick these or other small, soft-bodied animals from the film of water or mud surface. When feeding in tidal areas and the tide is well out, many Leasts can still be found in such places in the marshes, while others spread far across the broad mudflats with other shorebirds, the Leasts no doubt gleaning smaller prey. The flight of this species is usually erratic when the bird is alone, but flocks proceed more directly except for occasional wheelings (see Western Sandpiper account). Where sandpipers gather on roosts at high tide, the Leasts almost always mix with Westerns or Dunlins, though frequently only on the periphery of a large flock. *Voice:* a high, thin *kree* or *screet.*

Range. Breeds from c. Alaska, n. Canada, and Labrador south to se. Alaska, s. Yukon, sc. and se. Canada. Winters along coasts from Ore. and N.C., inland from n. Calif. and s. Utah, to Tex. and south to Peru and Brazil.

Occurrence in California. Common by July to Very Common or Abundant late Aug.-Apr. and dwindling to Uncommon by late May (Occ. few in June) on lakeshores, marshes, and wet fields, especially in se. Calif. but also widely through cismontane lowlands and to salt-marsh borders; fewer on tideflats proper, and still fewer on sandy and rocky coasts. Common in ne. Calif. Apr.-May, Abundant July-Sept. Occ. to Dec. Also at Sierran lakes July-Aug. and s. mountain lakes Apr. and Sept..

Baird's Sandpiper (*Calidris bairdii*) (fig. 36)

Recognition. L. 7-8 in. (18-20 cm.); WS. ca. 16 in. (41 cm.). A large "peep" about the size of a Sanderling, but with slenderer bill almost as long as the head. The chest and head are quite buffy (though not as much as the breeding feathering of the Sanderling and some others) and the back feathers of immatures are dusky with broad buffy edges, giving a distinctive "scaly" appearance. The legs and bill are black (compare with Pectoral Sandpiper, fig. 36). In flight the Baird's shows the white on either side of the upper tail coverts like most peeps, but the stripe along the base of the secondaries is faint.

Habits. Another primarily midcontinent migrant, of which only the most peripheral small groups reach the Pacific coast. Even so, most inland records of it in California are spottily distributed except near the Salton Sea, perhaps partly due to its predilection for sandy shores. Careful search of all sandpipers about the margins of large streams and lakes at the appropriate season, especially east of the Sierra, might well disclose that this species occurs there regularly also. In its feeding and flight behavior, the Baird's resembles most other peeps (see Dunlin and Western Sandpiper accounts). *Voice:* a slightly rolled, soft *kree* or *kreet.*

Range. Breeds from ne. Siberia across n. Alaska and Canada to w. Greenland. Winters on high plateaus from n. Ecuador to w. Argentina and Chile, rarely north to Central America (and Casual s. U.S., although some of the Dec.-Jan. occurrences reported in Calif. are suspect). Migrates primarily through mid-U.S., but immatures also in small numbers on both e. and w. coasts, especially in fall.

Occurrence in California. Uncommon to Locally Fairly Common July-early Oct., Occ. Rare Nov.-Jan., and Irreg. Rare to Fairly Common late Mar.-May on sandy shores of outer coast, river mouths and lagoons; fewer in nw. Calif., at Salton Sea, and Occ. on other inland shores, including Sierra and ne. Calif., Aug.-Oct.

Pectoral Sandpiper (*Calidris melanotos*) (fig. 36)

Recognition. L. 8-9½ in. (20-24 cm.); WS. ca. 17 in. (43 cm.). A medium-sized sandpiper (or large "peep") with straight beak about as long as the head, or a bit shorter. The legs and bill are dull yellowish-green, or the bill mostly gray. It can be distinguished from other species of nearly the same size by this leg color and the rich buffy and dusky mottled back, on which the light feather edges usually appear as narrow streaks, the dark-streaked chest abruptly demarcated from the white belly. In flight, a Pectoral shows white on the sides of the upper tail coverts but only a faint white line, if any, on the upper wing surface.

Habits. A common sandpiper east of the Rockies, the Pectoral prefers the sparse, low sedges or grasses of muddy shores amid which to probe for food in the manner typical of its family (see Dunlin account). Even on bayshores they tend to stay near the edge of the salt marshes, as do the similarly colored but smaller Least Sandpipers (see fig. 36 and plate 9). When large numbers of the commoner "peeps" are in such places, careful scanning for the larger, dark-chested Pectoral is often successful, at the appropriate season. Sometimes, however, they spread with the other species far out onto open mudflats. When alarmed, the Pectoral flies away erratically more often than the other peeps do. Its characteristic voice, given at such times, is a vibrant *krrrik,* rather low in pitch but not so harsh as the call of the Dunlin.

Range. Breeds in the Arctic from e. Siberia across w. and n. Alaska and n. Canada to s. Hudson Bay. Winters in s. South America and in small numbers in sw. Pacific Islands (Casual elsewhere). More widespread coastally in fall migration than in spring, when most pass north through the Great Plains and Mississippi Valley.

Occurrence in California. Fairly or Locally Common Sept.-mid-Oct., Irreg. Uncommon or Rare late July-Aug., Nov., and Apr.-May on marsh borders and wet fields of coastal to Central Valley areas, somewhat fewer in nw. and s. Calif. in similar habitat, at mountain lakes, and in tidal areas anywhere.

Note: The primarily East Asiatic **Sharp-tailed Sandpiper** (*Calidris acuminata*) is a very close relative of the Pectoral, distinguishable from it by its bright rusty-brown crown and more prominent buffy wash on the neck and chest (less evident in fall), with few or no dark streaks in the mid-chest region. Individuals have been reported in California, usually coastally, a dozen times in Sept.-Nov. (see Appendix).

Rock Sandpiper (*Calidris ptilocnemis*) (fig. 36)

Recognition and Habits. L. ca. 9 in. (23 cm.); WS. ca. 15 in. (38 cm.). This species is the hardiest of the shorebirds of the rocky coast, with its main wintering area well north of California. It is slightly smaller than a Surfbird and has stout green-gray legs and a yellowish-based, slender beak that curves down slightly. The general color in winter is dark brown-gray above, fading to white on the belly and throat and line above the eye. In spring there is a richer brown above, a whitish face and foreneck, and a large dusky blotch on the breast (not the belly as in the Dunlin, fig. 36 and plate 10). Rock Sandpipers may forage alone or, frequently, with turnstones and Surfbirds. They tend to be rather tame.

Range and California Occurrence. Breeds from e. Siberia and cw. Alaska south to Kurile Is. and s. Alaska. Winters on coast from Aleutians and s. Alaska to n. Calif., where it is Uncommon Oct.-Apr., and Irreg. to c. and s. Calif.; stragglers early May and Sept. Recorded on bayshores a few times.

Dunlin (*Calidris alpina*) (fig. 36; plate 10)

Recognition. L. 8-9 in. (20-23 cm.); Bill ca. 1½ in. (3.8 cm.); WS. ca. 15 in. (38 cm.). A medium-sized sandpiper (large "peep") with a bill that is downcurved toward the tip and distinctly longer than the head (about equal to it in some immatures). The legs and bill are black. In fall and winter the feathering is brownish-gray above (darker than the Western) and whitish below, except for a variable gray-brown wash on the neck and chest and a white line above the eye. The breeding feathering (plate 10) is more distinctive, with a black belly patch and reddish back, the American subspecies having formerly been called the Red-backed Sandpiper from this feature. In flight there is obvious white on the sides of the upper tail coverts and as a line along the bases of the flight feathers.

Habits. In midwinter this species is usually the most abundant of the sandpipers on tideflats from San Francisco Bay north. Along with the Least Sandpipers, they are also the chief ones in the Central Valley. However, the first migrants do not arrive in California until about the time of the peak migration of the Western Sandpipers, which far outnumber the Dunlins most of the year on the coast.

Unlike most sandpipers, Dunlins complete their molt before leaving the Far North. When feeding with their usual associates on the tideflats, the Dunlins tend to segregate somewhat from the larger dowitchers or the smaller Western and Least sandpipers. As the receding waters expose more and more of the animals taken by these birds for *food* (polychaete worms, small crustaceans, small mollusks), each species tends to feed up to belly-deep in the shallow water. They move along with head down, probing rapidly in numerous places, and congregate closely when this random search method locates unusually abundant pockets of food. Although there is much overlap, the birds with different lengths of legs and bills compete less for food than would be true if all were similar. To see best this tendency to spread out differentially when feeding, one must visit a tideflat when the ebbing tide is just beginning to uncover it. In their flight and in use of high-tide roosts, Dunlins are very similar to the Western Sandpiper (see account), although they sometimes associate with dowitchers instead. *Voice:* a raspy, fairly low-pitched *greeg* or *tzeeep*.

Range. Breeds around world in Arctic (except e. Canada, w. Greenland) south to w. Alaska and n. Hudson Bay and to n. Europe. Winters along coasts from s. B.C. to n. Mexico, from Mass. to Fla. and Tex., and in w. Europe, sw. Asia, and n. Africa; also Locally inland (as in Central Valley of Calif.).

Occurrence in California. After first arrivals in Sept., Abundant to Very Abundant Oct.-Apr., declining to Irreg. Rare to Fairly Common by late May, on bay tideflats, roosting in nearby salt marshes, salt ponds, or other open lands with pools. Somewhat fewer at comparable seasons in lowland cismontane and se. Calif. on lake and marsh margins and wet fields; fewer yet on outer coast beaches, Occ. to rocky flats. Recorded in ne. Calif. Apr.-May and Sept., at Sierra Nevada lakes in May, and at s. Calif. mountain lakes in Dec.

Stilt Sandpiper (*Micropalama himantopus*) (fig. 33)

Recognition. L. 7½-8½ in. (19-20½ cm.); WS. ca. 16-17 in. (41-43 cm.). A long-legged, slender large sandpiper with a bill like a Dunlin (fig. 36). In size it is a little smaller than the Lesser Yellowlegs and in flight appears somewhat like it (fig. 32) and the Wilson's Phalarope (fig. 37) because of full white upper tail coverts and unmarked dark wings. The legs are light greenish. In nonbreeding feathering it is otherwise gray above and whitish below, with the usual sandpiper white stripe above the eye. In breeding feathering the underparts are pale greenish-buff with conspicuous short dusky bars everywhere, and the head has four reddish-brown stripes.

Habits. This species favors shallow ponds or pools with mud bottom, and feeds in close flocks with the same rapid, deep probings that dowitchers exhibit, often being found with them.

Range and California Occurrence. Breeds in arctic or sub-arctic Canada and ne. Alaska; winters in South America. Migration chiefly through c. U.S., and in fall to Atlantic coast and sparingly west to se. Calif. Recent intensive observation in the Salton Sea-Imperial Valley area has shown Stilt Sandpipers to be Uncommon to Irreg. Fairly Common there July-Sept. and Apr.-May, Occ. through winter. Recorded elsewhere in Calif., including on tideflats in wc. and nw. Calif. Feb.-Mar. and July-Oct., and several times Aug.-early Oct. in Central Valley.

Note: The **Buff-breasted Sandpiper** (*Tryngites sub-ruficollis*), a species migrating primarily in the Great Plains area and nesting in the Arctic, has been found a few times on or near the s. Calif. coast in Sept., and once in Death Valley in July. Some have been found on beaches, although their favored habitat is short grassy uplands, including golf courses and airports. The Buff-breasted is a medium-sized sandpiper (L. 8 in. = 20 cm.) of chunky build, with straight black beak about head length, very buffy overall feathering, nearly unmarked on the underparts, and white wing linings.

Note: The **Ruff** (*Philomachus pugnax*) is primarily an Old World shorebird about the size of a Lesser Yellowlegs, but somewhat stockier and with shorter, yellow-green to orangish legs and nearly straight bill a little more than head length. The female (known as a Reeve) has plumage brownish above to paler brown on chest and white on belly. Males in breeding feathering have elaborate ruffs of feathers of varied color patterns on the neck. Two oval white patches on either side of the dark rump and tail covert area are distinctive marks. There have been at least twenty occurrences Sept.-Apr. in Calif. since 1950, most on s. coast, almost all single individuals, and most in female or immature feathering. Several have stayed for months in one area (see Appendix).

30. FAMILY PHALAROPIDAE (PHALAROPES)

(Picture Keys A and C)

Red Phalarope (*Phalaropus fulicarius*) (fig. 37; plate 1)

Recognition. L. 7½-9 in. (19-23 cm.); WS. ca. 15 in. (38 cm.). When in breeding feathering, all three species of phalaropes are easily identified by their contrasting color patterns, the females being more brightly colored than the males. The Red and Northern phalaropes both show a white line along the extended wing and white sides of the upper tail covert area, like many sandpipers. In nonbreeding feathering the Red Phalarope has more smooth light gray on the back (entirely so in adult females), and often a more extensive white forecrown, than the Northern. Both species have a dark ear patch but are quite variable in head pattern otherwise. At all times the less needle-like (but still thin), yellowish-based bill distinguishes the Red Phalarope, if it can be seen.

Habits. This species is the most pelagic of the three in this family of "swimming sandpipers," and apparently migrates so far off California shores in some years that very few are detected, while in other years they are abundant along the coast (see Graphic Calendar in Appendix). Whether on the open ocean or in quiet waters of the coastal ponds or bays, however, they swim like activated corks and make quick jabs here and there for the swimming or floating tiny animals that make up their *diet*. Although they sometimes whirl, Red Phalaropes do this less often than the Northern, perhaps because of the rough water they so often occupy. In flight, phalaropes bunch irregularly, much like sandpipers, but seldom rise very far above the water. At the main migration

FIG. 37

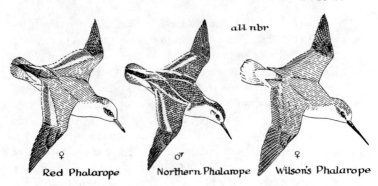

all nbr

Red Phalarope   Northern Phalarope   Wilson's Phalarope

periods, large flocks sometimes mass on the ocean to rest and feed. In the late fall such flocks are usually almost all Reds, though they mix with the Northern earlier. *Voice:* a short, light *kit*, similar to the call of the Northern.

Range. Breeds around the world in high Arctic, south in America to w. Alaska and n. Hudson Bay. Winters at sea, chiefly off South America, w. Africa, and New Zealand. Migrates widely over oceans near continents, rarely inland.

Occurrence in California. Irreg. Uncommon to Very Common Aug.-Dec., Occ. Jan. and July, and Irreg. Uncommon to Very Common late Mar.,-early June, most reliably on ocean, sometimes well offshore but in some years widespread on coastal waters of all sorts. Decidedly fewer (usually) reach salt ponds and other waters of inner bays and the Salton Sea. Recorded inland elsewhere in cismontane lowlands May and late July-Dec. and in ne. Calif. in Nov.

Northern Phalarope (*Lobipes lobatus*) (fig. 37; plate 1)

Recognition. L. 6½-8 in. (16½-20 cm.); WS. ca. 14 in. (36 cm.). This most numerous of the phalaropes is very like the Red Phalarope in nonbreeding feathering, but has more dark gray on the back, the light buff or whitish feather edges there often forming streaks lacking in the Red. This is most obvious in immatures and males, which are otherwise all dark on the back; but adult females often have a considerable mixture of light gray feathers, and thus resemble male Red Phalaropes in back pattern. The more slender, truly needle-

like, all-black bill of the Northern is then the only safe crite-
rion. In breeding feathering, Northerns have conspicuous white
throats and reddish-brown on the side of the neck and chest.
The wing and tail pattern are like the Red Phalarope.

Habits. At the end of summer, when the salt-evaporating
ponds of San Francisco Bay are teeming with brine shrimp,
and some also with tiny water-boatmen, these little birds
gather at times in flocks of 10,000 or more to swim and dab
daintily at the surface for the harvest. There is an even greater
concentration (to over 100,000) at the same season on alkaline
Mono Lake, where the brine shrimp, their favorite *food*, is the
largest swimming animal found. When these prey are deeper,
the phalaropes often whirl round and round in one spot,
creating a vortex of water from which they continue to snap
up tiny animals for minutes at a time. On waters with a more
varied assemblage of small swimming animals, the phalaropes
have a quite varied diet of fly and mosquito larvae, young bugs
and beetles, small crustacea, and sometimes even tiny fishes
and seeds. In flying from danger or merely to new feeding
grounds, Northern Phalaropes usually bunch together and may
engage in irregular twistings and turnings or even "close-order"
wheelings like the small sandpipers. *Voice:* a sharp, high-
pitched *kit* or *chik*, somewhat like a Sanderling.

Range. Breeds around world in middle and s. arctic regions,
south in America to sw. Alaska, n. B.C., and Labrador. Winters
at sea, especially off South America (Abundant near Peru),
and in Old World tropics to New Zealand. Migrates abundantly
inland in w. North America as well as coastally.

Occurrence in California. Common by late (Occ. early) Apr.
to Abundant in May (stragglers in June), and Very Abundant
July-Sept., declining to Uncommon by late Oct., on salt ponds
(especially of San Francisco Bay); somewhat fewer at same
seasons on ocean (but often Abundant there at peak migrations) and bays; fewer yet (Occ. Abundant Aug.) on cismontane lakes. Uncommon to Irreg. Common Nov.-Mar. in San
Diego Co. coastal area, very Rare in c. Calif. then. In ne., ce.,
and se. Calif., Fairly Common to Abundant Apr.-May and
July-Sept. and Very Abundant Locally in Aug. on alkaline

lakes, Occ. to Oct. Recorded in main migrations on Sierran and s. mountain lakes.

Wilson's Phalarope (*Steganopus tricolor*) (fig. 37; plate 1)

Recognition. L. 8½-10 in. (22-25 cm.); WS. ca. 17 in. (43 cm.). The largest and slenderest phalarope, with the upper wing surface solid gray and a white area all across the upper tail coverts. These features, and the feet extending beyond the light gray tail, make it appear very like the Stilt Sandpiper (fig. 33) and even a little like a yellowlegs (fig. 32) in flight. The phalarope, however, has a needle-like bill. In nonbreeding feathering, the color is otherwise white below and smooth gray above except for a white forehead and stripe above the eye. The chestnut-striped neck and back of the breeding feathering are distinctive.

Habits. In its inland habitat, this phalarope often feeds along muddy shores or in wet grass by probing like the smaller sandpipers or by "side-sweeping" like an Avocet. On open water they also feed in the fashion characteristic of the family. On migration, the Wilson's Phalarope favors alkaline lakes or salt-evaporating ponds where brine shrimp are abundant, usually mixing with the more abundant Northerns there, but with an earlier migration peak in late summer (see Graphic Calendars in Appendix). Female phalaropes are the more active partner in *courtship*, with which their brighter plumage is correlated. The male Wilson's defends a territory in low marsh growth or wet "hay meadows" in which he builds a grass-lined *nest* on the ground. The female lays the 4 (rarely 3) *eggs* which are pale buff variably and usually finely marked with dark brown or black. The male then performs all the incubation (20-21 days) and care of the young, although the female may participate in scolding of intruding predators until she leaves the area to molt. The *downy young* are orangish-buff with whitish belly, marked above with a blackish median stripe and lateral spots. They apparently remain in the marsh until well grown. *Voice:* softer than other phalaropes, as *check* or *chik*; also a low-pitched, nasal grunting *nya* or *unk* given by birds on territory or flying over it.

Range. Breeds from c. B.C. across s. Canada to Manitoba and south, east of Cascades, to ce. Calif. (and Locally in Central Valley) and across n. and c. U.S. to Ind. Winters chiefly in Argentina, a few rarely north to s. Tex. and s. Calif.

Occurrence in California. Common to Irreg. Abundant mid-Apr.-May, then few stragglers until Common to Abundant late June-Sept., at Salton Sea and on Central Valley lakes, and somewhat fewer (though Abundant at times) on salt ponds from San Francisco Bay south; Irreg. Rare to Uncommon Oct.-Mar. in far s. Calif. Locally Uncommon to Fairly Common as breeder on wet meadows in Central Valley late May-July. In ne. and ce. Calif. and n. Sierra Nevada, Farily Common to Common late Apr.-June, nesting May-July on wet grassland and marshy lakes, and Abundant to Very Abundant July-Aug. on open lakes, esp. alkaline ones such as Mono Lake. Migrants also recorded Apr.-May and late June-Sept. at s. mountain lakes, in n. coastal section, and Occ. inland in sw. and wc. Calif.

31. FAMILY STERCORARIIDAE (JAEGERS, SKUA)

(Picture Key C)

Pomarine Jaeger (*Stercorarius pomarinus*) (fig. 38)

Recognition. L. 20-23 in. (51-58 cm.); WS. 46-50 in. (117-127 cm.). A sea bird about the size of a California Gull, but with hooked beak, sharp curved claws on the toes, and more pointed wings and powerful flight — as is characteristic of its family. The central pair of tail feathers in adults extend beyond the others about 1 to 4 in. (to 10 cm.) and are twisted, with rounded ends. Immatures through their first winter and spring lack this feature, but can sometimes be told from other jaegers by their larger size, heavier bill, and more extensive white along the shafts of the primaries (on the five to nine outermost feathers). All immatures and some adults have the feathers otherwise dark brown with numerous rusty to white feather edges in the immatures and somewhat whitish underparts in second-winter birds. Some adults have a "light phase" feathering, with blacker cap and whitish underparts except for

FIG. 38

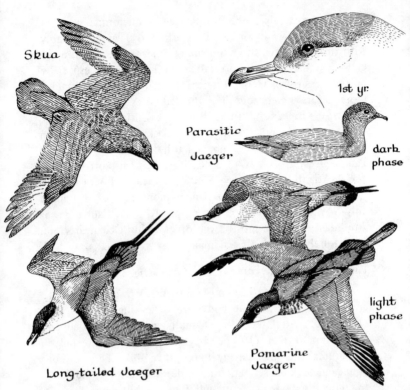

Skua

Parasitic Jaeger

1st yr.

dark phase

Long-tailed Jaeger

Pomarine Jaeger

light phase

a variable dark collar on the foreneck and a whitish or golden collar across the hindneck.

Habits. When breeding on the tundra, jaegers are predators of lemmings and various small birds. Most of the year, however, they are pirates of the seas, obtaining most of their *food* by robbing terns or gulls of their catches. Although their hooked beak and claws may help in attacking these birds, most of the supremacy of the jaeger is attained simply by close pursuit or disabling the flight of the other bird. With a sudden burst of falcon-like wingbeats and speed, they single out a food-laden victim and continue to close in on it even though it twists and turns in attempted escape. Usually the fleeing bird will disgorge its food, which the jaeger then snaps up quickly; but I have seen a Pomarine Jaeger drive the larger Western Gull into the water in a burst of thrashing wings, beaks, and feet.

Despite such tactics, jaegers flying along at normal speed or resting on the water or shore are often associated peaceably with the very species they pursue at other times. If a group of gulls or terns flying about erratically is investigated, especially on the ocean proper, one often finds that the cause is one or more jaegers. *Voice*: squealing whistled notes.

Range. Breeds in arctic tundra, essentially around the world, in w. America south to w. Alaska and nw. Canadian coast; nonbreeders in summer south to B.C. Winters in e. Pacific Ocean from s. (Occ. n.) Calif. to Peru; also off Australia and in Atlantic near se. U.S., South America, and Africa.

Occurrence in California. Uncommon to Irreg. Common July-Nov., Occ. through winter (mostly in s. Calif.), fewer Apr.-May, on ocean mostly well offshore but Occ. near shore, Very Rarely on bays. Casual inland.

Parasitic Jaeger (*Stercorarius parasiticus*) (fig. 38)

Recognition. L. 16-21 in (41-53 cm.); WS. ca. 44 in. (112 cm.). A jaeger about the body size of a Ring-billed or Mew Gull. The central rectrices of adults are sharp-pointed and extend up to 4 in. (10 cm.) beyond the rest of the tail; in immatures they are scarcely longer than the others. In both light- and dark-phase feathering, older birds are very similar to the Pomarine (see account and fig. 38), but somewhat less white shows in the spread primaries — the shafts of three to five outer ones being noticeably white from above (compare with Long-tailed, fig. 38). The bill and feet of adults are black; first-year birds have the tarsi and basal part of the toes and webs more or less light blue to pink.

Habits. When in California waters, the Parasitic Jaeger is similar in its piratical food-getting habits to its larger relative, the Pomarine (see account); but of course it is more likely to be seen pursuing the somewhat smaller birds, especially terns. However, each species is capable of obtaining *food* from birds larger than itself because of the jaeger's superior powers of flight and great maneuverability. On occasion a jaeger may catch a fish or other sea animal of suitable size for itself, usually using its hooked beak. Too little is known about the relative numbers of the three species of jaegers off our shores, partly because immatures viewed at a distance cannot be identified to species. It seems certain, however, that the spring migration of all of them is either farther out at sea or accomplished so rapidly that it is usually "missed" by the relatively few observers offshore at that season; a spectacular migration of both Parasitic and Pomarine jaegers was noted forty miles off San Clemente Island in Nov. 1964 (see Appendix). The Parasitic seems the most prone to spread into bays and inland, and one even persisted for several winters in traveling with gulls over San Jose, enroute to and from a reservoir in the hills. *Voice*: dry *tik* or *kuk* notes; also wails and mews, but usually silent in fall and winter.

Range. Breeds in or on borders of arctic tundra around the world, in w. America south to cs. Alaska (and probably Aleutians), s. Mackenzie, and n. Manitoba. Winters in e. Pacific from s. Calif. to Chile, in sw. Pacific, and in Atlantic from Maine to n. Argentina and from Great Britain to South Africa and near Arabia.

Occurrence in California. Uncommon to Irreg. Common Aug.-mid-Oct., fewer to late Nov., Rare Mar.-July and in s. Calif. Dec.-early Feb., on ocean both offshore and at times also along shore, even to beaches and river mouths; Very Rare in bays and at coastal lakes. Migrants recorded inland, mostly Aug.-Oct., in ne. and wc. Calif., Central Valley, L. Tahoe, and at Salton Sea.

Long-tailed Jaeger (*Stercorarius longicaudus*) (fig. 38)

Recognition and Habits. L. 14-23 in. (36-58 cm.); WS. ca. 38 in. (97 cm.). With their main migration apparently passing far offshore, very few Long-tailed Jaegers are ever detected even from boats chartered for bird trips off California shores. They are abundant in parts of the Arctic, however, and should be watched for at the proper seasons here. A major reason for the scarcity of records (all embodied in the Graphic Calendar in Appendix) is the great difficulty of distinguishing this species from the more common Parasitic Jaeger. Adult Long-tailed Jaegers have only a light phase without a darker collar on the foreneck, with a broader white or yellowish collar across the hindneck and a paler back than the Parasitic. The central rectrices are sharply pointed and project 3 to 10 in. (7.6-25.4 cm.) beyond the others, but birds in the lower half of this variable range cannot be safely identified as Long-tails on this basis alone. In both immature and adult Long-tails the tarsi and part of the toes and webs are blue-gray (as in immature Parasitics), and only the outer two or three primaries show noticeably white shafts from above. Immatures sometimes lack the rusty feather tips of the larger jaegers, but these may fade or wear away by spring in any species.

Range. Breeds in parts of the arctic tundra around the world and Locally in alpine areas (as at Mt. McKinley, Alaska). Nonbreeders summer commonly in s. Alaska. Winters in tropical oceans and temperate s. Atlantic and se. Pacific.

Occurrence in California. Rare Aug.-Nov. on ocean, nearly all well offshore.

Skua (*Catharacta skua* and *C. maccormicki*) (fig. 38)

Recognition and Habits. L. 20-22 in. (51-56 cm.); WS. ca. 58 in. (1.5 m.). Like a large, heavy-bodied jaeger with somewhat rounded tips on the broader wings and a conspicuous white "window" in the wings at the base of the primaries, which also show white shafts. The feathering is otherwise dark brown with small whitish markings or somewhat rusty below, and the central tail feathers barely project beyond the others. The general shape is somewhat like a *Buteo* hawk, but the flight when pursuing birds it victimizes is strong and powerful. Because of its great size (body nearly the size of larger gulls), it does not hesitate to pirate *food* from almost any sea bird. The Skua was formerly considered very casual in the north Pacific; but so many records have now accumulated, through frequent visits by parties of observers to the necessary offshore areas, that it should be termed an erratic transient. Most of our birds appear to be the variably pale to dark brown, less "rusty" South Polar Skua (*maccormicki*).

Range. Breeds in Iceland, near Great Britain, in s. South America, New Zealand, Antarctica, and various islands in southern oceans, ranging nearby seas in all seasons. Occurs somewhat farther, including North Pacific, when not breeding.

Occurrence in California. Irreg. Rare Sept.-Nov. (twice to Fairly Common level in Oct.), and recorded Casually Jan.-Aug., on ocean off c. and s. Calif., mostly well offshore. Casual off n. Calif. and on bays. Two species often not distinguished.

32.-33. FAMILY LARIDAE (GULLS, TERNS)

32. SUBFAMILY LARINAE (GULLS)

(Picture Keys A, B, and C)

Glaucous Gull (*Larus hyperboreus*) (fig. 39)

Recognition. L. 26-32 in. (66-81 cm.); WS. ca. 60 in. (1.5 m.). A large to very large gull (size of Western or larger). Adults have a light gray mantle fading to full white on all but the basal parts of the primaries, pink feet, and a bill similar to that of other large gulls but often heavier. The iris is light yellow and a *narrow* ring of yellow (edge of eyelids) encircles the eye (this ring is bright red in the slightly smaller but similarly plumaged Iceland Gull). First-winter Glaucous Gulls are mottled pale brown and white, white predominating in some, while other individuals are about as dark as the first-year Glaucous-winged Gull (fig. 39). Such birds are best distinguished from that abundant species by the *two-colored beak*, the black terminal third to half *abruptly* separated from the pinkish base. The primaries of immature Glaucous Gulls are paler than the rest of the upper wing surface, but so are those of a first-winter Glaucous-winged in the worn plumage so often seen in spring. From below, the primaries appear nearly white in both species. Second-winter Glaucous Gulls have much of the black still in the outer half of the bill, but the plumage is decidedly whiter, often almost entirely white; this is the only gull in California which is normally so, though albinos may occur in any species. In the third winter they are between this and the adult pattern, with a partial light gray mantle.

Habits. The central California coast is just about the normal southern limit for wintering of this far northern species, and most of those seen are in the first- or second-winter feathering. Although immature (first-year) Glaucous-winged Gulls are frequently misidentified as Glaucous by the overly hopeful, a few real Glaucous Gulls can be found every winter by careful search among the thousands of gulls at our bayside garbage dumps and other points of concentration of the larger species. Farther north, this species is said to be an important predator on various sea birds, or to pirate food from other gulls. It also ranges far offshore there, and can thus be expected to do so in our waters when in transit.

Range. Breeds on arctic shores essentially around the world, in w. America south through Bering Strait to Pribilofs. Winters from Bering Sea south to s. Calif. and Japan, and from Greenland to N.Y. and Europe, occasionally to Fla. and to Mediterranean.

Occurrence in California. Rare to Uncommon Nov.-Apr., in cw. Calif. at garbage dumps, fewer on open bays, tideflats, and ocean, Occ. to nearby lakes; stragglers recorded May-Oct. More Irreg. on s. coast at same seasons. Casual inland.

Glaucous-winged Gull (*Larus glaucescens*) (fig. 39)

Recognition. L. 24-27 in. (61-69 cm.); WS. ca. 54 in. (1.4 m.). As adults, all our large gulls have pink legs and feet, a yellow bill with a red spot at the angle on the lower edge back from the tip, a fully white tail, white trailing edge of the wing, and white underparts, neck, and head (speckled with brown in nonbreeding feathering in most species). The Glaucous-winged shows all these features (see plate 12) and has a light gray mantle (upper surface of wings and intervening back) continuing to *light gray wing-tips* with white subterminal spots (mirrors) on the outermost or longest feathers. The iris is medium but variable brown in adults, dark brown in first-year birds. The first winter feathering is uniformly light grayish-brown, mottled or flecked with pale buff or whitish except on most of the flight feathers and tail; the bill is then all black or nearly so. Second-winter birds are mostly light gray on the back and the head, the underparts are brown and whitish mixed, and the

FIG. 39

Large Gulls

Glaucous Gull

ad br
white primaries

2nd winter

1st winter

all adult legs flesh colored

1st winter

ad br

Glaucous-winged Gull

2nd winter

1st winter
Herring Gull

ad br

yellow eye-ring
ad br

ventral

ad
lavender eye-ring

Thayer's Gull

1st winter

tail is white basally, but the wings are still mostly brown. The bill is light basally, but most of the terminal third is black. In the third winter, a pattern very like the adult is evident, but with some light grayish-brown in the wing coverts and often subterminally on the tail. Vestiges of this subadult pattern, and even more regularly some of the dusky mark in front of the red on the bill, may persist for several more years, even though the bird breeds.

Habits. At garbage dumps, fishing centers, and other spots with regular and sizable food sources on the larger bays and outer coast in the northern half of California, this gull is usually one of the most abundant among the five or six wintering species. They also, of course, forage in smaller numbers along beaches and over open waters in typical gull fashion. After feeding for a period, gulls usually assemble on open flats or shores or along piers, etc., to rest and preen. Most species also visit fresh water daily for drinking and bathing, the Glaucous-winged being among those that commute from the Bay to such places as Lake Merced in San Francisco and the reservoirs in the East Bay hills. In late afternoon, small groups or large straggling flocks of gulls converge from miles away on roosting areas, which in this species are often on the roofs of large bayside warehouses or docks, or on breakwaters or salt-pond dikes. Many Glaucous-winged Gulls are banded in Washington and British Columbia, and observers in California should watch for these marked ones (including those found dead). With a telescope, it is even possible to read the band number on a live gull at times. *Voice*: many calls similar to the Western (see account), and a softer, mellow *kowk* that is possibly distinctive.

Range. Breeds from w. Alaska and islands of Bering Sea south and east along coast to nw. Wash., where it hybridizes with the Western Gull. Winters chiefly along coast from se. Alaska to s. Calif., a few to s. Baja Calif. and Sonora, Mexico, and in major valleys of interior Calif.; also south to n. Japan.

Occurrence in California. Abundant Nov.-Mar., fewer from Oct., Apr., May, and a few stragglers all summer on n. and c. Calif. bays and nearby garbage dumps; somewhat fewer at other bayside and various outer coast locations, even to far offshore in migrations. Fewer in similar habitats in s. Calif., and usually fewer yet well inland on wet fields, at dumps or at lakes or rivers in cismontane lowlands and at Salton Sea, though Occ. Common there Dec.-Mar., and Locally in Central Valley.

Western Gull (*Larus occidentails*) (plate 12)

Recognition. L. 24-27 in. (61-69 cm.); WS. ca. 54 in. (1.4 m.). A large gull with characteristics common to all such (see previous two accounts and fig. 39 as well as plate 12), except that the adult is fully white-headed all year (and one subspecies of the Gulf of California has yellow feet). The adult Western has a pale yellow iris, a mantle that is either medium-dark slate-gray (a shade darker than the California Gull) or very dark and appearing blackish-gray (in the subspecies breeding from Monterey Co. southward). The primaries are black or mostly so, with small white tips except for large white tips on the outer one or two (sometimes with a dark mark). In their juvenal and first-winter feathering, Western Gulls are darker brown than all the similar gulls, mottled with whitish everywhere except over most of the long wing and tail feathers. The rump and upper tail coverts show greater contrast of white and dark markings than in the Herring and California gulls, and the bill is solid blackish, the feet dark pinkish-brown. At each molt during the next three and a half to four and a half years these feathers are replaced (or partly so in spring) by new ones that are more and more like the adult, major stages in the complicated sequence shown in plate 12. As in the Herring Gull, some third-year birds may resemble typical second-year, some four-year-olds still show some dark in the tail, etc., because individuals vary in the rate of progress of these changes.

Habits. Since this is the only species of gull that nests along the California coast, a visit to one of their breeding areas will enable the patient observer to see many acts not performed by gulls so commonly seen at other seasons and places. This species closely resembles the Herring Gull (much studied in England by Tinbergen and others) in that the males establish small territories in which they walk about slowly with head high and bill somewhat down (mild threat), or sit on potential nest sites by the hour. If another male intrudes, he is threatened with lowered, extended head and fought off if need be by use of both wings and beak. Two neighbors often display at their common territory boundary by "sham battles" that involve violent pecks at the ground, pulling up of vegetation, and slow retreat from one another. *Courtship* involves the female's acceptance of a fish disgorged by the male, usually preceded by a duet of quick vertically upward tosses of the bill by both birds, accompanied by a muffled musical call. *Voice*, at other times, includes quavering nasal notes, a louder *qua-ARR-ik*, trumpeting "long calls" in which each note is a loud *kyuk* or *kee-yuk*, and a mellow low-pitched *kuk, kuk-kuk* or series of same, and others. When food is found in abundance, this and other gulls give a quavering high-pitched series of wheezy notes which seems to serve the function of attracting others to the feast.

The general food-finding and roosting behavior is similar to that of the Glaucous-winged (see account), although the Western shows much more restriction to marine situations. Since it is the only gull present in numbers at the colonies of cormorants and murres in our area, the Western is most significant as a predator on any of their unguarded eggs or young. *Nest*: a low mound of soft plants, usually in loose colonies on tops or gentler slopes of islands, occasionally on ledges of cliffs even on the mainland, and even, rarely, on bridges, boats, or dockside roofs. *Eggs*: 1-4, buffy with many irregular blotches of brown; incubation 24-26 days, by female more than the male. *Downy young*: light buff to cream-colored, with numerous dark marks above, larger on the head; remain in nest a few days, then walk, run, and hide until they first fly, at about 6 weeks; are able to forage on their own at 8 weeks.

Range. Breeds along coast and nearby islands from nw. Wash. to Baja Calif. and Sonora, Mexico, large colonies on South Farallon I. and a number of the Channel Is., small ones on several small islands in San Francisco Bay, at Point Lobos, Morro Rock, etc. Spreads both north and south somewhat in fall and winter, and inland Locally along major streams in late summer or fall (at time of the salmon run in nw. Calif.).

Occurrence in California. Abundant all year on outer coast and over nearby ocean and outer parts of bays. Adults largely withdraw to vicinity of breeding colonies Apr.-July, so total numbers elsewhere are less then. First adults on nest territories by Feb.; eggs noted Apr.-July, mostly May-June, dependent young June-early Aug. Fewer at comparable seasons to inner parts of large bays (most evident late summer). Rare or Uncommon at lakes and major streams somewhat inland in c. and nw. Calif. Also recorded at Salton Sea (a few on Colorado R.) every month but Oct., Occ. Common late June-Sept. when yellow-legged subspecies prevails there.

Herring Gull (*Larus argentatus*) (fig. 39)

Recognition. L. 23-26 in. (58-66 cm.); WS. 51-57 in. (1.3-1.5 m.). One of the large gulls, with their usual characteristics (see Glaucous-winged account, fig. 39, and plate 12). The adult Herring Gull has a *pale* gray mantle, relatively small black wing-tips with white mirrors on the outer one or two primaries, light yellow iris, and (in breeding season at least) yellow-edged eyelids (compare Thayer's, fig. 39). Hybrids between the Herring and other large gulls overlapping it in breeding range are fairly often encountered — e.g., a few birds in the San Francisco Bay region annually seem to be Herring × Glaucous-winged or Western × Glaucous-winged, which resembles it. In first-winter feathering, Herring Gulls are medium brown mottled with buffy to whitish, the smaller individuals usually distinguishable from California Gulls of the same age (fig. 40 and plate 12) by the all-dark beak, which may fade to lighter basally by spring but without an abrupt separation from the dark such as the California has. Second- and third-year birds have comparable patterns of feathering and bill colors to same-age Westerns (see plate 12), but are recognizable because of the pale shade of gray on the back or mantle.

Habits. Being the commonest gull in the North Atlantic, this species has been more thoroughly studied than any other. It has long associated with man and has shown mushrooming populations as a result of the ever-increasing amount of *food* available at solid waste disposal sites. In the masses of gulls at such areas, each bird obtains food by merely walking up to it or by flying down amid the melee of gulls and grabbing with its beak, often close to the bulldozers used for compacting the trash. Both of these feeding methods are used also by gulls more widely dispersed, as when searching on mudflats or in flight over water or a beach. Experiments have shown that gulls are quick to investigate any shiny object on or falling into the water, but not the sound of a splash, particularly. When several gulls repeatedly execute the typical figure-eight route in turning and swooping down to food, other gulls within sight are quickly attracted, particularly if the quavering "food-call" (see Western Gull account) is given. Though omnivorous, all gulls prefer protein and fat-rich animal tissues. Herring Gulls may dive with difficulty (from 10-40 feet up), and have been noted retrieving fish from the bottom of water up to 4 feet deep. Two unusual food-getting techniques they occasionally employ are "puddling" or stomping rapidly in place on soggy mud or turf to drive worms out, and dropping clams or mussels 40-50 feet onto rocks or hard-packed sand to crack them open. *Voice*: loud *kyow* or *hyah* notes, higher pitched than comparable calls of the Glaucous-winged and Western; other calls similar to the Western, but long-call notes more often doubled.

Range. Breeds from c. Alaska, arctic Canada, and Greenland south to s. Alaska and c. B.C., Mont., Mich., n. N.Y. and on Atlantic coast to Va.; also across most of Eurasia. Winters from se. Alaska, sw. and se. Canada and nc. U.S. south to Central America and West Indies; in Old World to c. Africa, se. Asia, and Philippines.

Occurrence in California. Very Common to Locally Abundant Oct.-Mar., fewer May and Sept., Occ. July-Aug., and a few stragglers in June, at cw. and nw. Calif. garbage dumps and nearby bays and harbors. Somewhat fewer in sw. Calif., on similar habitat, and on bays and ocean generally, including to well offshore in migration; and to lakes and fields or in towns, especially near coast. Uncommon to Locally Very Common

Nov.-Mar. in Central Valley at dumps, lakes, and rivers, and on wet fields or grassland; fewer inland in cw. and sw. Calif. and on Colorado R.

Thayer's Gull (*Larus thayeri*) (fig. 39)

Recognition and Habits. L. ca. 23 in. (58 cm.); WS. ca. 53 in. (135 cm.). Averaging smaller than most subspecies of the Herring Gull to about the size of a California, this gull is colored very much like a Herring Gull and has only recently been designated as a separate species. Since it is so difficult to distinguish from the Herring Gull, most of the Graphic Calendar (in Appendix) is extrapolated from the few reliable records of numbers. The adult (and third-year?) Thayer's differs from the Herring Gull at comparable ages in having a brown or mottled brown and cream iris, a rose-*lavender* eyelid ring (instead of yellow), and smaller black area on the wing-tips, the outer two or three primaries typically having black only on their narrow outer webs and on the tips (beyond the mirrors), although there is some dark gray more basally. Particularly from below, the wing appears to have only a very small black tip with an arc to the narrow black leading edge (fig. 39). First-year birds are mottled like most gulls, but with sharper light edges to the back and covert feathers than in the Herring Gull. The darkness of the basic brown varies from medium to quite pale, and the wing-tips are usually darker than the brown elsewhere, but decidedly paler than in a Herring or California gull (fig. 39 and 40). The bill tends to be all or nearly all blackish. Even though breeding across much of the Arctic, this species oddly concentrates on the Pacific coast in winter and is suspected of migrating mostly *via* the coast — i.e., around Alaska. (One that was wing-tagged at a San Francisco Bay dump was seen the following season on the west coast of Alaska, just north of the Alaskan peninsula.)

Range. Breeds from nw. Greenland across most of high Canadian Arctic; nonbreeders summer from n. Alaska south to B.C. Winters on coast and in some coastal valleys from s. B.C. to s. Calif. and (possibly only irregularly) to se. Calif., s. Nev., and c. Baja Calif.

Occurrence in California. Fairly or Locally Common Dec.-Mar., Uncommon Nov. and early Apr., at garbage dumps near bayshores and to a few miles inland; fewer at Central Valley and inland sw. Calif. dumps, and widely on bays and outer coast elsewhere. Recorded also in se. Calif., chiefly Colorado R. area, Oct.-Mar.

California Gull (*Larus californicus*) (fig. 40; plate 12)

Recognition. L. 20-23 in. (51-58 cm.); WS. ca. 52 in. (132 cm.). A medium-large gull, the adult with mantle of medium neutral gray and outer primaries increasingly black subterminally with conspicuous white mirrors on the outermost two, legs and feet light grayish-green to yellowish, bill yellow with a red spot at angle of lower mandible and usually a narrow black mark in front of this, iris brown, and eyelids edged orange-red. In their first fall and winter, immatures are quite dark brown (though paler than the Western, see plate 12), mottled everywhere except on the long wing and tail feathers with light buff to whitish. The bill is dusky brown to black, abruptly dull pinkish basally (up to half); the legs are dusky, becoming pinkish and then bluish-white toward spring. The back is fully gray by the second fall, the body much whiter, the tail coverts fully white and the basal half or so of the rectrices also; but the wings show mixtures of brown and gray plus blackish tips with or without faint mirrors. The beak at this stage is bluish-white to pale yellow with a broad subterminal band of black (thus easily confused with a dull adult Ring-billed, fig. 40), usually also with at least a faint red spot. Third-year birds (and a few older) are intermediate between this and the adult coloration, retaining small blackish areas on the primary coverts and subterminally on the tail.

Habits. Coming from their inland nesting grounds in late summer, many California Gulls migrate west or southwest across the Sierra Nevada, while others go far northwest to British Columbia and then south along the coast — to become until spring one of the commonest and most widespread of gulls in both coastal and coast-slope lowlands of California. In Aug. and Sept. they are usually the most numerous species

FIG. 40

Medium Gulls

2nd winter

1st yr

black tail

2nd winter

ad nbr

1st yr

California Gull

legs greenish to gray

legs greenish to yellowish

complete ring

Ring-billed Gull

ad br

2nd yr

Mew Gull

greenish legs

1st yr

yellow bill

ad br

slight fork

ad

Black-legged Kittiwake

1st yr

237

about garbage dumps; but with the later influx of the larger gulls to the San Francisco Bay region, many of the California Gulls retreat to places where the competition is less intense, or move on southward. In the southern third of the state, the California is often the commonest gull even along the beaches, but relatively few get more than five miles offshore, and these, chiefly in migration periods. In most major cities, particularly those within thirty miles of the coast or a large bay, this species is also conspicuous except in summer as a daily visitor to schoolyards, shopping center parking areas, and large park lawns. The spring movement brings many flocks northward along the outer coast, where they pass close along any available seacliffs. Other flocks go via inland routes and flap in V formation or soar in circles to great heights over hills and mountains. A little patient watching of such birds soon discloses that they are bent on traveling a distance and not just moving to feeding grounds or nightly roosts, on which shorter flights they tend to straggle much more.

Along the coast the California Gull is an omnivorous scavenger much like the Herring Gull, but inland it often follows plows for the upturned worms and grubs or *feeds* on grasshoppers and other large insects, and even mice when these are abundant. The "sea gulls" that from time to time invade insect-plagued croplands in Utah are, in fact, mostly California Gulls from nearby nesting colonies. *Voice*: varied *kaks* and squeals, somewhat harsher yet more muffled than comparable notes of the larger gulls. *Nest*: a loose cup of plants or debris on open ground, on a mound in marsh, or occasionally in low shrubs; neighboring nests but a few feet apart. *Eggs:* 2-4 (usually 3), light grayish to buffy-brown, irregularly marked with various dark browns; incubation probably 24-25 days, by both parents. *Downy young*: light buff with varied browner areas and black spots and blotches above; remain in nest only a few days.

Range. Breeds in colonies on inland lakes from nc. Mackenzie and Manitoba south in Wash. and Ore., east of Cascades, to ne. and ce. Calif. (Mono and Topaz lakes), Utah, Colo. and N.D. Winters from s. Wash. to Guatemala along the coast, and Common inland in Calif., (Locally north to Idaho.)

Occurrence in California. Abundant, Locally Very Abundant, July-May, fewer in June, on bays, salt ponds, and through coastal towns and dumps; somewhat fewer at lakes and on outer coast, but few at any distance offshore. Fewer but Locally Common Aug.-Apr. at Salton Sea and at lakes and streams, in towns, and on wet fields widely through cismontane lowlands and se. Calif. In ne. and ce. Calif., present on lakes, fields, and marshes at least late Mar.-Sept.; Very Common near colonies, where nests with eggs present May-June, dependent young mostly June-July (first migrant young-of-year cross Sierra Nevada in late July). Also recorded on many lakes of Sierra Nevada May-Nov., Common July-Aug., and s. mountains July-Dec.

Ring-billed Gull (*Larus delawarensis*) (fig. 40)

Recognition. L. 18-21 in. (46-53 cm.); WS. ca. 48 in. (122 cm.). A medium-sized gull (about the size of a crow); the adult is white except for a *pale* gray mantle, black wing-tips with white mirrors in the outer two feathers, yellow bill with a black band across it, and yellow or greenish-yellow legs and feet. In fall and winter, there is some gray-brown streaking on the head and neck. Most Ring-bills attain these adult colors in their third fall, and even in the preceding year are very similar except that the outermost primaries and primary coverts are extensively dull black, making a wedge-like dark area from bend to tip of the wing with only faint or no mirrors. Usually there is also some scattered dusky elsewhere on the wing and subterminally on the tail. Compared to the larger species (fig. 39 and plate 12), first-year Ring-bills show more light gray in the wings and back and are whiter below and more coarsely mottled brown and white on the inner wings, upper tail coverts, nape, and part of the back. The tail is most distinctive at this age — grayish-white with a broad, even-edged subterminal band of blackish (but the second-winter feathering of larger species also shows a broad dark subterminal band). The legs and bill are similar to those of the first-year California (fig. 40).

Habits. The Ring-billed is like the California Gull in much of its behavior and choice of habitat (see account) and often flocks with that species; but it is more widespread inland in winter, particularly far up the Sacramento and San Joaquin valleys and in the Imperial Valley. It is very rare offshore, at least in central California. Ring-bills are inveterate plow-followers; they sometimes catch large insects high in the air, and even occasionally obtain berries or acorns by swooping or hovering beside the shrubs or trees and yanking them off with the beak. Most of the time, however, they stick to their preferred *diet* of miscellaneous animal materials as available from water or land surface. *Voice*: shrill squeals, a raspy *kaar-ik*, muffled *kowk* calls, plus others. *Nest*: in colonies similar to, and sometimes mixed with, California Gulls. *Eggs*: 2-3 (rarely 4), pinkish- or greenish-buff with many irregular dark brown or gray markings; incubation 21 days, by both parents. *Downy young*: similar to California Gull (see account), but frequently somewhat paler overall.

Range. Breeds from c. Wash., Alberta, and sc. Manitoba south through e. Ore. to Honey Lake area of ne. Calif., sc. Colo. and ne. S.D.; also from c. Que. and Newfoundland south to n. Mich. and N.Y. Winters from Ore. and Nev. and from Great Lakes and New Brunswick south through Fla., and Cuba, and Mexico.

Occurrence in California. In se. Calif. and Central Valley, Abundant Aug.-Mar., dwindling to Uncommon by June and Increasing again in July, on fields and wet grassland, and at lakes and dumps; somewhat fewer on bays and rivers, cw. and sw. lowlands, and at Salton Sea; still fewer on outer beaches (more in s. Calif.) and in marshes and towns widely through cismontane lowlands. Locally Common at dumps unless larger gulls are too numerous. In ne. Calif., Fairly Common late Mar.-Oct., Common to Abundant near colonies through summer — eggs May-June; dependent young late May-July — and Irreg. Uncommon to Locally Common Dec.-Jan. Common to Very Common in Bear Valley Mar.-Oct.; and recorded most months there and at other lakes in s. mountains and in n. Sierra Nevada to near Mt. Shasta June-July.

Mew Gull (*Larus canus*) (fig. 40)

Recognition. L. 16-18 in. (41-46 cm.); WS. ca. 38-40 in. (96-102 cm.). A medium-small gull with adult mantle medium gray, about the color of a California Gull, the wing-tips black with larger white mirrors than either California or Ring-billed gulls, and yellowish-green legs and feet. The relatively small bill is unmarked greenish-yellow in the adult, dusky with a pinkish base in first-winter birds, which also have pinkish-brown legs. The feathering of the immature appears very like a small California Gull (fig. 40) in the first fall, but gray appears on the back before midwinter. Young Mew Gulls become much whiter by spring, and in their second winter are similar to the adult, but with dusky areas near the bend of the wing and subterminally in the tail, and clouded black and white borders in the wing-tip. They wear adult dress in their third autumn.

Habits. In winter, Mew Gulls prefer to forage in areas of turbulent or mixing water — along river mouths, waste-water outlets, surflines near islands or headlands, etc. There they fly about actively in a more fluttery fashion than the larger gulls, picking up small floating food items or occasionally plunging or settling on the water for the larger ones. Toward spring, they also gather about ponds or follow plows in fields very close to the coast and rivers, thus beginning to display the inland habits that characterize them in the far north. When resting on bayside flats or open beaches, Mew Gulls usually keep somewhat apart from the larger gulls, and they rarely feed at garbage dumps.

Range. Breeds from nc. Alaska, s. Yukon, and n. Mackenzie south to s. B.C. coast and n. Sask.; also a larger subspecies in n. Europe. Winters from se. Alaska along coast and major rivers to s. Calif., and through much of w. Eurasia.

Occurrence in California. Increase from Irreg. Uncommon Sept. to Abundant Jan.-Mar., decline to Rare by mid-May, on bays, river mouths, and along outer coast south to Monterey area; fewer at same seasons well offshore and in same habitats of s. Calif. and on coastal lakes, streams, and fields; stragglers

Rare through summer. Also in Central Valley Sept.-Mar., chiefly in and near Delta area, where Fairly Common to Common in winter. Recorded inland in sw. Calif. Nov.-Jan., and in se. Calif. Nov. and Apr.-May.

Laughing Gull (*Larus atricilla*) (fig. 41)

Recognition. L. 15-17 in. (38-43 cm.); WS. ca. 40-42 in. (102-107 cm.). A rather small gull with dark gray mantle like a Western (plate 12) and similar conspicuous white trailing edge of wing, but without mirrors in the blackish wing-tip. In breeding feathering, the head is black except for white eyelids, the bill dark red, and the feet reddish-dusky. Adults in non-breeding feathering have only a dark gray patch on the hind-crown (by late summer). First-winter birds are rather smooth medium to dark brown fading to whitish on throat and belly, the secondaries white-tipped like the adults, tail and coverts white with a broad subterminal blackish band, and the bill and feet dusky or tinged with red.

Habits. This is a common species of the warm seas in its normal range, favoring aerial foraging over either water or land, but also feeding on animals or debris from the surface and even at times at garbage dumps. The southward-flowing cool water off the California coast is probably the reason its distribution is so limited here, in contrast to the east coast with the influence of the Gulf Stream.

Range. Breeds on Atlantic coast from Maine to Fla., Tex., and West Indies, and from Baja Calif. (formerly a few at Salton Sea) to Central or South America. Winters from S.C. and cw. Mexico to Brazil and Ecuador; occasionally wanders to s. and c. Calif. coast.

Occurrence in California. Irreg. Rare to Fairly Common Apr.-Oct., even Locally Very Common July-Aug., at s. end Salton Sea and nearby fields and streams. A small colony present May-July on sandy islands in the southern part of the Salton Sea (later inundated) was last observed in 1956, but in only a few years were any young known to have been raised. The reappearance in the area of over 300 postbreeding birds in July-Aug., 1970-74 gives some hope for reestablishment of the colony. Stragglers recorded Nov.-Feb. in se. Calif.; on or near

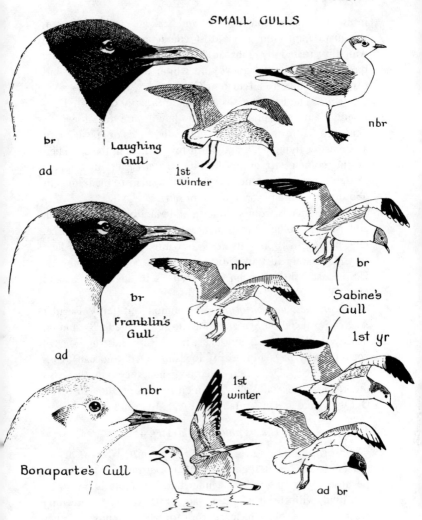

FIG. 41

SMALL GULLS

br

ad

Laughing
Gull

1st
winter

nbr

br

ad

Franklin's
Gull

nbr

br

Sabine's
Gull

1st yr

Bonaparte's Gull

nbr

1st
winter

ad br

sw. coast and bays nearly throughout year; and in cw. Calif.
May and Aug.-Sept.

Franklin's Gull (*Larus pipixcan*) (fig. 41)

Recognition. L. 13-15 in. (33-38 cm.); WS. ca. 35 in. (89
cm.). Between the Laughing and Bonaparte's Gulls in size, with
head coloration like the former, or the dusky nape area larger

243

in nonbreeding season. The feet and beak are also brighter red, the mantle medium gray. Most characteristic is the whitish wing-tip crossed by a broad black triangle. The first-winter bird is grayish-brown above and whiter about the head, but otherwise similar to a Laughing Gull.

Habits. The Franklin's is a species primarily of the midcontinent plains and prairies, where they forage for insects in dense flocks on the ground in cultivated fields or grassland, or at times by flying back and forth low or high in the air. Their nesting colonies are in marshes; but the stragglers that appear in California often associate with other small to medium gulls about waters of any sort — fresh or salt.

Range. Breeds from s. Alberta and Manitoba south to Utah and Neb. A colony established ca. 1957 at Malheur L. in ce. Ore. had up to 1000 nests in 1973. Winters from La., and Tex. to Chile, rarely in Calif. Migrants appear on both e. and w. U.S. coasts, though main movement is through c. U.S. and Mexico.

Occurrence in California. Uncommon but fairly Regular Aug.-Nov. at Salton Sea and along lower Colorado R., and on nearby fields, streams, and lakes; Irreg. Rare there and on sw. coast (mostly bays and nearby lowlands, Occ. to ocean) Apr.-May and Nov.-Dec., but recorded in every month. Also recorded in Central Valley all months except June-July and Oct.-Nov. and at s. mountain lakes Apr.-June and Aug.

Bonaparte's Gull (*Larus philadelphia*) (fig. 41)

Recognition. L. 12-14 in. (30½-35½ cm.); WS. ca. 32 in. (81 cm.). Our smallest gull of normal occurrence in California. Because of its size and its slender bill, it is more likely to be confused with terns (figs. 42 and 43). Adults in breeding feathering have black heads, except for white eyelids. At other times, there is a dark spot in the ear region on an otherwise white head. The mantle is light gray, and a conspicuous white triangle extends from the bend of the wing to the outer parts of the outer primaries, bordered narrowly by black around the tip of the wing. The bill is black and the feet bright red, or dull reddish in first-year birds, which also show more black in the outermost primaries, a dingy brown band along the middle coverts area, and a narrow black subterminal tail band.

Habits. Endowed with great powers of flight despite its small size, this tiny gull feeds mostly on the wing and favors as its winter foraging beat the surflines, areas of mixing tidal currents, and waste-water outlets. Tiny animals and bits of refuse are picked from the surface on numerous quick swoops even over rough water; or, where such *food* is abundant and the water calm, as on salt-evaporating ponds, these birds swim along pecking here and there much like slow, overgrown phalaropes. Such swimming, of course, is in sharp contrast to the terns; but the Bonaparte's Gull sometimes plunges head first into the water also, though not with the vigor of a tern. In April and May, when many have or are acquiring the neat black hood, migrant flocks of Bonaparte's Gulls appear widely on fresh waters along various major valleys in the state. There is no indication in their behavior even then, however, of their *tree-nesting* habit, unique among gulls. *Voice*: a low-pitched nasal *cherr*; also soft whistled and higher pitched, somewhat tern-like notes.

Range. Breeds from w. Alaska, s. Yukon and ne. Manitoba south to c. B.C. and east to Ont. Winters from sw. B.C. to s. Baja Calif., chiefly along coast, and from Ohio to Mass. and south to Fla. and se. Mexico.

Occurrence in California. Very Common to Abundant late Sept.-May, fewer in Aug., with stragglers to June-July on large bays and salt ponds; somewhat fewer over other bay, lagoon, ocean, and coastal waters (well offshore in spring and fall, and to various coastal ponds and fields in spring); still fewer inland throughout Central Valley and se., cw., and sw. Calif. on lakes, streams and marshes, and at times over fields. Also recorded in ne. Calif. or adjacent Nev. Apr.-Dec.; in the Sierra Nevada June and Oct., its w. foothills Dec.-Jan. and Apr.-May, and s. mountain lakes Apr.

Heermann's Gull (*Larus heermanni*) (plate 12)

Recognition. L. 18-21 in. (46-53 cm.); WS. ca. 46 in. (117 cm.). A medium-sized gull, so dark at all stages that it is like no other. Adults (third year or later) have a dark slate-gray mantle with blacker wing-tips (no mirrors) and a white trailing edge of the wing except at the tip. Otherwise they are smooth medium gray on the body, somewhat lighter below, and pale gray on the tail coverts. The tail itself is black with a narrow white tip, and the head is white, somewhat clouded with gray in fall and winter. The bill is red (or red with yellow tip), the legs and feet black. Second-year birds are similarly patterned, but the gray is slightly browner and the head mostly covered with dusky feather tips, and the bill is blackish terminally. First-year birds are dark sooty-brown throughout, mottled with slightly paler feather edges apparent only at close range, and have a mostly blackish bill. They are sometimes confused with Sooty Shearwaters (fig. 4), but lack the pale wing-linings and of course show a different flight silhouette and manner of flight (see Picture Key C).

Habits. This is the gull most strictly associated with the outer seacoasts and adjacent ocean waters. It even bathes in salt water, or occasionally in immediately adjacent freshwater ponds, and undoubtedly drinks sea water. As in other ocean birds, its nasal glands excrete the excess salt into the nostrils. The favored feeding grounds are along the kelp beds just offshore and about the sand coves and rocky promontories that are interspersed along much of the coast. Several Heermann's Gulls often gather around a fishing pelican, and are sometimes successful in grabbing some of the fish it is "re-arranging" to swallow. Along with the Western Gull, they may also hang around foraging sea otters for the bits of sea urchins or other prey they cast aside. When catching their own *food*, gulls of this species take small fish, shrimp, amphipods, small mollusks, and dead animals of almost any sort.

Since they spread from Mexico northward into California (as well as southward) after an early nesting period (Mar.-May), many are present in California waters around colonies of other sea birds, such as on Anacapa I., Pt. Lobos, and the Farallones. Without nesting duties themselves at this time, they are quick to shift location to any unguarded nest to sample its eggs or small young. *Voice*: a subdued cackling, a short *kow-uk*, and a higher, whining call, among others.

Range. Breeds on islands in the Gulf of California and Locally off w. Baja Calif., Sinaloa, and Nayarit, Mexico. "Winters" along the coast from c. Calif (Occ. Ore.) to Guatemala, following a large movement of postbreeding birds in midsummer and fall extending north to B.C.

Occurrence in California. Very Common or Locally Abundant late June-Oct., Irreg. Fairly Common through March, usually Uncommon Apr.-May on s. Calif. outer coast and ocean to offshore islands, fewer coming into river mouths and outer parts of bays or lagoons. Also in similar habitats of c. and nw. Calif., where Rare on inner parts of large bays, and few Dec.-May anywhere north of Monterey. Also recorded in se. Calif. chiefly Salton Sea, Apr.-Nov., and a few times inland in sw. Calif.

Black-legged Kittiwake (*Rissa tridactyla*) (fig. 40)

Recognition. L. 16-18 in. (41-46 cm.); WS. 36-38 in. (91-97 cm.). Adults of this oceanic gull can be distinguished from all other species by their pale gray mantle, "straight-across" black wing-tips without mirrors, white square-cornered to slightly notched tail, white body and underwing (except for black tip), and white head, somewhat clouded with gray on nape in nonbreeding feathering. The bill is yellow and the legs black in full adults, while second-year birds may have a dark tip on a greenish-yellow bill and some greenish-yellow on the tarsi, plus remnants of the brownish-black wing markings. In its first winter, a Kittiwake is among the easiest of gulls to identify because a dark, broad M-shaped band shows on the upper wing surface, made up of blackish outer primaries and their coverts and brown-black middle and some lesser coverts (see also New Zealand Shearwater account and fig. 4). At this stage there is a dusky spot by the ear, dusky bar across the nape, a black terminal band on the tail, and dingy brown-gray legs and bill.

Habits. A major wintering population of this species seems to be at considerable distances off California shores, but probably fluctuates from year to year. Since few observers take boat trips during that season, records are mainly of the even more erratic appearances on inshore waters. Most of the sightings of significant numbers of Kittiwakes from shore have been at Pt. Reyes, the Monterey and Palos Verdes peninsulas, and similar promontories. In their nesting habits, these birds are specialists at cliff-dwelling; and when they do come to shore in winter, rocks still attract them. There is no ready explanation for the occasional flocks remaining into summer so far south of their breeding area (see Graphic Calendar in Appendix). *Food*: pelagic mollusks (pteropods), small fish, and crustacea (especially euphausids or "krill"). *Voice*: mews, groans, and a raucous *kitti-waaak*, though usually silent in winter.

Range. Breeds on arctic and cold temperate coasts and islands essentially around the world, in w. America south to the Aleutians and se. Alaska. Winters in the ne. Pacific Ocean from off B.C. to nw. Baja Calif., in the nw. Pacific to Japan, and in the Atlantic south to N.J. and w. Africa.

Occurrence in California. Irreg. Uncommon to Locally Abundant Nov.-May, mostly offshore on ocean south through Monterey Co.; stragglers Occ. through summer, Common July-Aug. in a few recent years; fewer near or on shore except Locally, as at Monterey Peninsula, and on ocean off s. Calif., including islands. Rare at coastal lakes; vagrants noted a few times on bays Dec.-Apr., at Salton Sea Apr.-Sept., and inland in c. Calif.

Sabine's Gull (*Xema sabini*) (fig. 41)

Recognition. L. 13-14 in. (33-36 cm.); WS. ca 33 in. (84 cm.). A small gull with triangle of black from bend of wing to tips of all but the inner several primaries, and another triangle of white thence through the secondaries. The rest of the mantle is light gray in adults, somewhat darker brownish-gray and extending onto the hind-crown in first-year birds, which also have a black band at the tip of the shallowly forked tail. Adults have black legs, a yellow-tipped black bill, and in breeding feathering a slate-gray hood, which is replaced by a vague gray collar on the nape in fall.

Habits. A transient in our region except for rare vagrants, this species usually passes so far offshore that it is not observed every year. The migrating flocks are mostly small and rather compact. They fly with steady wingbeats low over the water, showing little interest in following boats for possible handouts as do the larger gulls. When far offshore in migration, however, Bonaparte's Gulls may behave similarly, and hence a close-enough view to see the different plumage pattern is necessary to tell these species apart. *Food*: Sabine's Gulls are plankton eaters, taking tiny animals from the water by quick dips of the bill while flying or hovering. On their breeding grounds, however, they do forage along beaches and even inland, chiefly on insects.

Range. Breeds around the world in high arctic areas, in America south only to w. Alaska, nw. Mackenzie, and Arctic Islands. Winters in Pacific Ocean off Peru (and elsewhere), and in Atlantic.

Occurrence in California. Fairly or Locally Common as migrant over ocean well offshore in May and Aug.-Oct., stragglers in all months; decidedly fewer reach the coast, but Occ. even to lakes there. Vagrants recorded in same seasons on bays, in se. Calif., and a few times elsewhere inland.

33. SUBFAMILY STERNINAE (TERNS)

(PICTURE KEYS B AND C)

Gull-billed Tern (*Gelochelidon nilotica*) (fig. 43)

Recognition. L. 13-15 in. (33-38 cm.); WS. 33-37 in. (84-94 cm.). Adults in breeding feathering appear nearly all white (actually pale gray above) except for gradually more dusky on outermost primaries and full black cap that is shaggy on the nape. The tail is shallowly forked and the wings relatively broad (for a tern). Most distinctive is the *all-black bill* that is relatively *heavier and shorter* than in other terns (figs. 42 and 43). Nonbreeding adults and first-winter birds have only a dusky stripe before and behind the eye, and fine brown streaks on the otherwise white head. The immature also has more brown on the upperparts, and a brown-gray subterminal band on the tail.

Habits. Gull-bills are versatile *feeders*, utilizing both of the fishing methods of other terns (see Forster's Tern ac-

count) but more frequently feeding on large insects obtained from the air or ground or plant surfaces. Their flight is
steadier than that of other terns of comparable size, thus
more gull-like. The near absence of sightings on the California coast is surprising, in view of their distribution (and
former abundance) well north on the east coast. Typically,
this species seeks the higher parts of salt marshes for *nesting*, but locally (as in the Salton Sea colony) they may
also make only a shell- or debris-lined scrape on a sandy
flat. *Eggs*: 1-4 (usually 3), cream to buff with sparse
brown markings. *Downy young*: pinkish-buff, streaked and
spotted above with dusky.

Range. Breeds from Salton Sea (erratically), Sonora, and
Tex. to Md. and south to Argentina; also Locally in
Eurasia, Africa, and Australia. Winters from Tex. to Fla.
and south to c. South America, and in Old World tropics.

Occurrence in California. Irreg. Uncommon (absent some
years?) to Common at s. end of Salton Sea and over irrigated fields (and streams and ponds?) in Imperial Valley,
late Mar.-Sept., Occ. to Dec. Formerly more numerous and
nested Apr.-June at s. Salton Sea until at least 1971.
Casual on s. coast and in Sacramento Valley.

Forster's Tern (*Sterna forsteri*) (fig. 42)

Recognition. L. 14-16 in. (35½-41 cm.); Tail 5-
8 in. (13-20 cm.), forked 2.3-5 in. (6-12½ cm.); WS. ca.
30 in. (76 cm.). Most widespread and numerous of our
terns, distinguishable as a tern and from small gulls by the
narrower wings, slender pointed beak, deeply forked tail,
and manner of foraging. Adults of all the similar-sized
terns have a light gray mantle, white body otherwise, a
mostly white tail, and in breeding feathering a full black
cap. In fall and winter, Forster's Terns show black on the
head only as a patch behind the eye (curving down a bit
at the rear), but the nape is sometimes partly gray and
toward spring may become quite black and thus yield a
pattern similar to the Common Tern. In fresh plumage the
primaries of an adult Forster's are silvery above, but when
worn may be quite dusky, as are those of the Common

FIG. 42

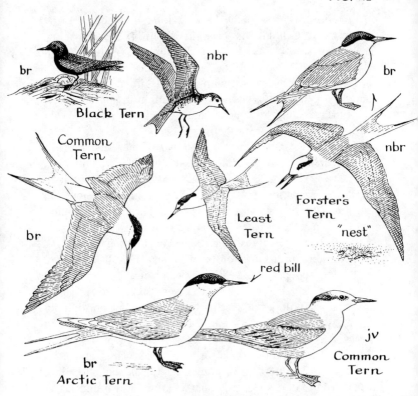

br
Black Tern

nbr

br

Common
Tern

nbr

br

Least
Tern

Forster's
Tern

"nest"

red bill

jv
Common
Tern

br
Arctic Tern

Tern. Both of these species have black-tipped orange-red to red bills (may be all black in winter) and red feet (to yellowish in a few). An absolute distinction, which can be seen under exceptional circumstances in the field, is the *white outer web* and gray inner web of the tail feathers of the Forster's, showing best on the outermost pair. Juveniles have brownish backs until early autumn.

Habits. The graceful beauty of the flight of the smaller terns has given them the popular name of "sea swallow" — and like the swallows, terns must fly to obtain food and so can be seen on the wing in nearly all kinds of weather. When fighting a strong wind, their wingbeats are very deep and quick. At calmer times, each downstroke of the wings, even of half-depth, produces a readily visible upward bounce of the course. When actually seeking food in flight over the water, a tern holds its beak at about 45 degrees below horizontal (see Picture Key C). An abrupt stop when available prey is sighted is followed by a vertical or steep plunge head first into the water, with the wings folded or the tips parallel to the body. The whole bird often disappears below the surface, only to emerge immediately into flight again with a few splashes from the wings.

Small fish compose most of the *diet* of the Forster's Tern in saltwater areas and on many lakes; but it takes crustacea, aquatic insects, and small amphibians also from the water. In shallow areas it often "scoops" these from just below the surface without plunging, and where flying insects are abundant often catches them in mid-air. When not feeding or nesting, these terns rest on pilings or floating objects or gather in compact flocks on open shores. Like all terns, they are seldom seen sitting on the water itself except when young and unable to fly well. *Voice*: sharp *kit* and nasal *kyarr* notes, both given irregularly during feeding and in mild alarm; also a longer, very harsh and rasping *zzree-arrr*, when more alarmed. *Nest*: a cup of grass or marsh plants placed on dikes, islets, or on platform of dead plants in marsh, sometimes floating (fig. 42). *Eggs*: 2-4, pale buff to olive, finely spotted or irregularly marked with brown and lavender, incubation 23 days. *Downy young*: light grayish to pinkish-buff, heavily spotted and streaked above with blackish; stay in nest only a few days, but do not reach flying stage for about 4 weeks.

Range. Breeds from sc. Alberta to Manitoba and south, east of the Cascades in Wash. and Ore. to ne. and sw. Calif. nc. Utah, e. Colo., and se. Wisc.; also locally along Atlantic and Gulf coasts from Md. to ne. Mexico. Winters from c. Calif. and Va., chiefly along coasts, to Fla. and Guatemala.

Occurrence in California. From Common Apr. and Nov. to Locally Abundant May-Aug. on bays, lagoons, and salt ponds from San Pablo Bay south through sw. Calif., nesting on salt pond dikes or low islands and marshes and foraging also over ocean (including well offshore in spring and fall);

fewer to coastal lakes, streams, and marshes; Uncommon to Locally Fairly Common in San Francisco Bay area, and Locally Common on s. coast Dec.-Mar. Inland on lakes, streams, and marshes (usually nesting in "old" marshes) – in ne. Calif. Uncommon to Fairly Common Apr.-Aug., somewhat fewer in Central Valley and s. Calif., esp. at Salton Sea, at same season and Occ. through Feb. Eggs, May-June; non-flying young, late May-July. Migrants recorded at Sierran and s. mountain lakes May-Sept.

Common Tern (*Sterna hirundo*) (fig. 42)

Recognition. L. 13-16 in. (33-41 cm.); Tail 5-7 in. (12½-18 cm.) with fork ca. 3½ in. (9 cm.); WS. 29-32 in. (74-81 cm.). See Forster's Tern account for characteristics possessed by both (also compare Arctic Tern). In fresh fall plumage, the dusky primaries of the Common Tern are fairly good clues to its identity, but the best clue is a complete blackish band around the hind-crown (late summer through winter) and a pure white tail except for the *dark gray outer web of the outer feathers*. This is visible at close range in the field when the spread tail is viewed either from above or below. Immatures are even easier to distinguish from Forster's in side view, as they show a brownish-gray area on the leading portion of the wing from the bend toward the body, where the Forster's is all light gray after the rusty-brown juvenal feathers are molted in late summer.

Habits. This species is much less common, even in migration, than the Forster's Tern in most areas of California visited by observers but it is suspected that many more pass undetected over the ocean well offshore. However, the irregular peak numbers shown on the Graphic Calendar (in Appendix) have been found in or near harbors and inlets of the southern California coast, whence most of the Dec.-Mar. records also come. In the eastern U.S., the Common Tern nests on sandy beaches, both on large inland lakes and along the coast. Several banded in Alberta, Canada, have been recovered in California and southward, so our migrants must cross to the coast somewhere, perhaps well north of the state. The best place to learn to distinguish Commons from other terns is among the mixed flocks resting on a beach or on bayside pilings or flats. Their feeding habits are similar to the Forster's (see account). *Voice*: short *kik-kik-kik*, a somewhat more trilled note, and a raspy *keee-yaaahr* of alarm (higher in pitch and usually more prolonged than the even harsher note of the Forster's).

Range. Breeds from c. Mackenzie east across s. Canada and south to Mont., S.D., and the Great Lakes, and on Atlantic coast south to N.C.; also in e. Tex., Fla., and Bermuda, and from w. Europe to Mongolia. Winters along coasts from s. (Occ. c.) Calif. and S.C. to s. South America; also in s. Africa, Indian Ocean, and sw. Pacific.

Occurrence in California. On s. Calif. coastal lagoons, beaches, river mouths, and the ocean to well offshore, Fairly Common to Common July-Aug., Common to Occ. Abundant Sept.-early Oct., Fairly Common to Rare Nov.-Jan., and Irreg. Rare to Fairly Common mid-Apr.-June. Similar habitat and seasonal pattern but fewer in cw. Calif, and still fewer on inner bays, at coastal lakes in nw. Calif., and at Salton Sea; Occ. elsewhere inland, chiefly July-Sept.

Arctic Tern (*Sterna paradisaea*) (fig. 42)

Recognition and Habits. L. 14-17 in. (35½-43 cm.); Tail 6½-8½ in. (16½-21½ cm.), forked 4-5 in. (10-12½ cm.); WS. 29-33 in. (74-84 cm.). Because of the difficulty in distinguishing this species from the Common Tern (and indeed, at any distance, from the Forster's), it may well be more numerous in California waters than the skimpy records indicate, especially well offshore. It presumably occurs in southward passage every year, but mostly beyond the range of the one-day charter-boat trips seeking birds. The even greater paucity of spring records remains unexplained. In breeding feathering the Arctic is similar to the Common, but the bill is somewhat darker and all red (blood-red rather than orange-red) and the underparts are *light gray* except for a white line next to the black cap. A few Common Terns may lack the black tip on the bill in late summer, and some Arctic Terns acquire a small black tip then. The legs (tarsi) of the Arctic are shorter, a feature visible only by direct comparison at fairly close range of the two species standing. In fall, Arctic Terns are quite white below but often retain a full or nearly full black cap through Sept. (as do a few Common Terns). For the experienced observer of terns, the Arctic at all seasons shows greater translucence of the whitish primary-outer secondary area of the wings (except for narrow dusky border) than either the Common or the Forster's, and the forehead is also slightly more abrupt.

Long proclaimed as champion of long-distance migrants, some Arctic Terns breed a full 11,000 miles by the shortest ocean route from their main "wintering" area (in the southern summer), and it has thus been computed that they must travel an average of 150 miles per day for some twenty weeks in order to make the round trip. However, banding recoveries show that their accomplishment is even greater than this, since New England breeders travel first to Europe and then southward to Africa.

Range. Breeds from n. Alaska, Canadian Arctic Islands, and n. Greenland south to Aleutians and nw B.C., and across c. Canada to Newfoundland, and south to Mass.; also in Old World Arctic and sub-Arctic. Winters on Antarctic and sub-Antarctic oceans around the world, in the e. Pacific north to c. Chile.

Occurrence in California. Migrates mostly over ocean well offshore, July-Oct., when Irreg. Rare to Common, and May-early June, when Rare to Fairly Common. Noted only Occ. in same periods along coast itself, and on bays Sept.-Oct.

Least Tern (*Sterna albifrons*) (fig. 42)

Recognition. L. 8½-9½ in. (21½-24 cm.); Tail ca. 3½ in. (9 cm.), forked ca. 1¾ in. (4½ cm.); WS. ca. 20 in. (51 cm.). The only tern in our region that is decidedly smaller than a Forster's and further distinguished by a white forehead and "eyebrow" area at all seasons. The rest of the crown is black in summer adults, the tail moderately forked, the mantle pale gray, the feet yellow, and the bill yellow with a black tip.

Immatures are darker on the wing coverts to blackish on the outer primaries, have black on the hind-head only, and their bill and feet are dusky.

Habits. Until recent years this tiniest of terns was common in southern California, and its northernmost known nesting area was on beach dunes near Moss Landing, Monterey Co. Now, though reduced there from perennial disturbance of beaches by people and dune buggies, it has begun to nest on salt-pond dikes or bare flats or sandfills about bays and estuaries, including San Francisco Bay, and may be staging a "comeback" thereby. Feeding is carried out both in the calm waters of narrow estuaries or large bays and for a short distance off the beaches in the open ocean. The hovering and plunging habits are conspicuous, as in the Forster's, but the Least Tern's small size makes its impact with the water appear rather weak. Small fish and crustacea are the almost exclusive *diet*, some being caught by quick dips of the bill on shallow swoops over the water surface. *Voice*: a distinctive, very high-pitched *kit* or *kit-ic* note; also a prolonged succession of these interspersed with excited, raspy squeals. *Nest*: a scrape in bare earth or in sand of high beach or low dunes, with at most some bits of shell or debris as a rim, usually in loose colonies. *Eggs*: 2-3, pale buff with darker, "sand" pattern of markings; incubation 20-22 days. *Downy young*: also sand-colored above; remain in nest only a few days, then move to various hiding spots nearby until able to fly, at about 4 weeks.

Range. Breeds along coasts and bays from c. Calif. to Baja Calif. and from Mass. to Fla., and Tex. and south to Brazil; also inland along the Mississippi R. and major tributaries, and over much of the Old World. Winters from Gulf of California and West Indies south, and in Old World in tropics and subtropics.

Occurrence in California. Uncommon to Locally Common Apr.-Oct., Rare Nov.-Mar., over lagoons, bays and ocean waters by sandy beaches and on some salt ponds; nests with eggs May-July; non-flying young June-early Aug. Also recorded in se. Calif., mostly at Salton Sea, late Apr.-Aug. and inland in sw. Calif.

FIG. 43

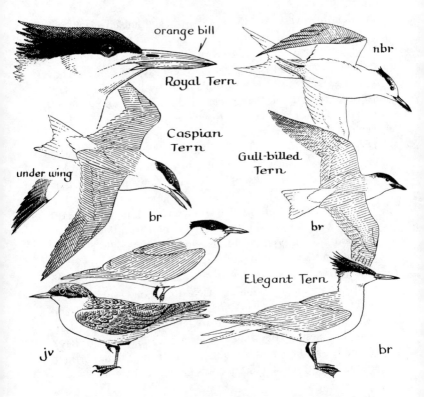

orange bill

Royal Tern

nbr

Caspian
Tern

Gull-billed
Tern

under wing

br

Elegant Tern

jv

br

br

Royal Tern (*Thalasseus maximus*) (fig. 43)

Recognition. L. 18-21 in. (46-53 cm.); Tail forked 3-4 in. (7½-10 cm.); WS. 42-45 in. (107-114 cm.). A large tern with moderately deep fork in the tail and a color pattern very similar to the still larger and more widespread Caspian Tern, from which a Royal can best be distinguished by the lack of any extensive dark area on the underside of the primaries, only a narrow blackish edge of the outer primary and general grayish elsewhere under the wing-tip being evident. The red to orange bill of a Royal is heavy but not as thick as that of a Caspian, and the feathers of the hind-crown project into a longer, shaggier crest. The crown itself is all black in spring and through the summer in some birds, but nonbreeders and some

257

breeders are white on the forecrown and mixed black and white behind this.

Habits. The Royal Tern is at all times a saltwater bird, usually nesting on sandy beaches or bayside flats. It has become less common in California than it was thirty years ago, perhaps because of the spread of the Elegant Tern (see account). The only breeding Royals known in the state have been near or in the midst of the Caspian Tern colony in San Diego Bay in 1959 and 1960. The *voice* includes high, clear *tsear* or *kreer* calls and a rolling whistle, both higher and less raucous than the notes of a Caspian, more melodious than those of the Elegant.

Range. Breeds from wc. Baja Calif and Sonora to islands off wc. Mexico, and Locally on Atlantic coast from Md. to Tex. and south to n. Venezuela; also in w. Africa. Winters from s. Calif. and S.C. to Peru and Argentina; also w. African coast.

Occurrence in California. In sw. Calif., from Uncommon to Fairly Common July-Jan. or Occ. Common, esp. in Sept., along outer coast beaches, ocean, and nearby lagoons and bays; Locally and Irreg. Rare to Fairly Common Feb.-June (and has nested May-June at s. end of San Diego Bay). Wanderers in late summer and fall have occurred north to Marin Co., but few c. Calif. records in recent years.

Elegant Tern (*Thalasseus elegans*) (fig. 43)

Recognition. L. 16-17 in. (41-43 cm.); WS. unknown. A medium-large tern with pale gray mantle merging to somewhat darker gray wing-tips (no real blackish below as in Caspian), and a slender, long-pointed, and somewhat downcurved bill which varies from orange-red to almost pure yellow (in some immatures). The feathers of the hind-head are even longer and form a more pronounced crest than those of the Royal Tern. Most breeding-season birds have the crown wholly black and a pronounced rosy-pink tint on the white underparts, some retaining this until Aug. or Sept. The feet are black to greenish-gray.

Habits. Elegant Terns are strongly gregarious when nesting, and still prominently so when resting on beaches or tideflats at other seasons, usually not mixing with Forster's or the larger terns on the same areas if large numbers of their own kind are present. They dive for fish mostly in the ocean beyond the breakers, occasionally in the bays or lagoons behind outer beaches. *Voice*: a rather clear *ke-e-er*, dropping slightly in pitch; also a rougher, nasal *karreek* or *ki-ki-k-kareek* similar to a call of the Least Tern, but louder and lower in pitch. *Eggs*: 1-2, white to pinkish-buff, marked with medium to dark brown or gray.

Range. Breeds on islands in Gulf of California, and Locally on w. Baja Calif. coast and on San Diego Bay. Moves north Irreg. following breeding to San Francisco and Bodega bays. Winters mostly from Peru to Chile, a few in s. Calif.

Occurrence in California. This species was formerly only an occasional post-breeding visitor to California shores from colonies to the south, though it sometimes appeared in numbers (as in 1926). Beginning about 1950, it has occurred on the southern California coast in quite large flocks in most years, and in 1959 a colony of nesting birds was discovered in the midst of the Caspian Tern nesting area on dikes between San Diego Bay salt ponds. About coincident with the establishment of this "beachhead," the late summer invasions began bringing flocks north along the central coast, where the species had been unrecorded for many years.

Irreg. Uncommon to Abundant July-Nov. along sandy beaches of s. Calif.; none to Abundant on c. coast, and Rarely to n. Calif., at same season. Fairly Common to Common Mar.-June in s. Calif., mostly near colony on dike between San Diego Bay salt ponds, where eggs found May-early June, young birds June-early July. A few also have wintered in recent years at Santa Barbara and San Diego, Rarely elsewhere.

Caspian Tern (*Hydroprogne caspia*) (fig. 43)

Recognition. L. 19-23 in. (48-58 cm.); Tail forked ca. 1½ in. (4 cm.); WS. 50-55 in. (127-140 cm.). The largest tern, with body size about that of a Ring-billed Gull, a massive bright red to orange-red beak, black feet, and shallowly forked white tail. The mantle is pale gray, and most of the underside of the primaries appears dusky to blackish (visible through careful observation even on the folded wing-tips, which extend well beyond the tail). Adults have a short crest on the hind-head, this and the whole cap being black in late spring and summer, while there is much white admixed on the crown at other times. The juvenile is mottled with light and medium brown on the back and upper wing coverts, shows dusky on the upper primaries and subterminal tail band, and has a duller bill but an all-black crown.

Habits. In straightforward flight this large tern is relatively steady, appearing much like a gull. But when seeking fish in the water below, the position of its beak is below the horizontal, and is typical of terns (fig. 43). Much of its foraging is done from heights of 50 feet or more, from which it plunges with a resounding splash. At other times it seeks *food* from lesser heights, but seldom feeds by the "bill-dipping" method. A variety of fish up to six inches or more in length are taken, some of which have been observed being carried for 8 miles or more to the young. In fact, a trout tag from a lake 16 miles away was found in a San Francisco Bay nesting colony. Caspians are often seen briefly at remote reservoirs and at many points along the ocean and bay shores where they do not nest. When the food supply is found to be good in such places, large flocks sometimes assemble there. In the fall, juveniles leave the nesting area with and continue to beg and receive food from the adults for weeks or months (see Graphic Calendar in Appendix).

Up to 1952, the only known colonies in the state were at Salton Sea, at Clear Lake Reservoir (Modoc Co.), and on a dike between salt-evaporating ponds northwest of Newark (Alameda Co.) – the last site used continuously since at least 1922, when only seven nests were found. This colony increased to about 500 pairs in 1954, then declined and was finally abandoned about 1969 when the dike was awash. By then two or three other locations on south San Francisco Bay salt ponds were in use, in or near the new San Francisco Bay National Wildlife Refuge which will now serve to protect them. A similar colony continues at the south end of San

Diego Bay, and smaller ones on islands of several lakes in Modoc and Lassen cos., plus possibly still at Salton Sea.

Voice: a loud, low-pitched *kowk*, or longer, grating *kraaaak*; when alarmed or aggressive, a piercing and very raucous *ka-k'-r-ROW*; begging juveniles give repeated wheezy, much higher-pitched calls. *Nest*: an unlined hollow on bare earth, or sometimes with lining or rim if in marsh border; spaced 2-10 feet apart in colonies. *Eggs*: 1-4 (usually 2-3), light buff or pinkish with varied brown markings. *Downy young*: grayish-white to buff, sparsely marked with blackish above; they remain in nest several days, then walk about the colony or run into water and swim away if disturbed — soon becoming so wet that it is doubtful if they can survive prolonged human intrusion into the colony.

Range. Breeds at scattered locations on all continents except South America; in North America from e. Wash. and c. Mackenzie to se. Canada and south to ne. and coastal c. Calif. and Baja Calif., Nev., Utah, Wyo., and on Atlantic coast from Va. to Tex. Winters from Calif. to s. Baja Calif., from N.C. through Caribbean, and from s. Europe to Africa and New Zealand.

Occurrence in California. Common to Locally Very Common Apr.-early Oct., and Irreg. Fairly Common to Rare rest of year (chiefly in s. Calif.) on bays and large salt ponds of c. and s. Calif.; somewhat fewer on ocean and beaches, esp. near and at river mouths, and at Salton Sea, and foraging at streams and lakes near any of these. Eggs Apr.-July, most in May-early June; nonflying young May-July, a few to Sept. Locally Uncommon to Fairly Common May-Nov. in Central Valley. In ne. Calif., Locally Common and nesting at least May-Aug.; migrants Occ. Mar.-Dec. elsewhere inland and in nw. Calif.

Black Tern (*Chlidonias niger*) (fig. 42)

Recognition. L. 9-10½ in. (23-27 cm.); Tail ca. 3¾ in. (9½ cm.) forked nearly 1 in. (2½ cm.); WS. 24-25 in. (61-64 cm.). Unmistakable in breeding feathering, with solid black head, neck, underparts (except white undertail coverts), bill, and feet. The wings and shallowly forked tail are medium gray at all seasons. Beginning as early as July, adults show white on the underparts which spreads until this includes the face and side of the neck, leaving an irregular boundary between the white and the black of the hind-head and gray of the upperparts. Juveniles are similarly patterned, but brownish on the head and body.

Habits. The deep wingbeats and bouncy course of a Black Tern hawking for flying insects over a marsh or grassy meadow are so slow that they are often referred to as "languorous." Yet it occasionally flaps more rapidly, as when pursuing an insect or when hovering before a more typical tern-like drop to the water. Its *diet* consists of large to medium-sized and abundant insects such as grasshoppers, dragonflies, cicadas, mayflies, etc., which are taken either on the wing or plucked from their resting places on plants, and small fish, aquatic insects, crustacea, and amphibians, which are caught in or on the edge of the water. When in its winter range on the ocean, or enroute to it, the Black Tern feeds mostly by shallow swoops at the surface rather than plunging into the water.

The drainage of former marshes caused a decline of this species. Later the expansion of rice culture in the Central Valley brought an increase because of the new "marshes" (albeit artificial) and the semi-protected nesting sites possible on the small check dikes. Recently, however, the population has dwindled again (except in Imperial Valley?), possibly due to pesticide accumulation. Over their nesting areas, pairs or whole flocks engage in spectacular long swooping glides over the plant growth. *Voice*: a short, metallic *kik*, or slurred to *k-klee-a*. *Nest*: a skimpy cup of dead plant stems, on matted or floating masses of marsh vegetation (fig. 42). *Eggs*: 2-3 (rarely more), rather dark olive or buff, heavily marked with dark brown; incubation 17 or more days. *Downy young*: drab gray and cinnamon.

Range. Breeds from se. B.C. across s. Canada and south, east of the Cascades, through the Central and Imperial valleys of Calif. to c. Nev., Utah, Colo., and across much of ne. U.S.; also in much of Eurasia. Winters on ocean off w. and ne. South America and on major rivers there and in Africa.

Occurrence in California. Fairly or Locally Common (formerly Very Common) mid-Apr.-Sept., a few Occ. to mid-Nov., about marshes and rice fields and foraging over nearby lakes, moist fields, and streams of se. Calif. and Central Valley; Locally also inland in sw. Calif. Eggs May-June; dependent young late June-mid-Aug. Coastally in c. and esp. s. Calif., Uncommon, Occ. Fairly Common, on bays, salt ponds, river mouths, and ocean (to well offshore spring and fall) Apr.-May and late June-Sept., but a few stragglers all other months. Recorded also in nw. Calif. Sept.-Oct., inland in cw. Calif. Apr. and July, and at Sierra Nevada and s. mountain lakes or meadows May-early Oct., formerly nesting in June at South Tahoe. Nests also in ne. Calif. where uncommon to Locally Common at marshy lakes, wet meadows and over nearby fields and grasslands May-July, and probably to early Sept.

34. FAMILY RYNCHOPIDAE (SKIMMERS)

Black Skimmer (*Rynchops nigra*) (fig. 44)

Recognition. L. 16-20 in. (41-51 cm.); WS. 42-50 in. (107-127 cm.). A "big tern"-sized bird. Blackish-gray above and white below and on the forehead, sides of the short notched tail, and tips of the secondaries. The small feet are red, and a most spectacular black-tipped red bill has each half flattened like a knife, the lower mandible decidedly longer than the upper. Immatures are dingy brown above with light feather edges that wear off; they molt to adult feathering in their first spring.

Habits. When foraging, skimmers often fly for several hundred feet at a time low over the water, with the long lower tip of the beak in the water, making a small wake. Supposedly this attracts fish to the surface, and the skimmers grab them by sidewise swings of the head or, in tern fashion, on a subsequent overflight. They are loosely gregarious in such feeding, and much more so when resting on sandflats or beaches.

Range and California Occurrence. Breeds on Atlantic and Gulf coasts from Mass. to Tex. and south to Argentina, on

Pacific coast from nw. Mexico to Chile. The first sporadic occurrences in California were in 1962 on the Orange Co. coast, and in 1968-71 at Salton Sea. Then in 1972 a colony of five nests (one successful) was present late Apr.-early Sept. at s. end of Salton Sea,* and so a spectacular addition to the roster of California water birds is thus made. Winters from Carolinas and cw. Mexico south. Casual north to se. Canada and Irreg. Rare on s. and c. Calif. coast all months except June.

35. FAMILY ALCIDAE (ALCIDS)

(Picture Keys A, B, and C)

Common Murre (*Uria aalge*) (fig 45; plate 7)

Recognition. L. 16-17 in. (40½-43 cm.); WS. ca. 30 in. (76 cm.). Our most numerous alcid, the murres have rather a longer neck than most members of the family. The black, pointed beak is almost as long as the head, which feature, together with its lack of white areas on the back or wings, distinguishes it from all other alcids in our area except the nearly related Thick-billed Murre (see Note below). In breeding feathering, worn Mar.-June or by some birds as early as Nov. or as late as July, the throat and foreneck of a Common Murre are dark chocolate-brown and the rest of the underparts are white, including the underwing coverts. Elsewhere the color is dark slate at all seasons, except for narrow white tips on the secondaries. Some individuals have a narrow white eye-ring and line back from it. Adults in nonbreeding feathering have the white extending over the throat, foreneck, and side of the head, typically with a dark point extending back from the eye into it. Immatures are similar but with the dark portions including vaguely defined lighter gray areas.

Habits. Large numbers of Common Murres can be seen in a short time at their nesting areas or at the offshore rocks or few cliffs where nonbreeders assemble to rest. Healthy murres do not come ashore anywhere else, but when there they stand

*Condor, 76:337-338; American Birds, 26:906

FIG. 44

Black Skimmer

quite upright on narrow ledges or sloping rock surfaces, with their weight resting on the whole tarsi. The impression is somewhat penguin-like; but this is quickly dispelled when they launch themselves into the air and, after gaining speed by dropping altitude, fly off as rapidly as a diving duck. The food-getting of the alcids is quite like that of penguins, however, for they dive quickly and pursue their prey under the water by use of the wings for propulsion and their feet as rudders. The *diet* of murres consists largely of small fish and swimming crustacea, but worms and other bottom-dwelling animals are taken occasionally. Murres can dive deeply, some having been caught in fishnets set more than 100 feet down.

Many of this species move up and down the California coast, but little is known about the source of these apparently migrating birds which are irregularly noted from such headlands as Pt. Pinos and Pt. Reyes (see Graphic Calendar in Appendix). In late summer, nearly-grown young and adults in molt are flightless because they lose all the long flight feathers more or less at once, but they then scatter widely on the ocean. Among major breeding sites in the central California area are Southeast Farallon I. (Population 8-10,000) and a large rock just off Pt. Reyes; but the numbers are still far below those prior to the early 1900s, when commercial collection of their eggs for San Francisco markets (over 7600 dozen in 1896 alone) caused their near extirpation. A large colony formerly at Pedro Pt., San Mateo Co., was also exterminated. *Voice*: a hoarse purr (or "*murrrrr*") or growl, given chiefly in the colony. *Nest*: none. *Eggs*: 1 only, very pointed, making it likely to roll in a circle on the narrow rock ledge where usually laid; pale blue or greenish, with extremely variable brown and black markings; incubation 30 days. *Young*: at 2½-4 weeks, long before fully grown, take to water, where they are safer than on home ledge even though they have little or no flight feathers.

Range. Breeds from Bering Sea islands and the Aleutians south to coast of s. Monterey Co. (formerly to San Miguel I., Santa Barbara Co.) and to n. Japan; also in Atlantic from Greenland to Nova Scotia and nw. Europe. Winters from limit of ice to s. Calif., s. Japan, and Mediterranean Sea.

Occurrence in California. Very Common all year to Very Abundant near larger colonies Mar.-Aug., on ocean near to well offshore from rocky coasts and islands of c. and n. Calif.; fewer off sandy shores and far fewer in s. Calif. waters (chiefly in winter there). Eggs chiefly late May-July, but a few as early as Mar.; young reported at colonies late June-Aug. Also noted in outer parts of bays, chiefly late Aug.-Oct.; a few records in winter and in May.

Note: The **Thick-billed Murre** (*Uria lomvia*) (fig. 45) is a more northern species similar to the Common but with shorter, thicker bill more convex above, with a narrow whitish line along its cutting edge. In winter feathering, there is less white area on the side of the head also. This species has been detected in the Monterey area almost every year, beginning in 1964, and once in Marin Co. (see Graphic Calendar in Appendix).

Pigeon Guillemot (*Cepphus columba*) (fig. 46; plate 11)

Recognition. L. 13-14 in. (33-35½ cm.); WS. ca. 23 in. (58 cm.). The only alcid in California other than the murres (see Common Murre account above) with a relatively slender neck, distinguished in all plumages from others in the state by the large white patch on the secondary coverts or inner part of upper wing surface, which therefore shows well both in flight

FIG. 45

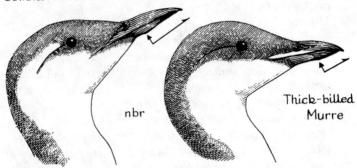

Common Murre

nbr

Thick-billed Murre

and with wings folded. In breeding feathering the adult is otherwise solid black with bright red feet (and mouth lining), while in fall and winter it is white below and has whitish mixed with dark gray above, especially on the head and neck. Immatures are similar to the winter adult but darker on the upperparts, sometimes with only a vaguely defined white wing patch.

Habits. This is usually the easiest alcid to see well from shore, since it nests at many places along the California coast, including the outer part of the Golden Gate and at state parks on the rocky coast from Trinidad to Montana de Oro. From the top of a seacliff in such places, one can often look down upon Pigeon Guillemots foraging in the clear waters of rocky coves, their red feet trailing behind them as they "fly" beneath the surface. When nesting, guillemots are more restricted to waters close to the coast than most alcids, their main *food* there being the small fish that lurk in crevices or along the bottom. Shrimp, crabs, amphipods, polychaete worms, and small or soft-shelled mollusks also are obtained in similar situations. The breeding colonies of Pigeon Guillemots are usually quite small, but an estimated 1500 bred on South Farallon I. in 1970. Adults often stand about on ledges of a seacliff, and may give their drawn-out, shrill, very sibilant call there or from within the nest crevice. *Nest:* none as such; birds nest within crevice in cliff or heavy talus slope, less often in a burrow. *Eggs:* 2, whitish with heavy dark brown markings;

incubation 28 days. *Young* go to sea by about 38 days, though cannot yet fly.

Range. Breeds from Bering Sea islands and the Aleutians south to n. Japan and along w. coast of America to Santa Barbara I., s. Calif. Winters in same area, except far north, but apparently more scattered or offshore; straggles south to San Clemente I. The very similar Black Guillemot is found in the Atlantic and along n. Asian coast to n. Alaska.

Occurrence in Califiornia. Fairly Common to Common late Mar.-Sept., Locally Very Common Apr.-June, along steep rocky coastline and to a few miles offshore. For balance of year Irreg. Rare to Fairly Common in same areas and farther offshore, esp. in s. Calif.; vagrants noted July-Nov. on San Francisco Bay.

Marbled Murrelet (*Brachyramphus marmoratus*) (fig. 46; plate 7)

Recognition. L. 9-10 in. (23-25 cm.); WS. unknown. A medium-small alcid with the short, thick neck and other features typical of the family. Its breeding feathering is unique — basically dark brown all over, with many short white bars or spots over the whole underparts and on the back. In nonbreeding feathering the underparts are pure white, extending upward as a partial collar on the side of the neck. In fall and winter the white scapulars form a distinctive white stripe just above the folded wing, and the upperparts are otherwise uniformly very dark gray. At a distance in these seasons, Marbleds are similar to the Ancient Murrelet but can be distinguished by the white scapulars and dark wing linings. The bill of the Marbled is also thinner and longer, and the chin is white.

Habits. Among all North American breeding birds, this is the only one whose precise nesting place long remained unsubstantiated by the essential evidence of collecting or photographing, in identifying fashion, an adult with egg or young *in the nest*. Finally a downy young was found in a high niche well up in a tree in Santa Cruz Co. in the summer of 1974, and became a specimen at the California Academy of Sciences. It probably also nests in burrows or niches amid the dense

FIG. 46

Pigeon Guillemot

nbr

Rhinoceros Auklet

nbr

Marbled Murrelet

br

nbr

Craveri's Murrelet

Cassin's Auklet

Ancient Murrelet

nbr

br

Xantus' Murrelet

Parakeet Auklet

imm

Tufted Puffin

understory of humid coastal forests, as its near relative the Kittlitz's Murrelet is known to do in northeast Asia. Marbled Murrelets are common about Puget Sound and northward in summer, fairly common where the redwoods are close to the coast in northwestern California, and can be seen there and locally in the Santa Cruz Mts. flying above the forested ridges at points up to 5 miles or more inland while "commuting" to and from the sea at dawn and dusk. They frequently give a shrill *scrrreeeee* or *meeer* call on such flights. Where the species nests so far inland, the young must make their first flight a successful one — all the way to the ocean or some large body of water if they are to survive. Yet there is no record on lakes or streams in the state, though young ones sometimes appear on the ocean near the coast while still not fullgrown. One was found in a campground at Big Basin State Park, Santa Cruz Co., in Aug. 1960, but no trace was found of its origin. Fledglings resemble the winter adult, but with some of the barring of the summer feathering also evident. Once on the ocean, the behavior of the species is very similar to that of other small alcids; but little is known of its food preferences or its numbers at any distance from shore. *Egg*: (taken from oviduct of female) pale greenish-yellow with small blackish-brown markings; probably laid singly. *Downy young*: soon to be described (see above).

Range. Apparently breeds on and near the coast from se. Alaska to wc. Calif.; occurs sparsely in summer to ne. Siberia. Winters on ocean, at least near shore, from s. Alaska south to San Luis Obispo Co., Calif.

Occurrence in California. Locally Fairly Common Apr.-Aug., Uncommon rest of year, on ocean off n. Calif. and Locally off c. coast, especially rocky areas; recorded flying to and from nearby dense forest areas (where probably nest) Feb.-Nov.; fledglings recorded late June-Aug., and one nestling found early Aug. 1974 (see Habits above). More Irreg. Sept.-Feb. on ocean, apparently because more dispersed.

Xantus' Murrelet (*Endomychura hypoleuca*) (fig. 46)

Recognition. L. 9-10½ in. (23-27 cm.); WS. unknown. A small alcid with solidly blackish upperparts and completely white underparts, including the side of the neck and head to just below the eye and most of the underwing coverts. This sharp black-and-white pattern without other markings is distinctive among our alcids, except for the nearly related Craveri's Murrelet (see Note below). The bill is rather short and slender.

Habits. More southern in distribution than most alcids, the Xantus' ranges beyond California's warm southern waters only during postbreeding dispersal, though less regularly in such northward occurrence then formerly. On their breeding areas they are colonial, the *nest* being merely a dark niche or crevice amid rocks or sometimes amid bases of plants. *Eggs*: 1-2, variably light blue to green to brown, sparsely marked with light to dark brown or reddish. *Young*: leave nest site very early, swimming well while still downy.

Range. Breeds on Anacapa, Santa Barbara, and San Clemente Is. off s. Calif. and south to Guadalupe I. off w. Mexico. Winters commonly on ocean, mostly offshore, from Monterey area south, Occ. north to B.C.

Occurrence in California. Irreg. Uncommon to Fairly Common Mar.-Sept. on ocean off s. Calif., Locally Common about islands where nesting, with eggs reported May-June (few in July), young July. Also Rare to Uncommon, Occ. Fairly Common, rest of year off s. Calif. and north to Monterey Bay, esp. July-Dec. Casual at Farallon Is. and off San Mateo to Humboldt cos. or dead on beaches there (Apr.-May and July-Nov.).

Note: The **Craveri's Murrelet** (*Endomychura craveri*) (fig. 46) breeds in the Gulf of California and occurs on the ocean west of Baja Calif. and north sparsely to s. Calif., Rarely to off Monterey. The July-Oct. records are all of one to six birds per day's encounter, except for thirty birds in Sept. 1972 between San Diego and San Clemente I. (see Appendix). This species resembles the Xantus' very closely, but has dark gray (rather than white) wing linings and a somewhat more sinuous junction of dark and white on the side of the neck and head. Some authorities feel this form should be considered a part of the same species as the Xantus'.

Ancient Murrelet (*Synthliboramphus antiquus*) (fig. 46)

Recognition. L. 9-10½ in. (23-27 cm.); WS. unknown. In nonbreeding feathering the Ancient Murrelet appears at a distance very similar to the Marbled (fig. 46 and plate 7), but lacks the white on the scapulars and has white underwing coverts. In good light the back shows medium gray, contrasting with the black of the crown and nape. In breeding feathering (which some individuals begin to develop by Dec.) the throat is black and a white stripe extends above the eye to the nape in the black part of the head. The pale bill of an Ancient is much shorter than that of a Marbled Murrelet.

Habits. Sometimes quite common as a wintering alcid, Ancient Murrelets can then be found foraging regularly in the deep coves close to rocky headlands. Others apparently spend the winter well out to sea, at least as far as waters around the Farallones and the Channel Is. off southern California. Few boat trips by experienced observers are taken Dec.-Mar., so these birds are probably more numerous then, at least in some years, than indicated on the Graphic Calendar (in Appendix). In late Nov. 1945 I saw numerous flocks of hundreds of this species and/or Marbled Murrelets passing southeastward over the area 3-8 miles east of the Farallones.

Range. Breeds on coast from e. Siberia through s. Alaska to c. B.C., Casually to Wash. Winters south to Taiwan and through Ore. and Calif. to n. Baja Calif.

Occurrence in California. Uncommon to Irreg. Common Nov.-Mar. and Rare May and July-Oct., on ocean both inshore near rocky coasts and well offshore and about islands of c. (and probably n.) Calif., fewer at comparable seasons off s. Calif., and Rare inshore there.

Cassin's Auklet (*Ptychoramphus aleuticus*) (fig. 46)

Recognition. L. 8-9 in. (20-23 cm.); WS. unknown. This smallest of our alcids appears all dark at any distance when on the water. At close range, the upperparts are seen to be blackish-gray, the face, foreneck, chest, and sides medium brownish-gray, the short bill dark with yellow base of the lower mandible, and the eye yellow with a tiny white spot above it. Sometimes the white belly is partly visible above the water line, and it contrasts with the gray chest when the flying bird is seen from the side or below. Immatures are similar but lighter gray on the chest, and have a whitish throat.

Habits. Although these tiny alcids are the most numerous breeding species on South Farallon I. (total estimated at over 100,000 in 1970 by Pt. Reyes Bird Observatory team), they are often difficult to locate on the nearby ocean because of the rough water they seem to prefer. On some circuits of the island in a chartered boat, flocks of several hundred were encountered, diving or flying off in alarm as the boat bore down on them, and one July trip discovered some 4000 at seven miles northwest of the island. Like all alcids, they must run along the surface to gain speed for flight, and in rough water they often strike several waves before becoming airborne above the crests — or else give up the attempt and dive even while flapping furiously. The full flight of this and all the small alcids gives a "buzzing" impression because of its speed and rapid wingbeats. Little is known of the food habits of this abundant species. They come and go from their nesting islands only during darkness (as do the petrels), occasionally colliding with objects or with persons. The advantage to small sea birds of appearing over land only at night is the safety thereby achieved from attack by the omnipresent hungry (and diurnal) gulls. On the water, of course, alcids can dive to escape. *Voice*: raspy *kwee-kew* or similar creaking notes, heard commonly at night on nesting grounds. *Nest*: wisps of vegetation or drift, or none, in a burrow up to 4 feet long, or in crevices among rocks or under debris on the surface. *Egg*: 1, dull white or pale blue or green; incubation probably about 30 days. *Downy young*: blackish-brown above, soon fading, and paler gray to white below, the down very long and thick; fly at about 40 days of age.

Range. Breeds from e. Aleutians east and south on coastal islands to cw. Baja Calif. Winters on nearby ocean north to Vancouver I., B.C.

Occurrence in California. Fairly or Locally Common all year, more Irreg. in fall and winter, on ocean off c. Calif.; to Very Abundant on Farallon Is. Jan.-Aug., Occ. to Dec.; fewer reported on ocean and about islands off n. Calif. and still fewer off s. Calif. Rare anywhere close to mainland, except as dead birds on beaches. Nest-digging extends from Jan. on; eggs reported Mar.-July, few even to Nov.; young in nests, May-Sept.

Parakeet Auklet (*Cyclorrhynchus psittacula*) (fig. 46)

Recognition and Habits. L. 10 in. (25 cm.); WS. unknown. An alcid similar in size to the Marbled and Ancient Murrelets, but with a short, bright red beak that is notably convex both above and below. The plumage is blackish above and all white below, or with blackish throat and dingy sides in breeding feathering. A single white streak or plume extends below the pale eye and backward. (The larger Rhinoceros Auklet has two plumes on each side of the head in full plumage, but may show only one transitionally; see account below). Parakeet Auklets are winter visitants, reaching their southern limit in California, the few records probably being in part a reflection of the lack of offshore observations in midwinter.

Range and California Occurrence. Breeds from ne. Siberia and Bering Sea islands to Aleutians and Kodiak I., Alaska. Winters from Bering Sea south to Japan and to Wash., Ore., and Rarely, late Nov.-early Apr., to n. Calif., and Casually s. Calif. Calif. records mostly of one or two birds (a number dead), although twelve reported once in Jan. (see Appendix).

Rhinoceros Auklet (*Cerorhinca monocerata*) (fig. 46; plate 7)

Recognition. L. 13½-15½ in. (34-39 cm.); WS. unknown. A moderately large alcid with the large head and short, thick neck typical of the smaller species. A "Rhino" at any season is dark brownish-gray above and somewhat lighter on the neck and sides, the white of the belly partly visible above the water line in some individuals. In full breeding feathering there are two narrow white lines of elongated feathers on each side of the head, and a short horn projects upward from the base of the rather thick yellowish bill. The horn is absent in fall and early winter, but one or even both white plumes may be evident then (compare Parakeet Auklet, above). Immatures are like winter adults without the plumes, with a dingier bill and a dark eye. They are most apt to be confused with immature Tufted Puffins (fig. 46), but have a less heavy, more pointed beak, lack the paler area above the cheek, and are whiter on the belly.

Habits. Small flocks of Rhinoceros Auklets are one of the common sights on the open waters of the continental shelf in winter, especially in areas of upwelling water or where river outflows leave flood debris in irregular lines. When pursued by a boat, they most often escape by diving, the wing-tips appearing on either side of the short tail, as in all alcids, as they go under. Their *food* includes fish and swimming crustacea obtained by "flying" pursuit beneath the surface. In air, the flight is speedy but usually more direct or broadly curving than in the smaller alcids, thus resembling puffins, as they do also in *nesting* habits by digging their own burrow. They come and go at their nesting colonies mostly after dark, however, like the smaller species.

Range. Breeds from se. Alaska to nw. Wash. and a few to sw. Ore. and c. Calif.; also from Sakhalin to Korea and n. Japan. Winters offshore from B.C. to Baja Calif., and off Korea and Japan.

Occurrence in California. From Uncommon or Fairly Common late Sept.-Nov. and late Apr.-May to Irreg. Very Common Dec.-Mar. on ocean well offshore of c. and n. Calif.; fewer on inshore waters and off s. Calif. Rarely recorded through summer, a few nesting on South Farallon I. off San Francisco in May. Over a century ago, the southernmost known colony was on South Farallon I. Then there was no breeding record south of Wash. until a 1966 discovery in Ore., followed by a sighting of courting birds at South Farallon in 1971 and a few pairs going to nesting sites there in 1972, and carrying food in 1974 — as witnessed by Pt. Reyes Bird Observatory personnel now on the island continuously.

Note: The **Horned Puffin** (*Fratercula corniculata*) has "occurred" at least fifty times along our coast, many as dead, sick, or injured birds probably carried southward by the main current. However, apparently healthy individuals were seen on or flying over the ocean off Santa Cruz in 1967; near the Farallon Is. and off Humboldt Bay in 1973; and since 1971 frequently (esp. off s. Calif.), May-July, up to 39 in a day. This species, which winters Regularly from Alaska to B.C., is similar in size and shape to the Tufted Puffin (fig. 46 and plate 11), but has white underparts up to a dark throat, and the whole face and cheek are white or (in winter) overlaid by gray.

Tufted Puffin (*Lunda cirrhata*) (fig. 46; plate 11)

Recognition. L. 14½-15½ in. (37-39,cm.); WS. ca. 27-30 in. (68-76 cm.). A large alcid with short thick neck and very heavy head and bill which, as in all puffins, is enlarged vertically and has colorful additional plates. Spring and summer adults are sooty-black except for bright red feet, red bill beyond its gray or yellow base, and a clownish white face with curving yellow plumes extending from behind the whitish eye. In winter the bill is less colorful (outer sheath is molted), the underparts are somewhat lighter, plumes are absent or vestigial, and the white face is obscured by dusky except in the plume area. Immatures (fig. 46) are like winter adults but still paler, sometimes almost whitish on the belly, and have dark eyes and a yellowish-dusky bill that is less high (but still somewhat more so toward the tip than the Rhinoceros Auklet, (fig. 46 and plate 7).

Habits. Puffins on the water or diving for food or to escape danger behave much like other alcids. In flight they have somewhat slower wingbeats, as befits their bulk, but progress as rapidly as others. Near their nesting or shore lookout stations they may be seen flying quite high, but over the open ocean they seldom do so. When on land, they are unlike all other alcids in that they walk about with their tarsi off the ground and consequently do not have the shuffling gait. They use both their strong feet and bill in excavating a nest burrow, to which they come and go in daylight – again unlike the smaller alcids. Rarely are more than a few puffins seen together when at sea. Often, though, they spend much time on lookout from rocky prominences, not necessarily near any nest; but presumably such places may develop into nesting sites if the birds are undisturbed there. The *diet* of the Tufted

Puffin is predominantly fish such as smelt, herring, and sea-perch up to ten inches in length, with small amounts of crustacea, mollusks, etc. Feeding is sometimes done 5 miles or more from the colonies, to which they return with fish held crosswise in the bill — sometimes a number in a row. *Nest*: A burrow in a sea bluff or deep crevice among rocks, with little or no added material. *Egg*: 1, dull whitish with faint markings of brown to lavender; incubation probably about 41 days (as in Atlantic Puffin). *Young*: may remain in or by burrow for 45-50 days, by which time it is full-grown.

Range. Breeds from Bering Sea area south to n. Japan and to islands off s. Calif. (Anacapa, Santa Rosa, and recorded present in June at or near San Nicolas and San Clemente islands). Winters in all but the northernmost part of this range.

Occurrence in California. Uncommon, to Locally Fairly Common at colonies and on nearby ocean, late Mar.-Aug., Occ. to Oct.; fewer in s. Calif waters. Eggs Apr.-Aug., most in May-June; young in nest July-Aug. Also Rare but more widely on ocean rest of year.

36. FAMILY ALCEDINIDAE (KINGFISHERS)

Belted Kingfisher (*Mergaceryle alcyon*) (fig. 47)

Recognition. L. 11-14 in. (28-35½ cm.); WS. 21-24 in. (53-61 cm.). A bird of very distinctive shape with doubly crested head, large straight pointed bill, small feet, short tail, and long wings (and accordingly powerful flight) — the king-fisher family traits. Since this is our only kingfisher, no further examination is necessary to identify it. The light blue to blue-gray above and white below is varied in the male by a band across the chest of the same bluish color, and in the female by this plus a rusty band or partial band below it and along the sides. There are numerous small white dots on the wings and tail.

Habits. A bird of such jaunty appearance, sitting regularly on exposed perches over or near open water, or at times hovering in place 10 to 15 feet above it, and flying with jerky wingstrokes while uttering a loud rattling call, is bound to attract attention. So this is one species that is usually known even to those who do not study birds. To obtain the small fish and occasional crustaceans or insects of their *diet*, these birds plunge headfirst into the water much after the fashion of terns — and they have a comparably smooth-edged beak in which they catch the fish crosswise. Some species of the family (in the Old World) are upland dwellers and catch insects, but ours is strictly a water's edge bird. It may be found at the appropriate season almost throughout California, even on desert streams and farm reservoirs in winter, along the outer rocky coast where they dive into tidepools, and in the better-watered sections of the state, including occasionally up to timberline in late summer. They are essentially solitary birds, so one has to cover considerable ground to see more than ten in a day other than in exceptional habitat — the east shore of Bolinas Lagoon, Marin Co., where some three miles of shore has utility cable over or paralleling it, normally supports from four to seven each winter. *Nest*: a burrow dug by the pair in a steep or vertical earth or sand bank, the opening usually within 2-3 feet of the top and the "nest" placed some 3-6 (even 10 or more) feet in from it, becoming lined with fish bones, scales, and other regurgitated items. *Eggs*: 5-8, white; incubation 23-24 days. *Young* are fed in nest burrow by both parents for nearly a month.

Range. Breeds across North America from c. Alaska and middle n. Canada south to Central America, but only Locally in southern areas. Winters from se. Alaska and B.C. to c. U.S., across Great Lakes to Mass. and south to n. South America.

Occurrence in California. Although widespread, usually only Uncommon all year, or Locally Fairly Common Oct.-Mar., along or over edges of all sorts of fish-bearing waters, fresh and salt, particularly those with many elevated perches near or over the water. Fewer in summer in s. Calif., and only Occ. in se. Calif. then; and few to none in mountains and on ne. plateau after freeze-ups in winter. Nests dug Mar.-Apr.; eggs Apr.-May; young May (+ June?).

FIG. 47

American Dipper

Belted Kingfisher

rust

♀

♂

37. FAMILY CINCLIDAE (DIPPERS)

American Dipper (*Cinclus mexicanus*) (fig. 47)

Recognition. L. 7-8½ in. (18-20½ cm.). The chunky build, short tail, and all slate-gray coloration except for white eyelids (frequently blinked) identifies the adult American Dipper. Juveniles are paler, especially on the belly, to nearly white on the throat. Their habit of repeatedly "dipping" or suddenly lowering the body for an instant as they stand is also characteristic, and their aquatic foraging is unique for a songbird.

Habits. This is the trout fisherman's companion — the only American songbird that swims below the surface of water and bobs cork-like to the surface again when its foraging effort of 10 to 60 seconds on the rocky bottom is over. An adult dipper's feathers are very dense and seem very water repellent, though captive birds get wet promptly if not exposed to water regularly. Much feeding is done in coves and eddies along the edges of the larger streams, sometimes merely by wading, but the swimming and diving enables them to use a much larger area. Under water, they often progress by a modified type of flapping of the wings, angled so as to aid in keeping their light bodies down, as well as by foot action. Reports of their "walking" on the bottom for any length of time without use of the wings are conflicting — usually the water they feed in is too rough for an observer to see them clearly. Occasionally, however, they do visit quiet pools or even lakeshores — most of these being birds away from their home areas.

Each dipper or pair ordinarily maintains a territory of from fifty yards to a half-mile or so of rocky streambed, which is their exclusive feeding area. Seldom do they deviate more than a few feet from the stream course, even in flying up and down it. New territories are established in the fall; but at higher altitudes, where streams freeze over, they are abandoned as their owners are forced downstream to temporary ones. In late summer some dippers move upstream to or above timberline. *Voice*: call a sharp, rough *zzeet*, sometimes repeated in series and then a little more liquid; song of variable melodious to shrill phrases in long sequence, loud enough to carry over much water noise, and most often heard in fall, winter, and early spring. *Nest*: a 10-12-inch domed-over structure of moss, with side entrance placed on bridge support or rock ledge above high water, but usually inaccessible to a person wading; sometimes behind waterfall through which birds may fly. *Eggs*: 4-5, white; incubation about 13 days. *Young* are fed in nest for about 18 days, then for a time afterward along stream banks; "dip" from the start.

Range. Resident except for local wandering in nonbreeding season, chiefly up and down streams, from Aleutians and c. Alaska to Alberta and sw. S.D., and thence south chiefly in mountains to s. Calif. east to N.M. and south to Panama.

Occurrence in California. Uncommon to Locally Fairly Common but widespread all year on turbulent streams in foothill and mountain areas west of deserts; most numerous in Sierra Nevada, and fewest, though Regular, in isolated mountains of cw. Calif. Nest-building by early Mar. in low-altitude areas; eggs in nests Mar.-May (a few in June, at higher altitudes?), young in nests late Apr.-early July.

ORGANIZATIONS

The following organizations are all actively involved in some way with study or conservation of birds in California. They can often provide guidance or services in their special areas of interest to those who want to learn more in that area.

Bleitz Wildlife Foundation, 5334 Hollywood Blvd., Los Angeles 90027. Many photos, tape recordings, dealer in bird-research equipment.

California Academy of Sciences, Golden Gate Park, San Francisco 94118. Museum exhibits of California birds; major research collection. Publishes *Pacific Discovery* (general natural history) magazine, and research papers in *Proceedings*. Monthly program for members.

California Department of Fish and Game, 1416 Ninth St., Sacramento 95814. Publishes *Calif. Fish and Game* (professional quarterly) and *Outdoor California* (popular monthly); various free pamphlets. Contact public information office.

California Field Ornithologists, c/o San Diego Natural History Museum, Balboa Park, P.O. Box 1390, San Diego 92112. Publishes *Western Birds*, quarterly emphasizing identification, distribution, and habitat aspects of birds. Offshore boat trips. (New name: Western Field Ornithologists.)

Cooper Ornithological Society, c/o Jane R. Durham (treasurer), P.O. Box 529, Tempe, Ariz. 85281. Publishes *The Condor* (professional quarterly), and *Pacific Coast Avifauna* (name to be changed) at occasional intervals for longer major papers. Local chapters meet monthly except summer at Berkeley and Los Angeles. Annual meeting in west, with research results presented.

Los Angeles County Museum of Natural History, 900 Exposition Blvd., Los Angeles 90007. Museum exhibits, research collection including many fossils; excellent exhibit of Pleistocene birds and mammals from La Brea tar pits.

ORGANIZATIONS

National Audubon Society, Western Regional Office, 555 Audubon Place, Sacramento 95825. Central point of information on conservation; 35 branches in various parts of the state, with some 35,000 members in 1976. Many branches have numerous field trips and programs dealing with birds. Also operate Richardson Bay Wildlife Refuge on Tiburon Blvd., Mill Valley, and the newly acquired Starr Ranch, Orange Co.; a consortium of branches owns and operates Audubon Canyon Ranch on Bolinas Lagoon.

Oakland Museum, Natural Sciences Division, 1000 Oak St., Oakland 94607. Exhibits include Hall of California Ecology, including many birds; educational program.

Pacific Grove Museum of Natural History, 165 Forest Ave., Pacific Grove 93950. Small, excellent exhibit, with marine emphasis.

Point Reyes Bird Observatory, 4990 State Route 1, Stinson Beach 94970. Research centered in Point Reyes National Seashore and Farallon Islands but also elsewhere, carried out by few staff and many volunteers. Publishes PRBO *Newsletter* and *Annual Report.*

Rotary Natural Science Center, Lakeside Park, Oakland. Small museum and display of live native birds; daily "duck feeding" (3:30 p.m.).

San Bernardino County Museum, 2736 Court St., Rialto 92376. Museum exhibits and research collection, including many eggs.

San Diego Natural History Museum, Balboa Park, P.O. Box 1390, San Diego 92112. Museum exhibits and research collection of birds.

Santa Barbara Museum of Natural History, 2559 Puesta del Sol, Santa Barbara 93105. Museum exhibits and research collection; educational program.

University of California Dept. of Zoology, Los Angeles 90024. Includes research collection (Dickey collection) of considerable size.

University of California Museum of Vertebrate Zoology, Berkeley 94720. Research museum on campus with large collection of birds; very small visitors' collection.

Western Bird Banding Association, c/o M. San Miguel, 409 Meadow Lane, Monrovia 91016. Professionals and amateurs who study birds through banding. Publishes *Western Bird Bander* (merged 1976 into *North American Bird Bander*), including annual report of bandings in all western North America.

Western Foundation of Vertebrate Zoology, c/o E.N. Harrison, Suite 1407, 1100 Glendon Ave., Los Angeles 90024. Research collection, many eggs and nests included, nest records, photographs.

In addition, various campuses of the University of California not listed above, and of the California State University and Colleges and several independent universities and colleges, have smaller collections of birds primarily used for teaching purposes. See *The Auk*, v. 90, pp. 145-147, for listing.

SOME REFERENCES

Books:

American Ornithologists' Union. 1957. *Checklist of North American Birds*, 5th ed. (New one in preparation.)

Bellrose, Frank C. 1976. *Ducks, Geese and Swans of North America.* 544 pp. Wildlife Management Inst. & Illinois Natural History Survey. (A "new and expanded version of the classic work by F. H. Kortright," 1943).

Bent, A. C. (and successors). 1919-1968. *Life Histories of North American . . .* (Birds—by families indicated in each title). 24 vols., as various numbers of the *United States National Museum Bulletins*; reprinted by Dover Publications.

Grinnell, J., and A. H. Miller. 1944. *The Distribution of the Birds of California.* 608 pp. *Pacific Coast Avifauna,* no. 27.

Hoffman, Ralph. 1927. *Birds of the Pacific States.* 353 pp. Houghton Mifflin Co. (Classification outdated, but text still good.)

Jaeger, E. C., and A. C. Smith. 1966. *Introduction to the Natural History of Southern California.* 104 pp. University of California Press.

McCaskie, Guy, and Paul DeBenedictis. 1966 (being revised, 1977). *Annotated Field List, Birds of Northern California.* 58 pp. Lucas Book Co., Berkeley.

Miller, Alden H. 1951. *An Analysis of the Distribution of the Birds of California.* University of California Publ. Zool. 50(6):531-644.

Ornduff, Robert. 1974. *An Introduction to California Plant Life.* 152 pp. University of California Press.

Orr, Robert T., and James Moffitt. 1971. *Birds of the Lake Tahoe Region.* 150 pp. California Academy of Sciences, San Francisco.

Palmer, Ralph S. (ed.). 1962. *Handbook of North American Birds.* Vol. 1, *Loons through Flamingos.* 567 pp.;1976. Vols. 2 and 3, *Waterfowl.* 521 and 600 pp. Yale University Press. (Details on all aspects, in condensed, easy-to-use format.)

Peterson, Roger Tory. 1961. *A Field Guide to Western Birds* (2nd ed.). 366 pp. Houghton Mifflin Co.

———. 1962. *A Field Guide to Western Bird Songs.* Three 33-1/3 rpm. monaural records. Houghton Mifflin Co. Voices of species in sequence following the above book.

———, and editors of *Life.* 1963. *The Birds* (a volume of Life Nature Library). 192 pp. Time, Inc.

Pettingill, Olin S., Jr. 1953. *A Guide to Bird Finding West of the Mississippi.* 709 pp. Oxford University Press.

Pough, Richard H. 1946. *Audubon Bird Guide, Eastern Land Birds.* 312 pp. Doubleday.

———. 1951. *Audubon Water Bird Guide* (eastern U.S.). 352 pp. Doubleday.

———. 1957. *Audubon Western Bird Guide.* 316 pp. Doubleday. (Does not give details for the continent-wide species treated in the other two volumes.)

Pyle, Robert L. (revised by Arnold Small). 1961. *Annotated Field List, Birds of Southern California.* 64 pp. Published by Otis Wade for Los Angeles Audubon Society.

Robbins, Chandler S., Bertel Bruun, and Herbert S. Zim. 1966. *Birds of North America, a Guide to Field Identification.* 340 pp. Golden Press.

Small, Arnold. 1974. *The Birds of California.* 310 pp. Winchester Press. (Summarized seasonal status and distribution of all species, plus chapters on bird habitats of the state.)

Smith, Arthur C. 1959. *Introduction to the Natural History of the San Francisco Bay Region.* 72 pp. University of California Press.

Stebbins, Robert C. 1959. *Reptiles and Amphibians of the San Francisco Bay Region.* 72 pp. University of California Press. (Material on field notes.)

Yocom, Charles F., and Stanley W. Harris. 1975. *Birds of Northwestern California.* 74 pp. Humboldt State University Bookstore, Arcata. (Provides habitat and status information for each species.)

Journals and Magazines*

American Birds (bimonthly). National Audubon Society.

Audubon (bimonthly). National Audubon Society.

The Auk (quarterly). American Ornithologists' Union, c/o Smithsonian Institute, Washington, D.C. (Professional.)

Bird-Banding (quarterly). Northeastern Bird-Banding Assn., South Londonderry, Vermont. (Professional & national in scope.)

California Fish and Game (quarterly). California Dept. of Fish and Game.

The Condor (quarterly). Cooper Ornithological Society.

National Geographic (monthly). National Geographic Society, Washington, D.C.

National Wildlife (bimonthly) National Wildlife Federation, Washington, D.C.

Natural History (10 times a year.) American Museum of Natural History, New York.

North American Bird Bander (quarterly). Eastern and Western Bird Banding Associations, c/o Eleanor Radke, co-editor, P.O. Box 446, Cave Creek, Ariz. 85331.

Pacific Discovery (bimonthly). California Academy of Sciences, San Francisco.

Western Birds (quarterly), formerly *California Birds*. California Field Ornithologists.

The Wilson Bulletin (quarterly). Wilson Ornithological Society, Morgantown, W. Va. (Professional.)

---

*Addresses of western-based organizations given in preceding section.

APPENDIX

Introduction to Graphic Calendars

Birds of some species migrate considerable distances between breeding and nonbreeding areas, while others move only a short way, and still others stay in one area the year round; the terms *permanent resident, summer resident* (for those present part of the year and breeding), *winter visitant,* and *transient* or *transient visitant* have long been used to describe seasonal status in a general way for a given area. In some species, however, some individuals migrate while others do not, or at a given locality the breeding individuals may leave, only to be replaced by others that have bred elsewhere. In addition, nonbreeding "stragglers" often remain through the breeding season far south of the breeding area of their species, this being especially true of subadults of the diving ducks, shorebirds, and gulls.

Consequently, a true picture of seasonal status must give an indication of total numbers on a much more detailed basis than "first and last" dates, or even these plus an estimate of abundance. The broader picture is portrayed in this section, for each species of water bird regularly occurring in California, by one or more horizontal bar graphs (Graphic Calendars) showing the relative numbers that can be found per half to full day of normally diligent general observing of birds (not an all-out search for the maximum number of one species, ignoring others) by an experienced bird student, in the preferred habitat of the species, on the date indicated. These graphs were constructed from and checked against many thousands of estimates of numbers in the author's own field records and those of other observers, the latter chiefly as published in *Audubon Field Notes* and its successor journal *American Birds*, plus other sources as available. The graphs are drawn to an accuracy in dates of about one week (7 or 8 days, depending on the length of the month). A circled numeral, preceding the first species in each family, corresponds to the number given that family in the species accounts.

Code for Numbers of Birds per day. From 1 to 5 lines are used, representing numbers as follows:

‒ ‒ ―――― = under 10	☰ ☰ ☰☰☰	= 250 to 999
= = ===== = 10 to 49	☰ ☰ ☰☰☰	= 1000 or more
☰ ☰ ===== = 50 to 249		

These categories are thus somewhat arbitrary standardized intervals encompassing the relative abundance often (and in the text of this book) indicated by the terms Uncommon, Fairly Common, Common, Very Common, and Abundant; but they are applied in this book to all species on a uniform basis. Thus those species not sufficiently numerous anywhere to be found in large numbers in one day by ordinary observing means (e.g., rails, jaegers, kingfishers) are not rated as Very Common or Abundant, even though they may be among the most numerous of any species of their family, and of regular occurrence. An x is inserted for published records of occurrence in which no mention was made of numbers present, and one to several ^X_X's at the approximate position on the Calendar when more accurate dates were not available.

Regularity of Occurrence. Continuous lines in the graphs indicate fairly regular to yearly occurrence at that level of abundance, and are extrapolated over gaps of as much as two weeks in available records. For occurrence which fluctuates from year to year in date or level of abundance, dashed lines are used, and these are not so extrapolated; thus many irregularities appear. Many of these may indicate nothing more than an incomplete record, but where the species probably occurs in numbers in the intervening periods comparable to those shown for an adjacent period, a ? is inserted. For occurrences of still less frequent sort, the following entries are made:

   o = from 1 to 5 records, at least several of 3 or more individual birds, but not "usual"

   Δ = 3 to 5 records of 1 or 2 birds (sometimes more, if record is old)

   ● = 1 or 2 records only, of 1 or 2 individuals each

In addition, where recent records are lacking for any particular date or level of abundance,

   + = records dating from before 1940 only (or other date if so specified).

Migration Periods. Particular spans of dates when distinct migratory movements have been noted are further marked for some species by a row of asterisks below these entries, or by a double row for peak periods: ******
                                                                          **

Multiple Calendars. In many species, the seasonal status is quite different in certain districts of the state compared to others, for water birds particularly so in and east of the Sierra Nevada and southern California mountain ranges, where there is the greatest contrast in climate to that of the coast-slope (cismontane) lowlands. Where sufficient information is available on the dates and numbers of a species of this sort, a separate graph is drawn for each district (or several districts) where its occurrence is notably different. The districts and habitats to which each graph pertains are indicated in the associated right column (see Biotic Districts and Habitats below). When only a few records exist for certain districts, and they indicate a seasonal pattern different from any of the graphs, numerals representing the districts are entered at the appropriate date position as another calendar. An underline of such a numeral indicates two or more records in a particular 7-8 day period (of any one or more years).

Breeding Phenology. For each species known to nest in one or more districts represented by one graph, the normal span of dates for each *phase of the nesting cycle* is indicated by a first and last capital letter (connected by dots if there are intervening records). Unusually early or late records are shown by small letters. The letters used for the phase are:

B	and b	for nest-building
E	and e	for nests with eggs
Y	and y	for nests with young or for nonflying young of precocial species
F	and f	for fledglings (young able to fly in limited fashion, but still under care of adults)

289

C and c   for courtship behavior
J and j   for independent juveniles still in the nesting area
N and n   for "nesting" records without mention of phase
T and t   for birds taking up territory
P and p   for pairs in evidence on breeding areas

Since these entries have been placed on the calendar entirely from data in the literature or in my own notes, and these sources are often incomplete, the irregularity of many of the entries merely underscores the gaps in the record of these basic aspects of the nesting cycle of many of our water birds. Users of this book can help to fill in those gaps by reporting in print upon their findings!

DESCRIPTION OF BIOTIC DISTRICTS

To best represent the varying distribution of birds across the state, California is divided into eleven different *biotic districts*, each of which is characterized by one to several ecological communities that are widespread within it, but not present or not widely distributed in adjacent districts. While these districts differ more in their upland vegetation than in wetland types, and are thus more important in explaining land-bird distribution, the associated differences in topography and distance inland from the major climatic influence of the sea do produce noticeable differences in the seasonal occurrence of water birds also. This is most apparent in many species when comparing their dates of occurrence and numbers for the coast (District 1), Central Valley (2), Sierra Nevada (7), mountains of southern California (7), Imperial Valley-Colorado River (8), and the elevated northeastern plateau area (9).

These and the other biotic districts of the state recognized here approximate those defined from the evolutionary standpoint for breeding birds of California by Miller (1951), but are more closely bounded by breaks in the continuity of major vegetation types as checked in the field. Also, the Monterey Peninsula and the bulk of the Santa Lucia Range south of it (recognized as the South Humid Coast District by Miller) are here lumped with the adjacent San Francisco Bay uplands, since the differences between these two areas, at least for water birds, seem much less than those between other adjacent districts. Also, all of the relatively warm deserts of southeastern California (Mojave, Colorado, etc.) are grouped as one district (8); and the interior drier parts of the north and central Coast Ranges in the woodland belt (recognized as Clear Lake and San Benito districts by Miller) are here included with the comparable belt of the Sierra foothills as District 6. Each district includes the minor valleys of the same areas. However, the Central Valley of California's interior, north of the Tehachapis, is so extensive, and as such affects so many kinds of birds, that it is recognized as a separate unit (District 2), a most important one for many of the species treated in this volume.

The eleven biotic districts into which California is divided are as follows (and shown on accompanying map):

1. Ocean, islands in it, all seacoasts, bays, and intertidal areas; also, for some species, immediately adjacent freshwater areas. Northern (n), central (c) and southern (s) parts distinguished for some birds.
2. Central Valley (Sacramento, San Joaquin, Kern Basin) and contiguous open grassland of lower foothills; Carrizo Plain and marshes around Suisun Bay are included.
3. Coastal-slope (cismontane) lowlands of southern California (San Luis Obispo to San Diego cos.) including the lower hills and slopes of the mountains up through extensive chaparral areas, but excluding shady broadleaf or mixed evergreen forests of major canyons and cool slopes, wherein the bird life is more like 7, (many such areas are too small to show on the map).
4. Semi-humid Coast Ranges (Monterey and n. San Benito co. outer ranges) and San Francisco Bay area lowlands and its more moist hills. Major forest areas of Santa Cruz Mts. and north are included in District 5, while the drier hills well interior in the San Francisco Bay region (e.g., Mt. Diablo) are in District 6.
5. Humid Coast Ranges and valleys— the main forested, southern part of the Santa Cruz Mts.; small areas in Marin Co.; and broadly from near the lower Russian R. northward to the Oregon line.
6. Interior foothills—the main woodland and chaparral belt, both in inner Coast Ranges and the Sierra Nevada, plus the areas of these same vegetation types across the north end

MAP OF CALIFORNIA

MAP LEGEND

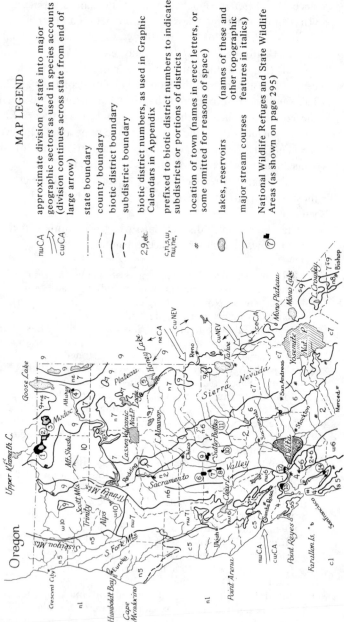

nuCA cwCA (arrows)	approximate division of state into major geographic sectors as used in species accounts (division continues across state from end of large arrow)
.....	state boundary
‒ ‒ ‒	county boundary
((	biotic district boundary
| |	subdistrict boundary
2,9, etc.	biotic district numbers, as used in Graphic Calendars in Appendix
c,n,s,w, nu,ne,	prefixed to biotic district numbers to indicate subdistricts or portions of districts
#	location of town (names in erect letters, or some omitted for reasons of space)
(oval)	lakes, reservoirs (names of these and other topographic features in italics)
(line)	major stream courses
⑦	National Wildlife Refuges and State Wildlife Areas (as shown on page 295)

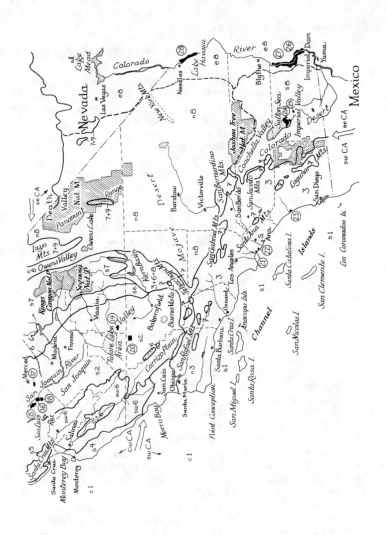

of the Sacramento Valley and interruptedly along the Tehachapi Mts.

7. Sierra Nevada montane forest (largely coniferous) zones and above, extending north to the Pit R. and south to northern Kern Co., plus outlying units of similar vegetation in the higher inner north Coast Ranges, Trinity Mts., etc. (or these could as well be considered part of District 10). Normally with snow cover and frozen shallow water in winter.

7. Southern California montane zones—forests, shady woodlands of some of the lower north-facing slopes and cool canyons, high-altitude chaparral, and semi-barren slopes. For the present volume, the bodies of water included are Big Bear and Baldwin lakes (where a majority of the bird records are reported) and Lakes Hughes, Elizabeth (on the desert border), Arrowhead, Hemet, Cuyamaca, and smaller ones of comparable altitudes. Snow cover and freezing over of shallow water variable but frequent.

8. Southeastern California deserts, including all vegetation and geological types on the north or east sides of the District 3 or District 7 types of vegetation in the Mojave and Colorado deserts, Owens Valley, Death Valley, and east to include the lower Colorado R. valley (both sides of the river, plus some occurrences north to about Las Vegas, Nev.). The bird life of the Salton Sea area in this district has been intensively surveyed only since 1960.

9. Northeastern plateau, or northern desert district including the Mono tableland (and Topaz and Mono lakes), Sierra Valley, and from Honey Lake area northward to Oregon, east of the main forest belt of Districts 7 and 10, plus outlying smaller valleys west to near Yreka. High counts of water birds are usually from the Tule Lake-Lower Klamath refuges, but many other lakes and marshes are included. Where seasonal occurrence is poorly documented within California (as is true of many species here), records from similar lakes and marshes in extreme western Nevada are included (Pyramid Lake through Reno and Carson City area). Cold to very cold winter weather normal, with small or shallow bodies of water usually frozen by late December, but snow cover not deep.

10. Cascade Mts. montane zones and above—forest zones and included areas of other vegetation types (mostly shrubs) and lakes and meadows at the same altitudes. Very few records seem to have been published on water birds found in this area of California, so the designation is omitted for most species. Presumably most have similar distribution here as in District 7, especially in the northern part.

In the column to the right of each Graphic Calendar, the districts to which that calendar pertains are listed, together with the habitats in which that species is found. Punctuation between numerals is as for the habitats (see below). If users of this book will learn the numeral or numerals of the biotic districts pertaining to the areas in which they are regularly afield, they will then have to refer to this list or the map only for occasional trips elsewhere.

NATIONAL WILDLIFE REFUGES AND STATE WILDLIFE AREAS
(numbers refer to circled numbers on map)

1	Lower Klamath N.W.R.	15	Kesterson N.W.R.
2	Tule Lake N.W.R.	16	San Luis N.W.R.
3	Clear Lake R.	17	Merced N.W.R.
4	Modoc N.W.R.	18	Los Banos W.A. (State)
5	Honey Lake Wildlife Area (State)	19	Pixley N.W.R.
6	Sacramento N.W.R.	20	Kern N.W.R.
7	Delevan N.W.R.	21	Seal Beach N.W.R.
8	Colusa N.W.R.	22	Bolsa Chica W.A. (State)
9	Gray Lodge W.A. (State)	23	Buena Vista Lagoon
10	Sutter N.W.R.	24	Salton Sea N.W.R.
11	Spenceville W.A. (State)	25	Imperial W.A. (State)
12	Suisun and Grizzly Is. W.A. (State)	26	Imperial N.W.R.
13	San Pablo Bay N.W.R.	27	Cibola N.W.R.
14	San Francisco Bay N.W.R.	28	Havasu Lake N.W.R.

KEY TO HABITAT DESIGNATIONS

The habitats in which each species is regularly found are given by letter code, referring to the list and brief descriptions below. This list has been modified somewhat from the general accounts of biotic communities in the *Introduction to the Natural History* . . . volumes of this series: Smith (1959) for the San Francisco Bay region, Jaeger and Smith (1966) for southern California. It also agrees fairly well with the communities recognized by Ornduff (1974). For this water-bird volume, there has been more emphasis on water and shore habitats and somewhat less on the forest and shrub types, but the list given includes the major categories in those upland types, which will be further subdivided in the volumes dealing with land birds. Reference to the introductory volumes cited will provide further characterization and illustrations of many of the habitats used in these volumes on birds.

To the right of each Graphic Calendar for a particular water-bird species, the habitats occupied by that species are indicated by letter codes immediately following the numeral or numerals for the biotic districts to which they pertain. These letters are mostly first initials of the main word of the habitat title, so the reader should find after only a few references to this list that he can read the habitat preferences for all species without difficulty. When there is any doubt, however, verification should be sought below:

A = alpine
B = bay or estuary, lagoon (brackish or salt water); in District 8 = Salton Sea
B̂ = impounded salt water, such as the salt-evaporating ponds of San Francisco-San Pablo, Newport, and San Diego bays and Elkhorn Slough
C = chaparral
D = desert (all types below the woodland belt of mountains)
E = broadleaf evergreen woodland or forests
F = fields, either cultivated or fallow, but with main vegetation other than grasses, or else barren
G = grasslands, including grainfields, and meadows in mountains; where extensive wetness is important to the bird, the designation "wet G" is used
H = humid forests of the Coast Ranges, usually of mixed tree types (needle-leaf, broadleaf evergreen, and/or deciduous)
I = islands
K = rocky areas, cliffs
L = lakes, ponds, sloughs, large ditches; the depth and extent of open water is often of importance, and mentioned in words for some species; the type of shore, if important, is also indicated
M = marshes, implying fresh to brackish water, including cattails, tules, other bulrushes, the larger sedges, etc.;

    i.e., nonwoody vegetation of wet areas usually strong
    enough to support small nests

M̂ = intertidal salt marshes, including a lower cordgrass
    (*Spartina*) zone and upper pickleweed (*Salicornia*)
    zone, plus locally gum plant (*Grindelia*), etc.

N = forests of primarily *needle*-leaf trees

O = ocean—the waters themselves, with alongshore, inshore,
    or well-offshore preference of some species indicated
    by words; commonly combined also as OS, OK, OI, OV
    (see below)

R = riparian (streamside) or lakeside woodland or shrubs
    such as willows, cottonwoods, maples, alders, syca-
    mores; but for water birds, the kind of tree is usually of
    little importance as long as it provides the needed perch
    or nest site

S = sandy beaches and flats, both on the outer coast and in
    bays, sometimes separately shown as OS and BS

T = tideflats, of mud or muddy sand, mostly regularly
    covered and uncovered by tides on bayshores or along
    river mouths, but also including similar shores of sea-
    sonal or fluctuating nontidal lagoons along the coast

U = urban areas, as residential, business, industrial, piers and
    buildings of harbors, etc.

V = rivers, creeks, and canals—with flowing water; subtypes
    given in words as necessary

W = woodlands—with openly spaced, usually round-crowned
    trees of short to moderate stature such as oaks, laurels,
    pinyon pines, junipers, and (especially in District 6) the
    taller Digger pine

Z = subshrub areas, the low shrubs usually openly spaced
    and with either herbs or bare ground between them

These habitat designations are combined without punctuation (as VR, ML, OV, or even OKI) when the edge between
the two types or a combination of them in a given area is the
essential feature for the bird species concerned. Underlined
letters indicate the habitats in which that species nests. If this
applies to certain biotic districts only, the numerals of these
districts are also underlined.

The preference shown by each species of bird for both the
habitats listed and the biotic districts in which these are found
is indicated through separation of symbols by commas, if little
or no difference in preference exists; or by > indicating a
greater preference for the type or types *preceding* that symbol

than for those following it; or by $>>$ when the preference is a conspicuous or decided one. Similarly, the symbol $<$ (used in few instances) indicates "less than."

When two district numbers are separated by the symbol $>$ followed by a single set of letter symbols, similar habitat preferences in the two districts are indicated, but with lesser abundance in the district following the punctuation (and thus often less than the Graphic Calendar indicates). If, instead, the district numerals are separated by commas, it indicates the author's estimate of roughly equal levels of abundance.

In the less well-worked parts of the state, however, much remains to be described on the details of habitat distribution and numbers of birds. Particularly needed are censuses in both summer and winter of all the widespread types of habitats, and in each of the biotic districts. When data of this sort are available, many alterations of the subjectively determined habitat preferences listed in this book will no doubt be in order. In the meantime, it is hoped they will be found useful to anyone interested in the habitat relationships of water birds.

GRAPHIC CALENDARS AND
HABITAT DISTRIBUTIONS

GRAPHIC CALENDAR   HABITAT DISTRIBUTION

1.

COMMON LOON (*Gavia immer*)

= n,c>s 1 O,B > 8 (esp.Apr.-May,Oct.-Nov.) L>B,V > 4,6, n,c>s 1 > 2,3 > 5 L; also 1 O well offshore in migrations (= ****).

= n 7 > 9 (+10?) L; "n" entries = old nest records (may still occas. nest).

??? = Occurrences reported on 7, c,s 7 L.

YELLOW-BILLED LOON (*Gavia adamsii*)

(↠ = individuals remaining)

= c,n 1 O,B. Also, individuals near Los Coronados Is.(collected 24 Nov. 1968), at L.Tahoe, 6 Jan.1973, & at Pt.Mugu,Ventura Co., 5-9 May 1976.

ARCTIC LOON *(Gavia arctica)*

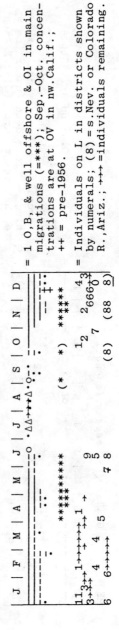

J | F | M | A | M | J | J | A | S | O | N | D

= 1 O>B, including OI & well offshore
(esp.in migrations = ***) >> c,n 1 L;
few are normally found far within
large bays.

= 8 (esp.Dec.-Apr.) L>B, 4,nw 6 L.

= Occurrences on L in other districts
shown by numerals; underline = more
than one record; (9) = w.Nev.; "0"=10;
→→ = individuals remaining.

2(cripple?)

(long dead)}(9)

RED-THROATED LOON *(Gavia stellata)*

J | F | M | A | M | J | J | A | S | O | N | D

= 1 O,B, & well offshore & OI in main
migrations (=***); Sep.-Oct. concentrations are at OV in nw.Calif.;
++ = pre-1956.

= Individuals on L in districts shown
by numerals; (8) = s.Nev. or Colorado
R.,Ariz.; →→→ =individuals remaining.

GRAPHIC CALENDAR HABITAT DISTRIBUTION

RED-NECKED GREBE (*Podiceps grisegena*)

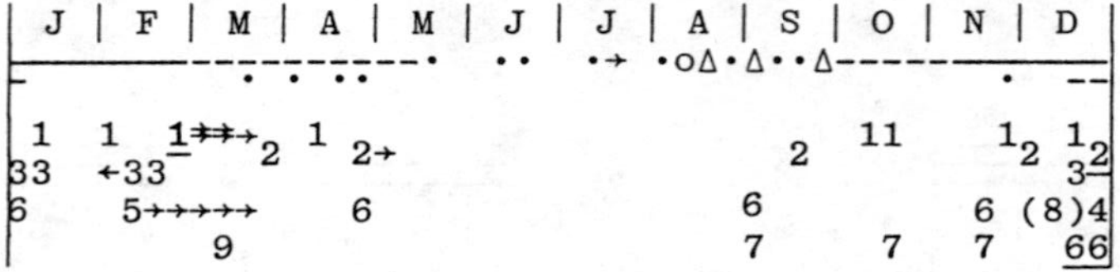

J	F	M	A	M	J	J	A	S	O	N	D
		. .	. .	. →	. °o∆·∆··∆––––––			.			—
	. . .	. .				2	11			1₂ 1₂	
1	1 1⧝⧝→₂ 1 2→										
33	←33									6 (8)4	
6	5→→→→→ 6						6	7	7		66
	9						7				

= n,c>s 1 B>O (+OI); no Apr.–Sep. record in s 1.

= Occurrences on L in districts shown by numerals; →→→ = individuals remaining; ← = long dead; underline = more than one record; (8)=s.Nevada.

HORNED GREBE (*Podiceps auritus*)

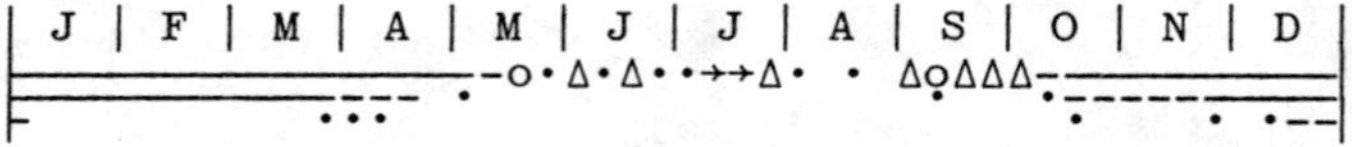

J	F	M	A	M	J	J	A	S	O	N	D
		–o·∆·∆··→→∆· ·			∆o∆∆∆· ·			∆o∆o· ·		––––––	
								.	.	.	

= 1 B>O>OI>>irreg. 1,2,6 > 3,4,5 L.

J	F	M	A	M	J	J	A	S	O	N	D
. . . . .		. → . · · · o––			. .	. °oo·•·o·		.	7 7 7₇₇		0
	7 7			7	7		0	7 0	0→		
9 (9)	9 9	9999	9								

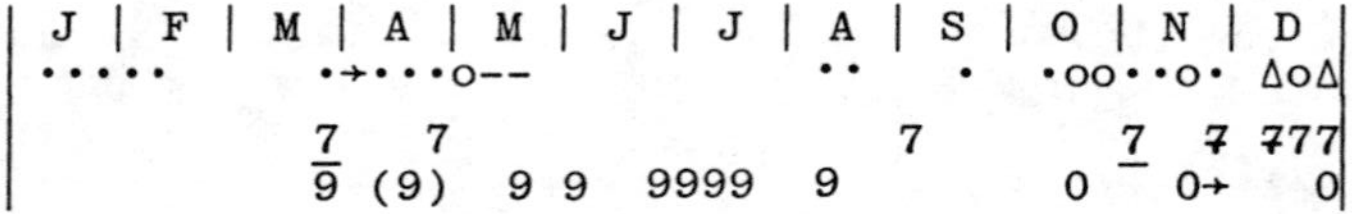

= 8 L,B.

= Occurrences on 7,7,9,10 L; "0"=district 10; (9)=w.Nev.; underline = more than one record.

EARED GREBE (*Podiceps nigricollis*)

J | F | M | A | M | J | J | A | S | O | N | D

= 8 B > (or < in June–Oct.) 1 B̂>>B,
e 8 L >1 O(+OI) >>1,4,5,6, w 8 L,
ML>V; peak nos. on Salton Sea to
300,000 (aerial estimate).

J | F | M | A | M | J | J | A | S | O | N | D

= 2,3 L, ML (but breeding colonies
now mostly depleted).

= Breeding phenology.

J | F | M | A | M | J | J | A | S | O | N | D

= 9 (chiefly Mar.–Oct.) > 7 > 7,10 L, ML;
fall peak based chiefly on records
at Mono L., where up to 300,000 can
be seen from shore; high Dec. nos.
at San Bernardino Mtn. lakes.

= Breeding phenology.

LEAST GREBE (*Podiceps dominicus*): Recorded along the Colorado R. (Calif.–Ariz.border) as follows: Imperial Dam or vicinity--at least 5 adults and 3 young
(adult male & flightless young collected) on 18–23 Oct.1946, 3 seen on 5 Jan.1951, 1 on 23 Nov.1954, 1 on
14,15, & 24 May 1955, and 1 on 9 Nov.1974; "near Yuma"--15+ seen on 5 Jan.1951; Havasu L. (near Headgate
Rock Dam)--2 on 28 Sep. & 1 on 8–9 Oct.1947. Among several reports from localities to the west, the only
one that seems probably valid is one seen on 20 Dec.1959 in Mission Bay, San Diego, at range of 50 feet.

GRAPHIC CALENDAR HABITAT DISTRIBUTION

WESTERN GREBE (*Aechmophorus occidentalis*)

J | F | M | A | M | J | J | A | S | O | N | D

= s,c 1 O,B, 8 L>B > n 1 O,B, non-nesting locations in 1,3 > 4,6,2(+5?) L; occas. to 1 OI in spring & fall.

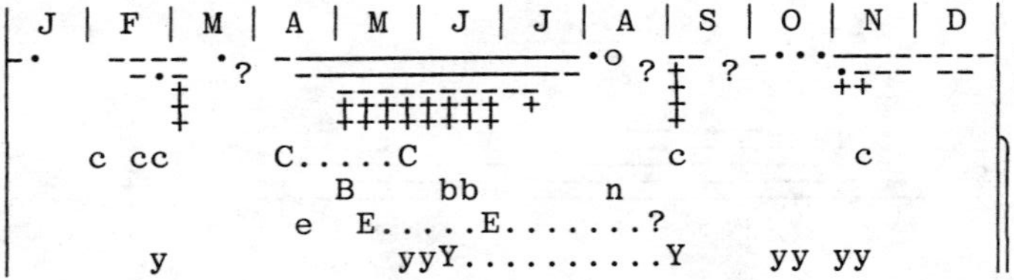

= nesting locations in 2,3 > 6 L, LM, & casually in 1,8 LM (++ level records =pre-1960, mostly on Tulare and/or Buena Vista Lakes, s2 L, now drained).

= Breeding phenology; some early courting is seen in non-breeding areas.

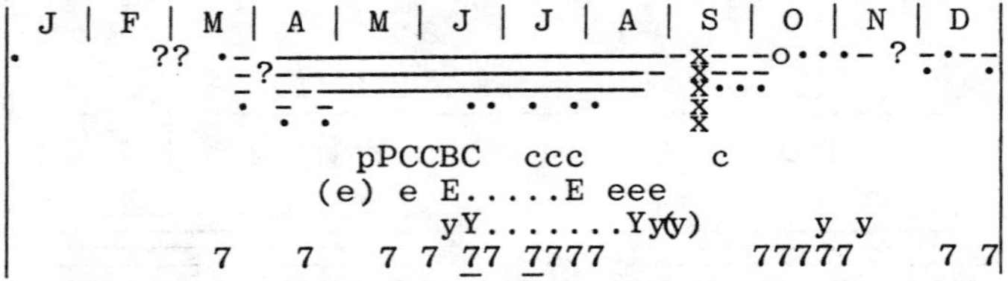

= 9 L, LM; X entry = 1971 record for Klamath Basin without specific date.

= Breeding phenology; ( ) records = from w. Nevada or s. Oregon.

= Occurrences reported on 7 L.

PIED-BILLED GREBE (*Podilymbus podiceps*)

```
J | F | M | A | M | J | J | A | S | O | N | D
.                                          ---- 
          X +  X}                     . 
          X +  X}  =Tulare L.1958  .   ----
  b  C?  b
 e  c   E..........E    ee
      y  yyyy Y........Yyyy y yy    yy y
                      j J.........Jj. . .  .
```
= 2,3,9 > (9=May-Sep., < other mos.) 1,5 > 4,6 > 8 L,ML,M (only local May-July in 8). Also on 7,7,10 L,ML (but fewer Dec.-Mar.), & usually fewer on 1 (+8?) B & esp. B,T̂ borders, Aug.-Apr. or early May & rare then on 1 O>>OI.

= Breeding phenology.

BLACK-FOOTED ALBATROSS (*Diomedea nigripes*)

```
J | F | M | A | M | J | J | A | S | O | N | D
..    ----.                  ..••o•o o
---...|...........        ---  ..•
    .|..........      ...•   ∆  ..•
     |.....         ...•     ..•
      |..          ...•      ..•
```
= 1 O, chiefly well offshore but occ. to few mi. of mainland, esp. near Monterey; one in San Francisco Bay, 14 Jan.

1931. Numbers beyond edge of continental shelf sometimes greater than shown by graph (from boat trips somewhat closer in). Peak numbers found off Eureka in spring (to 1000 in May 1976), summer, and fall, and off Monterey in summer.

LAYSAN ALBATROSS (*Diomedea immutabilis*)

```
J | F | M | A | M | J | J | A | S | O | N | D
o•  •∆o∆∆∆•∆•  •  •∆o
```
= 1 O, mostly farther offshore than the Black-footed; graph indicates records within 50 miles of shore, mostly off c. and n. Calif. A vagrant (probably after "entrapment" in Gulf of California) flew west near Desert Hot Springs, Riverside Co., 5 May 1976.

GRAPHIC CALENDAR   HABITAT DISTRIBUTION

SHORT-TAILED ALBATROSS (*Diomedea albatrus*):   Apparently was fairly common off California including inshore waters up to the late 1800's (dated records are from Jan.,Mar.,June, & Dec.). One recent record off our shores, 1 seen at about 70 miles off San Francisco on 17 Feb.1946 by Traylor (Condor 52:90). At least 7 records off Oregon & Alaska since then.

WANDERING ALBATROSS (*Diomedea exulans*):   One California record, a subadult on shore at Sea Ranch, Sonoma Co., 11-12 July 1967, & verified by photographs (details in the Auk 85:502-504, 1968). Normally found in oceans of Southern Hemisphere.

CAPE PETREL (*Daption capense*):   Individuals reported four times from California: undated specimen before 1853 from "opposite Monterey;" one seen on 9 Sep.1962 (details in Calif.Birds 1:39-40), and one on 13 Mar.1974, both in Monterey Bay; also one seen on 3 Sep.1965 near San Clemente I. Normally a Southern Hemisphere species.

NORTHERN FULMAR (*Fulmarus glacialis*)

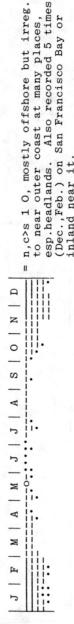

= n,c>s 1 O, mostly offshore but irreg. to near outer coast at many places, esp.headlands. Also recorded 5 times (Dec.,Feb.) on San Francisco Bay or inland near it.

STREAKED SHEARWATER (*Puffinus leucomelas*):   One collected on Monterey Bay, 3 Oct.1975 (American =(*Calonectris leucomelas*) Birds 30:118). Normal range in w. Pacific Ocean.

4.

PINK-FOOTED SHEARWATER *(Puffinus creatopus)*

| J | F | M | A | M | J | J | A | S | O | N | D |

= 1 O, mostly well offshore; one old July record in c 1 B.

FLESH-FOOTED SHEARWATER *(Puffinus carneipes)*

| J | F | M | A | M | J | J | A | S | O | N | D |

= c,s>n 1 O, mostly well offshore.

NEW ZEALAND SHEARWATER *(Puffinus bulleri)*

| J | F | M | A | M | J | J | A | S | O | N | D |

= c>n>>s 1 O, mostly well offshore, but occas. seen from outer headlands. Also one (debilitated) on 8 B on 6 Aug.1966.

SOOTY SHEARWATER *(Puffinus griseus)*

| J | F | M | A | M | J | J | A | S | O | N | D |

*********?************=observed mass movements

= c,n>s 1 O, mostly well offshore but irr.(esp.May-Aug.) close inshore; occ. outer parts of large 1 B (July-Sep.); occ. "crash" into land in dense fogs; one seen flying at 8 B on 14 Aug.1971.

307

GRAPHIC CALENDAR   HABITAT DISTRIBUTION

SHORT-TAILED SHEARWATER *(Puffinus tenuirostris)*

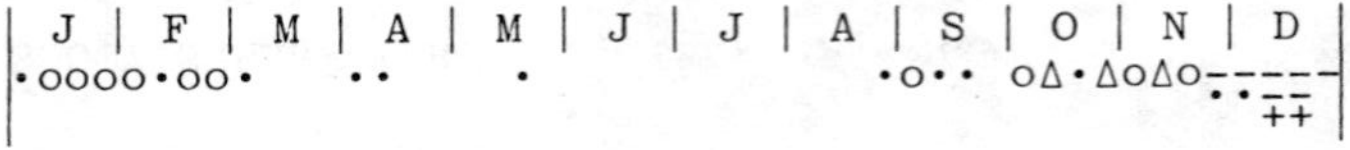

```
| J | F | M | A | M | J | J | A | S | O | N | D |   = 1 O, mostly well offshore; ++=
|.oooo.oo.    .       .           .o.. oΔ.ΔoΔo..---|     pre-1954.
                                              ++
```

MANX SHEARWATER *(Puffinus puffinus)*

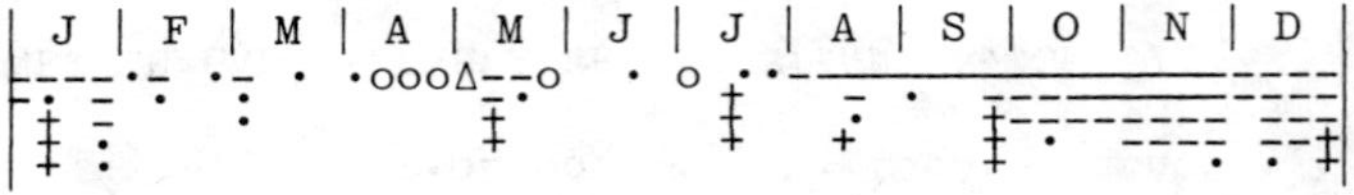

```
| J | F | M | A | M | J | J | A | S | O | N | D |    ---  = 1 O, mostly inshore waters, peak
|--.-.  . .oooΔ--o   . o   o..-.          ---        ---     numbers often a mile or two from
|--.   ++      ++     ++  ++  .          +++---       +++     shore; +++ = pre-1940 records,
|+++  ...                                 ---        ...     mostly near Monterey; recent large
```

numbers found only in s. Calif.; breeds in w. Mexico (Mar.-June).

SCALED, or MOTTLED, PETREL *(Pterodroma inexpectata)*:   Individuals found dead on shore near
Pt. Reyes on 25 & 26 Feb.1976, at Cayu-
cos and Cambria, San Luis Obispo Co., 28 Feb. & 31 Mar.1976.   Primary range is in South Pacific Ocean.

FORK-TAILED STORM-PETREL (Oceanodroma furcata)

```
  J | F | M | A | M | J | J | A | S | O | N | D
 •°oΔ •:          •:::: -:--  •+++++?•?? ?+-:--        -+••--o-•-Δ-•-
```

= n>c>>s 1 O, usually far offshore except near colonies on n 1 OI, but
close to land elsewhere irreg. after
strong onshore winds. Recorded in
S.F.Bay, Aug.-Sep. of several years;
& waifs at El Cerrito (31 Aug.1937)
& Hayward (late Feb.1971).

```
                                              n
 p  ppp  e ee E.E   e   e
         N y yY?.Y.?YY j                       n
(Breeding phenology; underlined letters = dates
 adults banded at colonies)
```

LEACH'S STORM-PETREL (Oceanodroma leucorhoa)

```
  J | F | M | A | M | J | J | A | S | O | N | D
 •    -:----•°•-•------- Δ•°°•°°
         15                    4     1
```

= n,s>c 1 O, mostly foraging far offshore
(graphed records = up to 100 mi. out).
Occurrences inland (districts 4,5) & on
San Francisco Bay (district 1).

```
  J | F | M | A | M | J | J | A | S | O | N | D
 •-   ::   •:::   ::•   ••+          -•   -•x-
```

= n>c 1 OI (breeding colonies) & at Los
Coronados Is., Mexico (near San Diego),
though few are seen on O nearby.

```
 bb  bbBPC.....CN B  p
      B.....B eE..../..E eee
             yyY......Y..y   y
             f   f f j   f
```

= Breeding phenology; egg dates to left
of the / are from c. & n. Calif. only.

GRAPHIC CALENDAR HABITAT DISTRIBUTION

ASHY STORM-PETREL (*Oceanodroma homochroa*)

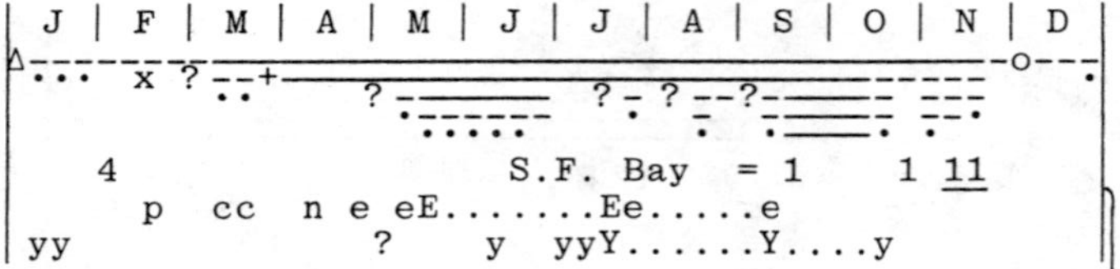

```
  J | F | M | A | M | J | J | A | S | O | N | D
Δ----- -x-?--+                            -o---
   ·:· ?      ?- ? --?            ?: ·    ---
          :·······   :·      ·    :·   ---
         4          S.F. Bay  =  1   1 11
      p  cc  n e eE.....Ee.....e
   yy         ?   y yyY......Y.....y
```

= c>>s 1 O,OI, usually foraging well offshore; also recorded in n 1 O (July, Oct.). Large nos. in Aug.-Nov. are from Monterey Bay, in Mar.-June from colony on S.Farallon I.; numerals indicate vagrants in other districts.

= Breeding phenology (mostly S.Farallon).

GALAPAGOS STORM-PETREL (*Oceanodroma tethys*): One Calif. record, immature female found alive in Carmel, 21 Jan. 1969 (specimen).

HARCOURT'S STORM-PETREL (*Oceanodroma castro*): One Calif. record, 1 reported seen some 25 miles west of San Diego, 12 Sep.1970 (Small 1973).

BLACK STORM-PETREL (*Oceanodroma melania*)

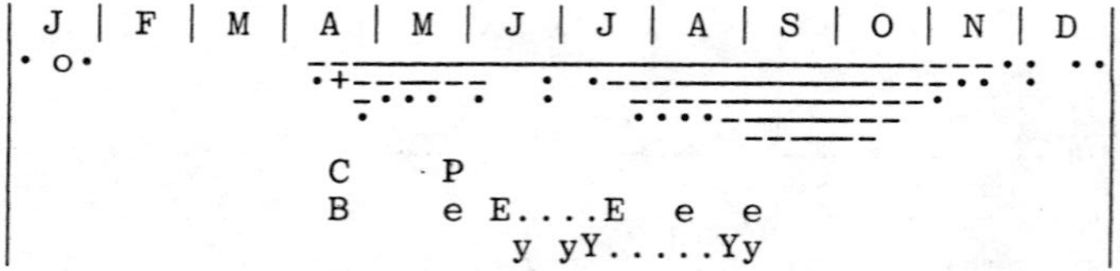

```
  J | F | M | A | M | J | J | A | S | O | N | D
· o ·     ·:+----- ·:      ·                 ---
              :·    ·    ·   :·····        ---
          C   P
          B   e E...E  e  e
                 y yY.....Yy
```

= s 1 O, + chiefly June-Oct. in c 1 O (Monterey Bay & few to Sonoma Co.), preferring continental shelf waters.

= Breeding phenology (on Los Coronados I., Mexico, near San Diego, Calif.).

BROWN BOOBY (*Sula leucogaster*)

| J | F | M | A | M | J | J | A | S | O | N | D |

= 8 B,L>V, with maximum of 8 birds; in addition one stayed 2 yrs. at a lake on the Ariz. side of the Colorado R. Also, 1 (?) at Pt.Reyes, Marin Co.,14 July 1973, photographed but not fully verifiable (see Amer.Birds 27:913).

= In s 1 O,OI (June-July entries=one bird in 1961 '65, '68 on an islet by San Miguel I., apparently on territory).

DOUBLE-CRESTED CORMORANT (*Phalacrocorax auritus*)

| J | F | M | A | M | J | J | A | S | O | N | D |

= c>s>n 1 B,$\hat{B}$ > s 1 O near K,B,OKI > 8 (except late July-Sep.), 1,$\underline{2}$,3 L>V,$\underline{LR}$,VR > c,n 1 OKI,OK, $\underline{4}$,5,$\underline{6}$ L,V,$\underline{LR}$ > 8 B; ++ = pre-1953 on c. San Francisco Bay; ** = reported migration movements.

= Breeding phenology; capital E egg dates left of "/" from district 8, to right of it = c. Calif. peak.

| J | F | M | A | M | J | J | A | S | O | N | D |

= 9 L(+V?),LR,LK > 7 L>7 V; ++= from Pyramid L, w.Nev., but also nests in Modoc Co. & elsewhere in ne. Calif.

= Breeding phenology (data incomplete).

LEAST STORM-PETREL (*Halocyptena microsoma*)

| J | F | M | A | M | J | J | A | S | O | N | D |

= s 1 O, mostly well offshore; mostly off San Diego, but including 8 records off Monterey (^=1, $\hat{}$=2-3 birds); also 1 reported off Humboldt Co., 1 Oct.1972, & "hundreds" at Salton Sea after hurricane, mid-Sep. 1976.

WILSON'S STORM-PETREL (*Oceanites oceanicus*)

| J | F | M | A | M | J | J | A | S | O | N | D |

= c,s 1 O, offshore; observed maxima of 2-3 birds in different years off Monterey, and 5 off San Diego (13 Sep.1975).

RED-BILLED TROPICBIRD (*Phaethon aethereus*)

| J | F | M | A | M | J | J | A | S | O | N | D |

= s 1 O, mostly well offshore (or near San Clemente I.). Also, 2 reported to have wintered off Santa Barbara in 1954-55, one adult seen on outer Monterey Bay on 14 July 1970, and one near Pescadero on 29 June 1973.

WHITE-TAILED TROPICBIRD (*Phaethon lepturus*): One adult was at Newport Beach, 20-23 June 1964, displaying to a radio-controlled glider! (Condor 67:186). Also an adult tropicbird, possibly of this species, was seen on Monterey Bay, 10 or 12 Mar.1970.

GRAPHIC CALENDAR HABITAT DISTRIBUTION

(7.)

WHITE PELICAN (*Pelecanus erythrorhynchos*)

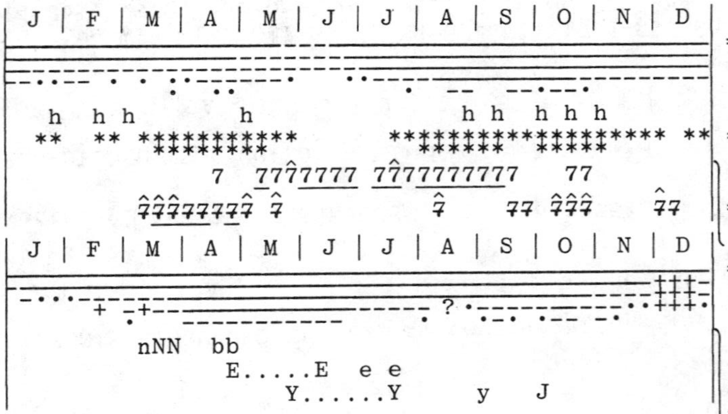

```
 J | F | M | A | M | J | J | A | S | O | N | D
                                        ** **
 h  h  h       h              h h  h h h
 ** ** ***********************************
            7 7777777 7777777777 77
                                    77 777
         7777777777 7                  77
```

```
 J | F | M | A | M | J | J | A | S | O | N | D
                                        ‡‡‡
                                        +++
       nNN  bb             ?  ?
              E......E  e  e
                 Y........Y     y     y     J
```

= c,s,1 B̂>B (esp. July-Dec. on primary salt ponds) >2,3>6>1,4 L>ML at non-nesting locations; "h"= on coast of Humboldt Co.

= Dates of migrations observed over upland habitats generally (largest nos. over s.Calif. mts. and passes). Occurrences reported on L in, or (as^ flying over mtn. districts as shown.

= 9 L,LI > (Apr.-Sep.15) or < (Sep.16-Mar with no Jan.-Feb.record in 9) 8 B,L (nested BI to at least 1952) > or < at nesting colonies in 2 L,LI (or dikes) but all now abandoned in 2?; ++= pre-1955 there.

= Breeding phenology, including at Pyramid Lake, w. Nevada.

BROWN PELICAN (*Pelecanus occidentalis*)

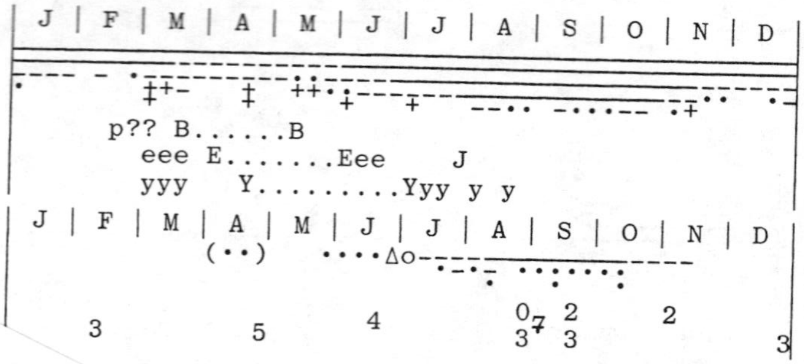

```
 J | F | M | A | M | J | J | A | S | O | N | D
    ‡‡-      ++ ‡‡‡-           -+
 .  ‡‡-   -‡ ‡++-+          -•-•-•-‡
       p?? B.......B
       eee E........Eee      J
       yyy  Y...........Yyy y y
```

```
 J | F | M | A | M | J | J | A | S | O | N | D
          ( ••)
                    -Δo-•-
    3         5          4        07    2     2
                                  37    3     3
                                              3
```

= 1 O,K,I>B,UB>S,T (land areas for resting) >>1 L,V, occ.B̂. In mid-Mar.-June the majority are near nesting colonies; in winter & spring usually rare in n 1.

= Breeding phenology.

= 8 B,LV; ( )= s.Nev. or w.Ariz., only.

= Records of vagrants at L>V inland in districts shown by numerals ("O" = 10).

...ED BOOBY (*Sula sula*): A "near adult" landed, was captured and released on S.Farallon I. on 26 Aug.1975; also, individuals (same one?) were seen there on 8 and Birds 30:118). Common in tropical oceans near coastlines and islands.

...a nebouxii)

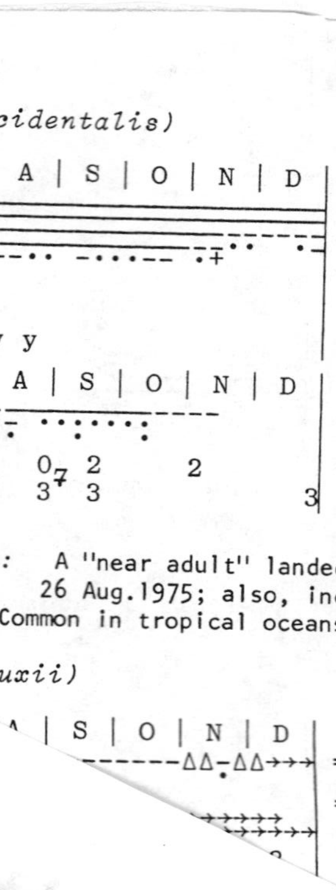

```
           | S | O | N | D
                  -ΔΔ-ΔΔ-++→
                          +→+
                          +→+
```

= 8 B>L>V (esp.Colorado R.Valley) > s 1 O.

= Records on 3 L (+→= same individuals remaining) & on 7 L (Big Bear L.). Additional more northern records: 1 at Tomales Bay, 8 Apr.1962; 4 to 2 at Pa-...alaveras Co., 15 Sep.-19 Oct.1976.

GRAPHIC CAL...

BROWN P...

OLIVACEOUS CORMORANT *(Phalacrocorax olivaceus)*: Individuals at West Pond, near Imperial Dam on lower Colorado R.: 13 Apr.1971 (ad.); 22-23 Apr.1972 (im.) & 7 Apr.1973 (details in Calif.Birds 2:134, Amer.Birds 26:792, & West.Birds 6:140). Normal range in s. Tex., Mexico & southward, occasionally in s. Ariz.

BRANDT'S CORMORANT *(Phalacrocorax penicillatus)*

J	F	M	A	M	J	J	A	S	O	N	D

```
                                         ?-            ?
        c cCCbbB...Bbb   bccb                            = c,n>s 1 O,OK,OKI > 1 outer B >> 1 inner
                     e?e?E.....E eee J J                   B.
                      y Y.......Yyy y
                                                        = Breeding phenology, c. California.
  P  B    e    E.......Eee
     e    yy   Y.......Y y     5                        = Breeding phenology, s. California.
     3                         5
                                                     bb = Records inland in districts shown
                                                        (in 5 = on V during salmon runs).
```

PELAGIC CORMORANT *(Phalacrocorax pelagicus)*

J	F	M	A	M	J	J	A	S	O	N	D

```
                                    ?
                                                        = c,n 1 O,OK, esp. OKI > s 1 OKI > s 1
                                                          OK, 1 outer >> inner B.  Essentially
        bnbB B   b                                        resident although birds wander when
        e eE....E eee  e                                  not nesting (1 on 5 V, 29 Sep.1973).
          y Y......Y  y   y                             = Breeding phenology.
```

* ANHINGA (or "SNAKEBIRD") *(Anhinga anhinga):* Recorded twice in Calif.: one on 9-12 Feb.1913 on lower Colorado R., & a probable escaped captive on 28-30 May 1939 at Searsville L., Stanford University, & (probably same individual) on 2 June to 16 July 1939 at L.Merced, San Francisco. Normal range is in cs. and se. U.S. and southward; casual in Arizona.

⑩ MAGNIFICENT FRIGATEBIRD *(Fregata magnificens)*

J	F	M	A	M	J	J	A	S	O	N	D
.									+	.	.	+
							˄	˄		˄		

= s 1 O (& over any nearby terrain), 8 B, L>V> c 1 O>B >>n 1 O +/or B (=˄). Invasion in 1972 with 26 records of 1-4 birds each; in 1976, 18 records, s.Calif.

⑪ GREAT BLUE HERON *(Ardea herodias)*

J	F	M	A	M	J	J	A	S	O	N	D
		‾‾	‾‾								┼·
		‾									‾┼
b	bB...B	b									
	cCdE......Eee	c	e								
e	yy	yyY......Y	yyy	yy	y						
		fF.F.	jjJ..J								

= 1 B̂, T > (July-Oct., < other mos. except near nesting colonies) 1 M̂, UB, OK, 2, 3 ̲8̲ > ̲1̲, ̲4̲, ̲5̲, ̲6̲ M, L, R (or other trees)>V̲; ++ entries = in 1952-3 at drying Buena Vista L. (s 2 L).
= Breeding phenology; earliest dates in s. California only.

```
 J | F | M | A | M | J | J | A | S | O | N | D
7   77  77  7   7   ? 77 77   ?   7  7 ? 77 ? 777
 7 7 7   7     7  777777777777777777777 ? 777
―――――――――――――――――――――――――――――――――――――――――――――
9       9     99   9? 9  9 999999999999999 ?  99
/
```

= Occurrences reported in 7 L,G (up to 20 at a location in Oct.), 7 L>V>M,G (mostly <10 at a location but 40 at a Lassen Co. nesting colony, 13 May 1964).

/ = Occurrences reported with dates from 9 L,LM,V,R̲,G, for most of which numbers of birds were not reported. However, up to 200-450 per day reported on surveys of major nesting colonies on Klamath Basin refuges; Christmas Counts include up to 30 in mild winters there.

GREEN HERON (*Butorides virescens*)

```
 J | F | M | A | M | J | J | A | S | O | N | D
                     .   .    .                  .
        ********       ***********
      bb BB  b    b
       eE....Eeee
       yy  yyY...Y  y
                       J
          9 (9)9  9   9999(9) 9→        9)
                       77      8   88
 8        888888       8         88       888888_8
```

= 5 > (except <,Nov.-Feb.) 2,3 > 1,4,6 VR, LR>M (locally L or V + other trees); occ. 1 M̃,I in migration; ***=reported arrivals, departures, transients.

= Breeding phenology.

= Occurrences reported in 7,7,9 L,V; (9) = records in w. Nev. only.

= Occurrences reported in 8 V>L. mostly where trees or shrubs are adjacent; (8) = records in w.Ariz.,s.Nev. only.

LITTLE BLUE HERON (*Florida caerulea*)

```
J | F | M | A | M | J | J | A | S | O | N | D    = s,c 1 T,M>L;→→→= same individuals.
OOOO→→→·Δ→→→→→→→ΔΔΔΔ··· ·· ··→ΔΔΔ·ΔΔ·· ····ΔOOOOO  = Breeding activity (incipient or sus-
              t      t           y?              pected, San Francisco Bay area).
                                                 = Records in 2,9 L or M, 8 L,M,V(+B?);
→→→→→→→    2 2  8 (98→→→8→88→→→→88 (88  (8) 8→→→→→    (8) = s. Nev.; (9) = s. Ore.
```

CATTLE EGRET (*Bubulcus ibis*)

```
J | F | M | A | M | J | J | A | S | O | N | D    = 8 pastured G, irrigated F,G, nearby R
──────────────────              ?        ?--       (+R at s. end of Salton Sea) > M,B,L.
·─?·?───────────────    ?    ?           ··:
───                                      │--     = Breeding phenology (reports incomplete).
       N    y       eN
J | F | M | A | M | J | J | A | S | O | N | D    = s 1,3 > c 1,2,4,5 in habitats as for 8;
·· ·───────────────              ··         ·      will probably expand to 1 T,UG, & gar-
·· ·───────────────    ··OOΔΔ····:·         ─      bage dumps.  In Nov. on San Nicolas I.,10 were seen in 1972, 1 in 1975; 1 was at
─────────              ·:· ·········         ·      Nat'l Wildlife Refuge, 14 Nov.1975; & 4 on San Clemente I., Mar.1975.
```

REDDISH EGRET (*Dichromanassa rufescens*)

```
| J | F | M | A | M | J | J | A | S | O | N | D |
|→→→→→→→→→→Δ→→→→→→→→ . →→→→→→→→→→→ .  . →→→→→→ . ·ΔΔ·-ΔΔΔΔΔ→→→Δ→·|
```
= s 1 T,B̂,M;→→= same individual remain-
ing. Also one on Otay L.,near San
Diego, 20 Sep.1943; and one at Elkhorn Slough, Monterey Co., late Aug.-8 Oct.1967.

```
| J | F | M | A | M | J | J | A | S | O | N | D |
|·→→→→→→→→}=Ariz.side of| . ·→·→·Δ→→→→·· . ? .  . |
                Colorado R.)
```
= 8 L,B shores (+M?,R? for roosting);
→→→ = same individuals remaining.

GREAT EGRET (*Casmerodius albus*)

```
| J | F | M | A | M | J | J | A | S | O | N | D |
||
||
++  --+++++                              ++          +
   +++++++++++++++++‡
```
= 1 B̂,T> in fall or near colonies in R,
BI (& < other times or places) 1 Ω, 1̲,2̲,
3̲,5 > 4 > 6 M,L,R (or W,H̲ locally) > G,F,
V̲; ++ = pre-1960 records. Also recorded occ. on 1 OI (May,July,Sep.).
= Breeding phenology.

```
| J | F | M | A | M | J | J | A | S | O | N | D |
‡‡‡    ? +?+‡‡+++++++++‡‡‡‡‡ ? + + ? +  ++? ‡‡‡
‡‡‡        x                              ‡‡‡
      pbB.B b
      CeE......Ee....eee
        yyY.........Yyyy yyf
```
= 8 LM,L,V,R, irrigated G>B shores; ++ =
pre-1960 records, present numbers probably to two-line level all year.
= Breeding phenology (data incomplete).

```
| J | F | M | A | M | J | J | A | S | O | N | D |
·                  N NN*   NYY (*= in Owens Valley)
```
```
| J | F | M | A | M | J | J | A | S | O | N | D |
 - ·   - ?x ? o ·?·  .    -  · - -·· ?   ?  ? o ··
                    N       - ·   .
            7      7 7 7777777
       7    b77         7     7
```
= 9 LM,L,V,R>G; nests (regularly?) but
only date reported = late May.
= Occurrences reported at 7,7̄ L; includes
a few nest-building at Mtn.Meadows Res,
Lassen Co., in 1964 (= b entry).

GRAPHIC CALENDAR

HABITAT DISTRIBUTION

SNOWY EGRET (*Egretta thula*)

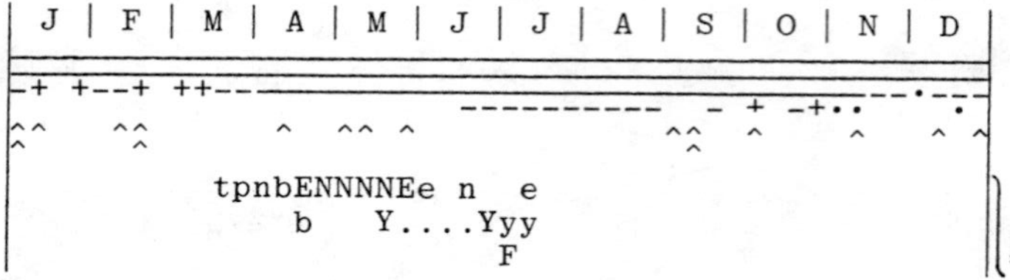

| J | F | M | A | M | J | J | A | S | O | N | D |

```
  +  +--+ ++--              -------------       <
  <      <   <^<  <                        ^<     <
  <      ^<                            ^<  ^<
                 tpnbENNNNEe n  e
                     b   Y....Yyy
                           F
```

= c, s 1 $\hat{B}$, T, $\hat{M}$ (+W at one colony), c, s 1,2,
8 > 3 M, F under irrigation, L, R > 4 > 6 M,
$\underline{L}$, c $\underline{T}$, 2, 3, 8 (+4, 5, 6?) V; occ. to $\underline{n}$ $\underline{1}$ T,
BI, L, ($\hat{}$ $\wedge$ records); ++= pre-1950.   Also,
undated record at S.Farallon I., & over
ocean between s.Calif. Is., 17 May 1961.

= Breeding phenology (data incomplete).

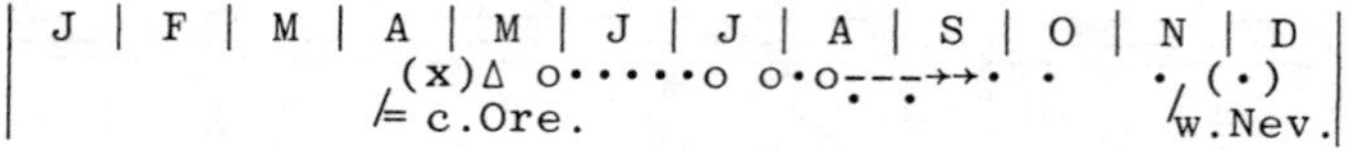

| J | F | M | A | M | J | J | A | S | O | N | D |

```
                      (x)Δ o•••••o o•o----→•   •     •(·)
                     ∕= c.Ore.                      ∕w.Nev.
```

= 9 > 7, $\overline{7}$ M, L(>V?); also an unverified report of nesting at Tule Lake, Modoc Co.

LOUISIANA HERON (*Hydranassa tricolor*)

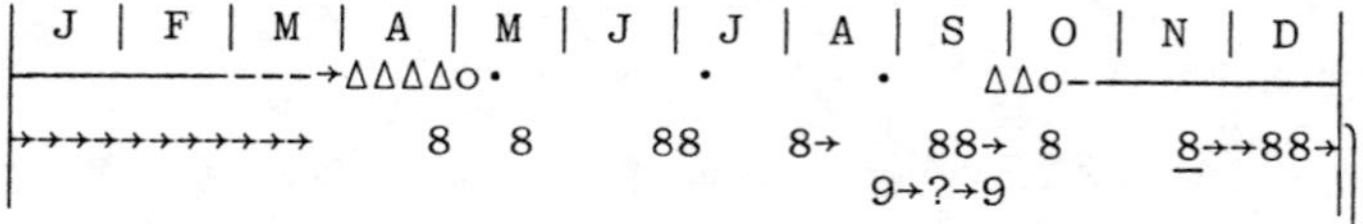

| J | F | M | A | M | J | J | A | S | O | N | D |

```
  ---→ΔΔΔΔΔo•        •     ΔΔo---
+++++++++++++   8   8   88   8→   88→ 8         8↔↔88↔↑
                             9→?→9
```

= s 1 (chiefly San Diego & Orange Cos.) T,
shallow B>L (+M?).

= Reported occurrences in 8 B,L & 9 ML
(latter near Honey L. in 1971); ↔↔↔ =
same individuals remaining.

BLACK-CROWNED NIGHT HERON *(Nycticorax nycticorax)*

J	F	M	A	M	J	J	A	S	O	N	D

= 1 B̂,M,T̂ > (near colonies, spring-summer)
or< 2 M (winter roosts) > 1,2,5,8 > 3,4 >
6 L,M̲,V,R̲ (or other trees), irrigated
F,G, 1 B̲U̲,OU,OK; also occ. to 1 OI
(Aug.-Sep.)

= Breeding phenology.

J	F	M	A	M	J	J	A	S	O	N	D

7 = Occurrences reported at 7 L (has bred
at Big Bear L., eggs in early June), 7
L, 9 L,M̲,R̲ (hundreds reported in 9, including nesting, but most dates lacking); up to 25 have wintered in w.Nev.

9 /=c.Ore. Fledged young

YELLOW-CROWNED NIGHT HERON *(Nyctanassa violacea)*

J	F	M	A	M	J	J	A	S	O	N	D

= s 1 L,T,U; + one photographed in 1963
at Claremont (="3"), & 1 in w.Ariz.(8).

= Reported continuity of records of one
individual (apparently) at San Rafael
in successive years, 1968-73.

321

GRAPHIC CALENDAR   HABITAT DISTRIBUTION

LEAST BITTERN   (*Ixobrychus exilis*)

```
 J | F | M | A | M | J | J | A | S | O | N | D
-ΔΔ••Δ•  o•••ΔΔΔΔo------ΔΔΔ•Δ-ΔΔ••   •   ••  Δ--
          c   e   E.E e
                  y    YY
             7      9     7    (9)
                          e       /=w.Nev.)
```

= 2,8 > 3 M > 4, s>c 1 M; also recorded in 5 (early Jan.), presumably in M.

= Breeding phenology.

= Available dated records in 7 M (S.Tahoe) & 9 M (possibly breeds there also).

AMERICAN BITTERN   (*Botaurus lentiginosus*)

```
 J | F | M | A | M | J | J | A | S | O | N | D
 ‖                                          ‖
‖   •                                       ┆┆      •
s    S S
     E  EE E
              Y•••••Y
```

= 2 > 1,3,4,5 M > 1 M̂ (late Aug.–May), 8 M (no June record). One seen on Farallon I., 12 Oct.1970.

= Breeding phenology (s,S = "singing").

```
 J | F | M | A | M | J | J | A | S | O | N | D  699
99                                               9
 6         7  99
         7 7777777777 7̲ 7  ?  7
           S  yE
```

= Available dated records in other districts as shown by numbers; breeds in 7 M̲ at S.Tahoe, probably also in 9 M.

12. WOOD STORK *(Mycteria americana)*

J	F	M	A	M	J	J	A	S	O	N	D

△→→→→→
. ++

6 4 29 2   2 2 22

= 8 irrigated F or G,M,B,L > s 1 T,M,M,B̂
>> 3 L,M (once to Cuyamaca L., edge of
7 L); →→=same individuals remaining; ++
= pre-1960 only. Numerals=dated re-
cords in other districts. Most winter records are from s 1 T,M,M̂ including recent ones
north to Morro Bay. There are also old records (in 1800's) on San Francisco Bay (undated).

13. WHITE-FACED IBIS *(Plegadis chihi)*

J	F	M	A	M	J	J	A	S	O	N	D

△△

E.. EE
Y   Y

= 2 M, shallow L, wet F, G (chiefly Gus-
tine-Los Banos area & S.) > s 1,3 >> c
1,4 LM, wet F,G > 1 T. Also casual in
n 1 (Oct.,Dec.).

= Breeding phenology (May records = from
Riverside Co. in 1911).

J	F	M	A	M	J	J	A	S	O	N	D

= 8 M, wet F,G, shallow L (+B?).

? ??   ?   ?   ?

PN   YyJ
(9)(9)(9)(9)(9)   999   (9)
7

= Breeding phenology (data incomplete).
= Dated records for 7,9 L,ML; possibly
nests in 9 M, as it does in w.Nev.(=()).

323

GRAPHIC CALENDAR

HABITAT DISTRIBUTION

WHITE IBIS *(Eudocimus albus)*: A species with normal range in s. Baja Calif., Tex., & se. U.S. to Peru. At least 4 individuals reported in Calif.: 1 in March 1914 at Palo Verde on the Colorado R.; 1 collected on Pt. Loma, San Diego, 20 Nov. 1935; an adult at Bolinas Lagoon, Marin Co., 14-19 May 1971, and apparently the same bird on 26 June to 15 Sep. 1971 near San Rafael, where it roosted at least at times in the heronry on West Marin I.; and an adult at n. end of Salton Sea, 10-24 July 1976, and at s. end of Salton Sea, 5 Aug. 1976. Also, an adult was seen on the Ariz. side of the lower Colorado R., 4-5 Apr. 1962.

ROSEATE SPOONBILL *(Ajaia ajaja)*

J	F	M	A	M	J	J	A	S	O	N	D

●──────○─······─○─·······─ o ·──

1111→→ ? → 11 22 2
3

● = 8 shallow L,V,R(+M?,B?); →→= same individuals remaining (Colorado R. Valley).

○ = Reported occurrences in s 1 (all = 1973) and districts 2,3. Undated records in 1849 and before to near San Francisco.

✱ FLAMINGOS *(Phoenicopterus* spp.): About 15 records in coastal Calif., all of birds probably escaped from captivity (see NOTE in text). Some individuals have remained in salt pond areas for months, and records extend over most of the year.

324

WHISTLING SWAN (*Olor columbianus*)

J	F	M	A	M	J	J	A	S	O	N	D

```
|---·|···ooo·|·  |   |   |   |   |   |   |·····|----|----|
|--?-|··     |   |   |   |   |   |  ·→→→→→→→→→··|----|·---|
        (ill?)
```

= 2 L,M, wet F >> loc.4,5,6 L,LM, c,n>s 1 L>B; & irr. to 8 L,LM, wet F(+B?), occ. V > 3 L; usually <50 at any one locality in s.Calif. The subspecies *jankowski* (E.Bewick's Swan) of ne.Asia was reported in 1975 & 1976 in the Sacramento Valley.

TRUMPETER SWAN (*Olor buccinator*)

J	F	M	A	M	J	J	A	S	O	N	D

```
|----|  ? o o|(:·)|   |   |   |   |   |   | ·  |----|----|
                  cripple/(w.Nev.)
********                              ********
7                                    77777777
```

= 9 L,LM, wet F,G >> V; winter numbers remain high until/unless extensive freeze-ups eliminate most open water.
= Main migrations (cold winters).
= Records with dates on 7 L (most at Lake Tahoe) and 7 L.

MUTE SWAN (*Cygnus olor*)

J	F	M	A	M	J	J	A	S	O	N	D

```
ooo→→ooooo→→                               ·o
3→→→→→→ 1(n.part)                       9 9
  6→       3
```

= c 1,4 L (families of 3-5 at Santa Rosa in 1967-69 indicated by ooo), c 1 B.
= Recent records on n 1 B (Humboldt Bay), 2,3, sw 6, 9 L; →→=same birds remaining.

MUTE SWAN (*Cygnus olor*): At least 6 records of birds living in wild state; native of Europe.

BLACK SWAN (*Chenopis atrata*): At least 4 records of birds living in wild state; native of Australia. These two species are commonly kept in captivity; neither has persisted for more than a few months in wild.

GRAPHIC CALENDAR

(15.) CANADA GOOSE (*Branta canadensis*)

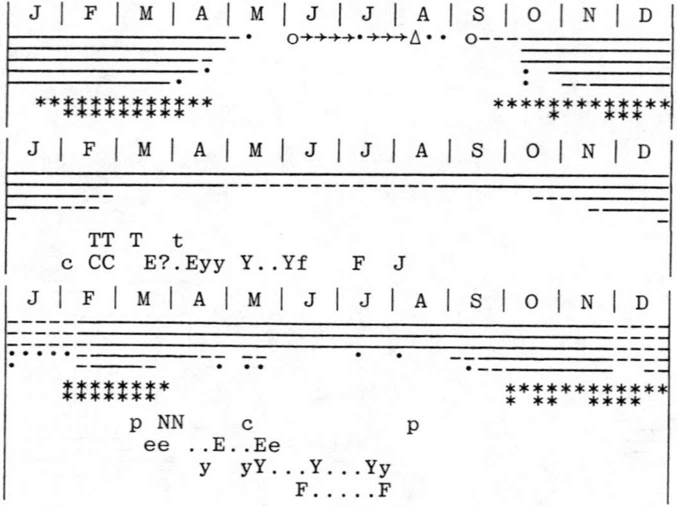

J	F	M	A	M	J	J	A	S	O	N	D

O→→→→→→→→Δ•• o---

= $2 > 4,5,6,8$ G,F,L,M > n 1 B,M̂,T,L,G > 3 G, F,L,M(+V rarely, various districts) > c,s 1 B,T,M̂, ₮ L,G; no s.Cal.record May–July.

= Main migrations; many flocks to s 1 OI also (in Mar.–Apr. at least).

J	F	M	A	M	J	J	A	S	O	N	D

TT T t
c CC E?.Eyy Y..Yf F J

= local at breeding areas in 4 (+5?) L, nearby G,M,W > c(+n?) 1 B,T,BI,L,M?,G.

= Breeding phenology (chiefly as noted in 4 UL,L).

J	F	M	A	M	J	J	A	S	O	N	D

p NN c p
ee ..E..Ee y yY...Y...Yy
F.....F

= $9 > 7$ (+10?) L,ML (or LG,LN),G>V.

= Dates of observed migration movements.

= Breeding phenology.

326

BRANT (*Branta bernicla*), primarily the BLACK BRANT (*B.b.nigricans*)

J	F	M	A	M	J	J	A	S	O	N	D

= 1 B,T>O>>OK,OS; includes occasional
  birds of light-bellied subspecies
  (recorded early Nov.- early May).
= Observed migrations, coastal and well
  offshore.

J	F	M	A	M	J	J	A	S	O	N	D
(∵)	.						·000-000000				.

= 8 B>L,V; ( )= s.Nev.+/or w.Ariz.

```
2 2       22      2                  2    2
     33    3                              3
    (9̲ 9) 7̲(9)                    99 9 9
```

= Records with dates in other districts
  indicated by numerals (includes one of
  light-bellied subspecies shot at Tule
  L., 20 Oct.1941). A number of additional records in 2,9 are without even a month given.

RED-BREASTED GOOSE (*Branta ruficollis*): At least 6 occurrences (Sep.-Feb.) in Calif. of birds in a "wild" state, all presumably escaped from captivity. The species is native of Asia, but is commonly kept free-roaming in zoos and animal parks.

EMPEROR GOOSE (*Philacte canagica*)

J	F	M	A	M	J	J	A	S	O	N	D
+o→Δ----Δ→→	···								···Δ-o-··---		

= n,c>s 1 B,KO > 2 L,M.

```
 0        9                    9999 9  9
```

= Records with dates in 9,10(=0)(L,M?);
  at least 7 other records of 1-9 birds each in fall and winter at Modoc Co.refuges, undated.

327

GRAPHIC CALENDAR

HABITAT DISTRIBUTION

WHITE-FRONTED GOOSE   (*Anser albifrons*)

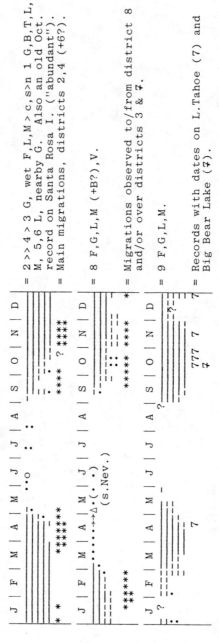

| J | F | M | A | M | J | J | A | S | O | N | D |
= 2 >> 4 > 3 G, wet F,L,M > c, s>n 1 G,B,T,L,
M, 5,6 L, nearby G. Also an old Oct.
record on Santa Rosa I. ("abundant").
= Main migrations, districts 2, 4 (+6?).

= 8 F,G,L,M (+B?),V.

= Migrations observed to/from district 8
and/or over districts 3 & 7.

= 9 F,G,L,M.

= Records with dates on L.Tahoe (7) and
Big Bear Lake (7).

BAR-HEADED GOOSE   (*Anser indicus*):   One seen & photographed at Lower Klamath Nat'l Wildlife Refuge,
Modoc Co., week of 18 Mar.1959, a possible escapee.   A species
native of c. Asia, commonly kept in zoos.

SNOW GOOSE *(Chen caerulescens)*, including "BLUE GOOSE"

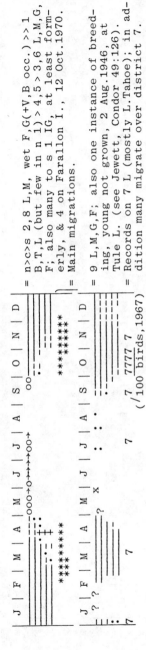

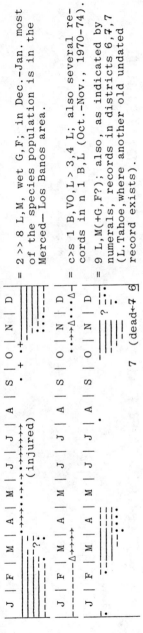

= n>c>s 2,8 L,M, wet F,G(+V,B occ.) >>1 B,T,L (but few in n 1) > 4,5 > 3,6 L,M,G, F; also many to s 1 IG, at least formerly, & 4 on Farallon I., 12 Oct.1970.
= Main migrations.

= 9 L,M,G,F; also one instance of breeding, young not grown, 2 Aug.1946, at Tule L. (see Jewett, Condor 49:126).
= Records on 7 L (mostly L.Tahoe); in addition many migrate over district 7.

ROSS' GOOSE *(Chen rossii)*

= 2 >> 8 L,M, wet G,F; in Dec.-Jan. most of the species population is in the Merced—Los Banos area.

= c>s 1 B,VO,L > 3,4 L; also several records in n 1 B,L (Oct.-Nov., 1970-74).

= 9 L,M(+G,F?); also, as indicated by numerals, records in districts 6,7 (L.Tahoe, where another old undated record exists).

BLACK-BELLIED TREE DUCK *(Dendrocygna autumnalis)*

J	F	M	A	M	J	J	A	S	O	N	D
									o→o→	·—ooooo	
						oo		1→		2	

= 8 M (+L,V?).

= Records in s 1 B, n(Sep.) & s(Nov.) 2.

FULVOUS TREE DUCK *(Dendrocygna bicolor)*

J	F	M	A	M	J	J	A	S	O	N	D
o·——————————————————————————	?ˣ?₂	———·—	———	·—ooooo
			n N(yEE) Y.Y y y J
						y			
11	1→→1→→→1→→1→→	11	1	<u>11</u>	1	1	1
33	3 <u>333</u>			37

= 8 <u>M</u>,L,V, wet F.

= Breeding phenology; ( ) = in 3 <u>ML</u>.

= Dated records in districts s 1,3,7.

J	F	M	A	M	J	J	A	S	O	N	D
o·——————	+	o—————	?	o·::	—————	·:·	·	
		ee E...E		y
		Y..Y		y 1 1→→1→→<u>4</u>→→ 1
1		9	1→		99 (9)
<u>4</u>→→→→→→→→44

= 2 M (formerly nested north to Los Banos area), L, wet F; additional undated records.in fall hunting season.

= Breeding phenology.

= Dated records in c (formerly nested), n 1 B,T,L, 4 L, 9 L,M; (9) = w. Nevada.

* RUDDY SHELD-DUCK *(Tadorna (Casarca) ferruginea)*: One wintered at Malibu from Oct.1961, and reappeared 17 Nov.1962; no doubt an escapee. The species is native of Eurasia, and is commonly kept in captivity.

16.

EGYPTIAN "GOOSE" (*Alopochen aegyptiaca*): Another Eurasian species commonly kept in captivity; 2 were shot by hunters in Merced Co., Oct. 1964.

(17.) MALLARD (*Anas platyrhynchos*)

```
J | F | M | A | M | J | J | A | S | O | N | D
_____
                                      c c C...
                 _._
                 _._
_____
..C....Cc cccc    F.........
  e E.......Eeee ee
  yyY............Yy.....y  yy
```

= 2 > 1, 3, 4, 5 > 6, 8, M,L, wet G,F,LG > 1 T, M̂, TG,B,B̂, even locally 1 UB.

= Breeding phenology; extremely early and late dates are from urban parks.

```
J | F | M | A | M | J | J | A | S | O | N | D
_____
__?__?_____
    ? ?
_____
         E.....Ee    F
         y  yyY.....Yy    y
```

= 9 >> 7, 7̂ L,ML, wet G,F,LG, even LN (but few to none in winter where shallow water is frozen).

= Breeding phenology.

BLACK DUCK (*Anas rubripes*): A species of c. to e. U.S. and Canada, but recorded four times in Calif. hunting areas: 11 Feb.1911 near Willows; 17 Dec.1960 in Butte Co.; and Nov. 1962 and Sep. 1963 on Lower Klamath National Wildlife Refuge, Modoc Co. There is a strong likelihood that some or all of these birds came from game farm stock or "plantings" to the north.

GRAPHIC CALENDAR

HABITAT DISTRIBUTION

GADWALL *(Anas strepera)*

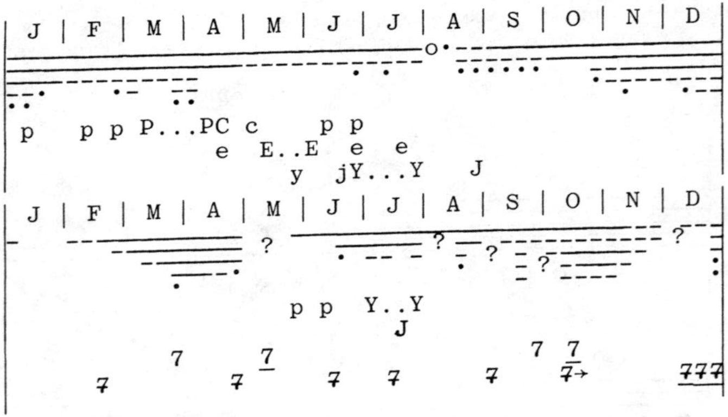

J | F | M | A | M | J | J | A | S | O | N | D

p   p p P...PC c p p
        e E.E e e
        y   jY...Y   J

J | F | M | A | M | J | J | A | S | O | N | D

            p p Y..Y
                  J

7   7   7   7   7   7   7
       7   7   7   7   7   7→
7   777

= 2 > 8 (except irreg. & few Apr.-Oct.) >
  4,6> 3,5, n, c+s ± 1 ML, M, L>G, F(+V?) > 1 T,
  B, B, M.

= Breeding phenology.

= 9 M, L, G, wet F.

= Breeding phenology (reports incomplete).

= Occurrences reported in 7, 7 L, M.

332

PINTAIL (Anas acuta)

J | F | M | A | M | J | J | A | S | O | N | D

= 2 > 8 M, L, wet G, F > 1 T M̂, B, B > 5, 4, 3, 1 L, M, wet G, F, irreg. ♀ L, G, LG or M; even to 1 O near shore when disturbed; in May -July, most regular in district 1.
= Observed migrations; also well offshore over 1 O & to 1 OI (Apr., Aug.-Dec.).
= Breeding phenology.

= 9 L, M, G, wet F (nesting commonly).

= Breeding phenology.

= Occurrences reported in district 7.

GREEN-WINGED TEAL (Anas crecca)

J | F | M | A | M | J | J | A | S | O | N | D

= 2, 8 (no record in 8, early June-mid Aug.), 9 (few in mid-winter) > 1, 4, 3, 5 ML, M, L>V, F, G, R > 1 T, B, B̂; sometimes to s 1 O when disturbed; also occas. to 1 OI.
= Breeding phenology.
= Occurrences reported in 6, 7, 7 L, M, V.
= Seasonal spread of records of A. c. crecca.

GRAPHIC CALENDAR HABITAT DISTRIBUTION

FALCATED TEAL *(Anas falcata)*: A male present in Golden Gate Park, San Francisco, 5 Apr.-21 May 1953, and one at Newport Bay, 2 Jan.-21 Feb.1969, both probably escapees. A species native of c. and e. Asia.

BAIKAL TEAL *(Anas formosa)*: Records of 5 individuals in Calif., all shot by hunters: e.Contra Costa Co., 13 Dec.1931; near Calipatria, 29 Dec.1946; near Riverside, 12 Jan. 1974; at Honey Lake Wildlife Area, 1 Dec.1974; and Gray Lodge Wildlife Area, 9 Jan.1975. All were possibly escaped captives although living in the wild. A species with normal range in e. Asia.

GARGANEY *(Anas querquedula)*: One male on 15 Mar.1972, 4 Apr.1974, & 19 Mar.1975 all at same location in Long Beach; possibly an escapee returning after migrating. A species with normal range in Eurasia, migrating south to Africa in winter.

BLUE-WINGED TEAL *(Anas discors)*

J	F	M	A	M	J	J	A	S	O	N	D
··	··	·· ---	--- ··	--- ··	o o·	--- ··	·· ---	·· ---	·· ·	·	·
pp p p p P.........Pc e P} =Breeding phenology)

C   Y.Y
I   ♂   ♂ ♂ ♂♂ ♂

= 1 L,M>M̂,T, 2,3,4,5,6 L,M, 8 (except late summer), 9 (except winter) L,M, wet G,F (+V?).

= Records in s 1 I ("I" entry), 7,♂ L.

CINNAMON TEAL *(Anas cyanoptera)*

J | F | M | A | M | J | J | A | S | O | N | D

= 2,3,4 M,L>G,F > 1,5,6,7 L,M (but few winter in 5,7) >1 M,B,T,B.

= Breeding phenology.

= 8 M,L > wet F,G; & has nested, with young out on 13 July.

= 9 > 7 M,L,G, wet F.

= Breeding phenology.

NORTHERN SHOVELER (*Anas clypeata*)

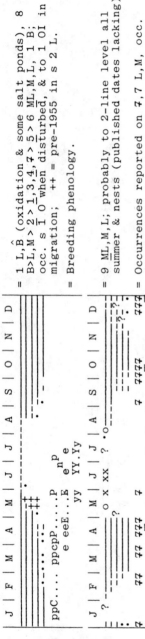

= 1 L,B̂ (oxidation & some salt ponds), 8 B>L,M>2>1,3,4,7̲>5̲>6 ML,M,L, 1 B; occ. s1̲0 when disturbed, & to 1 OI in migration; ++ = pre-1955 in s 2 L.

= Breeding phenology.

= 9 ML,M,L; probably to 2-line level all summer & nests (published dates lacking).

= Occurrences reported on 7̲,7 L,M, occ. abundant in Oct.

EUROPEAN WIGEON (*Anas penelope*)

= 1,2 L,M, nearby G > 3,4,5 > 6 L,M,G.

= Occurrences reported with dates in 7, 8,9 L,M (+8 B?).

AMERICAN WIGEON *(Anas americana)*

```
 J | F | M | A | M | J | J | A | S | O | N | D
                                -·····ΔΔ·····oo
 77    77  7 777   7   7   7   7  77777  7  777
 |·· |   |   |   |   |   |   |   | - - |   |   |
 |   |   |   |   |   |   |   |   |     |   |   |
7                                        y  7777777
                                            7
```

= n 1,2,8 > c, s 1,3,4,5, L,G,ML>F > 1 T>B>M̂ (+occ. nearby O when disturbed) > 6 L,M; nested at least once in 2 M.

= Occurrences reported on 7 L (sometimes to 1000 or more in Apr., Oct.-Dec.)

= 9 L,M,G.

```
 J | F | M | A | M | J | J | A | S | O | N | D
|| ? ?  ? --  ? .-x++-- - ?
|
|··

   C(dist.2)        pE..HYY)
                  7
```

= Breeding phenology; (YY) = in w. Nev.

= Occurrences with dates on 7 L,M.

WOOD DUCK *(Aix sponsa)*

```
 J | F | M | A | M | J | J | A | S | O | N | D
|-+-- -·+++-              ?···?·-- o · ·
                                      ·
           pp  P·····.P···     p      7  7777
           c         E.E....  e?        (9) 99  99
                 Y.......Y  yy  y
```

= 56 > 46, n 6 > 4 > 2,5, c,s 6 LR,VR, or W near L̲,V>M,L; rarely to n,c 1 BV̲,OV. _

= Breeding phenology.

```
 J | F | M | A | M | J | J | A | S | O | N | D
|--ΔΔ→-Δ---ΔΔΔΔooo-Δ-ΔΔoΔΔΔ ?  · ·Δ·----
 7      7     7  7077  77     7   7  7777
        99    9  9 99  0            (9)  99   99
 ppP..Pp→→p    p pn   n
               en         yy  y   Jy
               yy  y
```

= 3 > 8, s 1 LR,VR>L,M,V; nested at Riverside (1950s) & on Santa Ynez R. (1964).

= Occurrences reported with dates on 7,9 L,V(+R?,N?), 10 L ("O"=10; (9)=w.Nev.).

= Breeding phenology: 1st row=district 3; others in Yosemite Valley, Sequoia National Park, and ne. Calif.

337

REDHEAD (*Aythya americana*)

J | F | M | A | M | J | J | A | S | O | N | D

= n 1 B(winter), 8 B(chiefly winter) > 3, 4, 6, 7, 8 > 2 > 1 > 5 L, LM > c, s 1 B (few on S. F. Bay, although occ. unsubstantiated reports of many; also undated "winter" reports of up to 5000 on Humboldt Bay.

= Breeding phenology.

= 9 L, LM (including extreme w. Nev.).

= Breeding phenology.

= Occurrences reported with dates on 7 L, ML (formerly few bred at S. Lake Tahoe, young out in late June, late Aug.).

RING-NECKED DUCK (*Aythya collaris*)

J	F	M	A	M	J	J	A	S	O	N	D

= 4,6 > 2,5,7,8 > 1,3 L>ML > 1 (+8?) B.

J	F	M	A	M	J	J	A	S	O	N	D

= 9 L,LM; & reported nesting in the Tule-Klamath Lake Basin in June 1954.
(Occurrences reported on 7 L,LM (five

7   7   7   7   7   7   7   7   7   7   777

= nests at Mtn.Meadows Res., Lassen Co.,
in 1956; also partly grown young at S.
(Lake Tahoe, 24 Aug.1926).

TUFTED DUCK (*Aythya fuligula*)

J	F	M	A	M	J	J	A	S	O	N	D

(1)¹ →→→→→

4→→→→→ }=lakes near S.F.Bay in 1971 & 1972)

3→→→

Published records with dates, by districts as shown by numerals. →→ representing apparently same individuals staying: upper 1's = Arcata sewage pond, 1968–70; other 1's = L.Merritt,Oakland, 1976, and Rodeo Lagoon, Marin Co.=( ), 1975–76 (a hybrid); 3's = male at L.Sherwood, e.Ventura Co. in 1973–76 (successive rows). In addition, the first record for Calif., near Pleasanton, Alameda Co.; and the

a male shot "between 23 Dec. and 8 Jan. 1948–49" near
author saw a male on Berkeley Aquatic Park, 20 Feb.1953.

CANVASBACK (*Aythya valisineria*)

J | F | M | A | M | J | J | A | S | O | N | D

= c>n 1 > 8 B, 2,7,8 L>M > 1 $\hat{B}$ > 1,3,4 > 5,6 L>M, s 1 B (+ rarely 2,3,4,8,+?5 V).

= (Breeding records, possibly by partial cripples, in 2 LM except (y) in s.Nev.)

= 9 L,LM.

= Breeding phenology (reports incomplete).

= Occurrences on 7,10(=0) L; may nest at Grass L., n. of Mt.Shasta (June).

GREATER SCAUP (*Aythya marila*)

J | F | M | A | M | J | J | A | S | O | N | D

= n,c>s 1 B>>$\hat{B}$>L,O (Nov.-Mar. numbers are rough estimates since most scaups in large flocks are not identified to species.

= 2 > 4,5, n 6 > 3,8 L, 8 B.

= Occurrences reported with dates in 7, 9 L; (9) = w. Nevada.

LESSER SCAUP (*Aythya affinis*)

= 1 > 8 B > 1 B̂, 3, 7 > 1, 4 > 2, 5, 6, 8 L>>V;
e,y = breeding records in 4 LM, probably of wounded birds unable to migrate.

= 9 L,LM (sometimes nests commonly, but few published reports give any dates).

= Breeding records with dates.

= Occurrences with dates on 7 L; also old records at L.Tahoe "throughout winter."

N = several nests at Mountain Meadows Res., Lassen Co. in 1956.

COMMON GOLDENEYE (*Bucephala clangula*)

= n,c 1,8 B>B̂,M̂, 8 L,V > s 1 B, 1,2,4,6 > 3, 5 L>V,M; also, one at S.Farallon I. on 6 June 1968.

= Occurrences with dates from 7,7,9 L (in 9 to 3-line level, at least in Mar.,Dec.) Also "common" through winter at L.Tahoe, at least in the 1920's.

GRAPHIC CALENDAR HABITAT DISTRIBUTION

BARROW'S GOLDENEYE (*Bucephala islandica*)

J	F	M	A	M	J	J	A	S	O	N	D

= c>n 1 B,B̂ (brackish diked lagoon, as at
  L.Merritt, Oakland) > L.

= 10, n>c(+s?) 7 L,V, adjacent N,R; one
  also "wintered" at Tahoe Keys in 1975-76.

= Breeding phenology.

= Occurrences in other districts as shown
  by numerals (on L,V?); "1" = s 1 B; (8) =
  s.Nev. only (up to 57 birds at Davis Dam);
  other winter & spring records on 6 V are
  undated.

BUFFLEHEAD (*Bucephala albeola*)

= c>s,n 1 B̂,B > 8 B,L,V > 4,6 > 2,3,5 L>V.

= 9 L (+V?).

= Occurrences with dates reported from 7,
  7̲ L>V; & breeding data from n 7 L̲N or
  L̲R (Mtn.Meadows Res., Eagle L., etc.).

342

OLDSQUAW *(Clangula hyemalis)*

J	F	M	A	M	J	J	A	S	O	N	D
•		---ΔΔ-ΔΔ-Δ↔ΔΔΔ	↔	•	• •	• •				252	222
3	3		79	(8)					9	999995	

↔↔↔↔↔↔↔→   7→6↔↔6↔

= n,c>s 1 O,B,OI > 8 B,L, 1 L.

= Occurrences reported on 5 V, and on L in other districts as shown by numerals; ↔ = same individuals remaining.

HARLEQUIN DUCK *(Histrionicus histrionicus)*

J	F	M	A	M	J	J	A	S	O	N	D
•	-	---oΔ↔••↔Δ----Δoo ΔΔ↔↔Δo↔↔------							•	•	•
2		4		6		........7 7			2 6	7	

nested from Mar. onward, with young out as late as July (no recent nesting evidence).

= n,c>s 1 O (espec.near K points > OI >> 1 B.

= Occurrences reported on 2,4,6 L,V; and seasonal spread of known occurrences on c 7 turbulent V, along which formerly

KING EIDER *(Somateria spectabilis)*

J	F	M	A	M	J	J	A	S	O	N	D
Δ↔↔↔↔↔-↔-↔↔↔↔↔•↔-↔-↔•↔↔•↔		↔	•↔•?-↔?-?•↔-↔	↔	•				•	•••••↔↔↔o•	

(↔↔ = apparently same individuals remaining)

= c,n 1 O near K, harbors with breakwaters or many pilings > B; also an im.male at Malibu, 22 Nov.-28 Jan. 1973-74.

SPECTACLED EIDER *(Lampronetta fischeri)*: One Calif. record-- a male shot in Feb.1893 at Bitterwater L., San Benito Co. (partial specimen, see Condor, 42:309).

The normal range of the species in winter is from e. Siberia to the Aleutians.

343

WHITE-WINGED SCOTER   (*Melanitta deglandi*)

J | F | M | A | M | J | J | A | S | O | N | D

= 1 O,B>OV>OI, 8 B>L; higher summer numbers are primarily at c, n 1 OV.

= Observed coastal migrations.

= Occurrences reported with dates on L in
  other districts as shown by numerals; "O"
  = 10. There are also other undated winter records from L.Tahoe & Imperial Val.

SURF SCOTER   (*Melanitta perspicillata*)

J | F | M | A | M | J | J | A | S | O | N | D

= 1 O,B>OV>OI>>L; high summer numbers are
  chiefly at OV or n 1 OK,OS.

= Migration movements reported (along
  coast or offshore).

= 8 B>>L (nearly all = on Salton Sea).

= Occurrences reported on L in other districts as shown by numerals; (9)=w. Nev.

BLACK SCOTER *(Melanitta nigra)*

J	F	M	A	M	J	J	A	S	O	N	D

= n,c>s 1 O,B>OI>L.

= Occurrences reported on 2, 9 L, 5 V, 8 B; additional undated winter records for 2 L and 5 V not shown.

(9)=w.Nev.

RUDDY DUCK *(Oxyura jamaicensis)*

J	F	M	A	M	J	J	A	S	O	N	D

= 1,8 B, 1 $\hat{B}$ > in late Sep.-Mar. (<< other times) 1,2,3,4,7,8 (except few & irreg. in 8, June-Aug.) > 5,6 L,LM>>V; rarely a few to 1 O in winter or migration.

= Breeding phenology (y = in 8 ML, where only irreg. as breeder).

= 9 L,LM; large winter numbers in very mild years only.

= Breeding phenology (reports incomplete).

= Occurrences reported with dates on 7 L; & breeding dates from S.Tahoe marsh; a few also reported from L.Tahoe in winter but without dates.

19.

345

20.

HOODED MERGANSER (*Lophodytes cucullatus*)

J | F | M | A | M | J | J | A | S | O | N | D

= 6 > 5, 4, 9, 2 > 1 > 3, 8 L>V > 1, 8 B; "y" = fe-male and young, L.Earl, Del Norte Co., 1974.

= Seasonal distribution of records on 7, 7 L (up to 20 at L.Almanor in Feb.); also two pairs + female with nest & eggs, 13 May 1964, at Mtn.Meadows Res., Lassen Co.

COMMON MERGANSER (*Mergus merganser*)

J | F | M | A | M | J | J | A | S | O | N | D

= e 8 L,V (chiefly Oct.-Mar.) > non-breeding locations in 7 > 2 > 3, 4, 5, 6 L>V > (Sep.-Mar.) or < (Apr.-Aug.) 5 > 4 > 6 > n 2 V, L with adjacent trees with cavities > or = (winter) 1, w 8 L>V > 1 B.

= Breeding phenology; "2" = dates in n 2.

J | F | M | A | M | J | J | A | S | O | N | D

= 7, 9 L,V, nesting mostly in nearby tree cavities but also occas. amid rocks.

= Breeding phenology.

346

RED-BREASTED MERGANSER *(Mergus serrator)*

J	F	M	A	M	J	J	A	S	O	N	D
							•·°°°∆·°°∆°°∆°°∆°—				

= 1 B>O > 1 L,OI >> 3,4,5 L,V.

J	F	M	A	M	J	J	A	S	O	N	D
?					O→→·•		••				••
22		722	2				7??		6	66
6→6→	9	9			7	9			7→7	9?

= 8 L,V>B (high numbers along Colorado R.).
Occurrences reported on L,V in other districts indicated by numerals; additional
records in winter at L.Tahoe & in spring
in Mono Co. are undated.

SANDHILL CRANE *(Grus canadensis)*

J	F	M	A	M	J	J	A	S	O	N	D
					•x +x		* **************			
		******					î î 1 11 1̂			
1→→→	1		1			î î 1̂ 3̂ 1̂			
3̂ 3 333 3̂			4			3 4̂4̂ 3→3			
6 6666̂6̂6̂6̂			5→→→→→→→→→			4̂4̂ 3→3 4̂		
			7{=in 1800's)	77	6	7̂ 6			

= 2 G, certain F types, adjacent M, + shallow L or water in M for night roosting;
includes Carrizo Plain, the southernmost
major wintering area.
= Observed migrations.
= Occurrences reported in other districts
indicated by numerals, or (^) migrating
overhead; 6's in Jan.-Feb. = up to 1000
in upper Salinas Valley; →→= same individuals remaining.

J	F	M	A	M	J	J	A	S	O	N	D
‡‡‡‡‡‡		‡‡‡‡‡‡			•		xx·‡‡			
‡‡‡‡‡‡		‡‡‡‡‡					‡‡			
‡‡‡‡‡		‡‡‡					••			
••••							••			

= 8 F (e.g.,alfalfa), G>V, & M at least for
night roosting (s.end of Salton Sea);
++ entries = pre-1953 only.

21.

SANDHILL CRANE (*Grus canadensis*), continued

```
 J | F | M | A | M | J | J | A | S | O | N | D
 =                            ?                   o
 =(xx) xxx ?  ..o .+. o  o.--x
 _|_ _                        ?.?
 _ -  .                       (..)
   :  .
        (ne)EE (n)ppp p
                  y   Y YYJ)  (j)
```

= 9 G, <u>M</u>, LM>L; ( ) = c. to s. Oregon.

= Breeding phenology; ( )= c.Ore.,w.Nev.

CLAPPER RAIL (*Rallus longirostris*)

```
 J | F | M | A | M | J | J | A | S | O | N | D
 ..:. -   +   +                        .        ..:
                                                ---  ..:
          eE.....E eeeeE.Ee
              y   Y.....YyyY..Y                  .
```

= c>s 1 $\hat{\underline{M}}$ (occas. T,M of immediate vicinity; up to 3-line level can be found
all year by special techniques.

= (Breeding phenology (2 broods, with peaks
before & after extreme high tides of
early June).

```
 J | F | M | A | M | J | J | A | S | O | N | D
 ?   ?  -x   ?        ΔΔΔ... ... o   ?   ?  ? oΔ
                 x
                   eEE        YY
```

= 8 <u>M</u>, dense riparian shrubs; most are
probably absent from Ariz. & Calif. Oct.-
mid Apr.(see Condor 75:177-183).

= Breeding phenology.

```
 4   1   ^^       3        ^   ^^^^   4   4   1
                               ^^^   ^^^   ^I   3
```

= Occurrences reported in M (or as grounded
waifs) in districts 3,4 & in 1 M away
from any salt marsh, + I = on Farallon I.
in 1886;^ = in Humboldt Bay salt marsh.

VIRGINIA RAIL (*Rallus limicola*)

```
J | F | M | A | M | J | J | A | S | O | N | D
 .                                    . . . . .  ---
        eE.....E
        yy  Y.......Yy
                 JJ  j
```

= 1,2,3,4,5 > 6, 8 M; + (Sep.-Apr.) 1 M̂; & recorded on Farallon I., 2 Sep.1968.

= Breeding phenology

```
J | F | M | A | M | J | J | A | S | O | N | D      7
2   (9)   00  7                  777         7          99
          99 (9)  99  9  999 (9)       9
              e  n   n  nny
          yy  n
```

= Seasonal distribution of records in 7,7, 9,10(=0) M, with reported breeding data shown by n̲, e, y; (9) = w.Nev., c.,s.Ore.

SORA (*Porzana carolina*)

```
J | F | M | A | M | J | J | A | S | O | N | D
 .    ...Δx  .x  .                 .              ---
     E.....Eee
          777 7   7   7 7
       9  9    (9) 7 7
          eee e   j   j
              y
```
                                                      7
                                                       9

= 1,2,3,4,5(few in winter),8 (no June-July records) > 6 M; + (Sep.-Apr.) 1 M̂.

= Breeding phenology (reports incomplete).

= Occurrences reported with dates in 7,7,9 M, & breeding data indicated by e,y,j; (9) = w.Nev. Vagrants also reported on c, s 1 I in fall & winter, records undated.

YELLOW RAIL   (*Coturnicops noveboracensis*)

J	F	M	A	M	J	J	A	S	O	N	D

(Quincy)(Mono Co.)
1 1 ̂   1̂62 1̂ 11 1̂111
1⁄2 33¹   77   9 9 9 9   1⁄2 242 1̂242 242
                 N? e   N) old breeding data) 4 3

= Seasonal distribution of all dated records, by districts as shown by numerals;
preferred habitat = wet G or short M, but
occ. also in 1 M,M̲; ^=including since 1960.

BLACK RAIL   (*Laterallus jamaicensis*)

J	F	M	A	M	J	J	A	S	O	N	D

-o--o••Δ•••Δ•• •••   -------Δ→Δ Δ••oo----Δ • -•
                      ┼   I   I
     c   e Ncc Ecc cccc c
              (e)y e   Y Y j j
33   4   3 2 4 4 3   33 2   4   3
              9   33

= 1 M̲, 8 M(probably breeds); ++= estimated
breeding population on s.San Diego Bay in
1908; I = Farallon I., 1905,1909.

= Breeding phenology; c = calling; (e) =
near Chino in 1931.

= Occurrences in other districts (numerals).

COMMON GALLINULE   (*Gallinula chloropus*)

J	F	M	A	M	J	J	A	S	O	N	D

•   -•   •   •
-:
•-:
          e E..E e
            yy Y...Y   y
5   5   J.....J j j
          (9)9 I 99 9(9)

= 2 > 3,4,8 > 6, c, s 1 M,LM; also one recorded in n 1 (M?), in winter 1973-74.

= Breeding phenology.

= Occurrences reported with dates from 1 1
(Farallon I.,1975), & in districts 5
(near Arcata) and 9 ;   (9) = w.Nevada.

PURPLE GALLINULE *(Porphyrula martinica)*: One Calif. record: one flew into a wire at San Diego on 1 Oct.1961. Others have been seen and some collected at various points in N.M., Ariz., and on 7-8 & 20 Sep.1966, n.w. of Las Vegas, Nev. Normal range of the species is from se. U.S. to South America.

AMERICAN COOT *(Fulica americana)*

```
J | F | M | A | M | J | J | A | S | O | N | D
                              ? = ?
                                            tt...
t ....T....T t       c      eeee
bBeeE.....E^n   e
                yyyY.............Yyyy
                    jj J..............Jj
```
= 2 > 1,4,5,7 > 3,6 L,M,LM, wet G,F > 1 M̂>B, 1,2,3,4,5(+6?) V >> 1 Ō,OI (esp.Sep.-Oct., but has bred on Santa Cruz I.).

= Breeding phenology.

```
J | F | M | A | M | J | J | A | S | O | N | D
? ?          ?-?-?     ?.•   ?
      ? ?    ? ?
   (y)  (Y)y     J
```
= 8 B>L, wet G,F, M,LM>V.

= Breeding phenology (few reported); () = (w.Ariz)

```
J | F | M | A | M | J | J | A | S | O | N | D
? ?  ? ?                       ?-..•.. ?
                                    ?
      ? ....PE....E e
          yy Y.Y Y y
              j jJ...Jj
```
= 9 >> 7 LM,L>G,M(+V?).

= Breeding phenology.

23.

AMERICAN OYSTERCATCHER *(Haematopus palliatus)*

```
  J  |  F  |  M  |  A  |  M  |  J  |  J  |  A  |  S  |  O  |  N  |  D
→?→+•→→→→→→ (•)  •ΔΔ•+ ? ?(o→+)· ?•Δ? ·•→→•Δ· •·•→ (•)
```

= s 1 OI>OK,OS; ( ) = on Los Coronados Is., Mexico, near San Diego. Additional un- dated records two recent winters; & one at Pt.Lobos, Monterey Co., 3 Apr.1954.

BLACK OYSTERCATCHER *(Haematopus bachmani)*

```
  J  |  F  |  M  |  A  |  M  |  J  |  J  |  A  |  S  |  O  |  N  |  D
•- - ·   •       •            •••-       ··•:       ^8
                 5                 <
         n  CeeE...Ee...e            <
            C yY....Y y   y     f
```

= 1 OKI > n,c 1 OK > s 1 OK; rarely to 1 OS. (Records on 5 V (2 mi. e. of Shelter Cove), n,c 1 B(=^), & 8 B.

= Breeding phenology.

24.

AMERICAN AVOCET *(Recurvirostra americana)*

```
  J  |  F  |  M  |  A  |  M  |  J  |  J  |  A  |  S  |  O  |  N  |  D
                                        --- ?     J
                 teE.....E eee
                 B y yY.........Yyyy
```

= 8 B > (or < Apr.-June) ML,L, c>s 1 B̂ and dikes, T,M̂ pools, nearby M,L >> n 1 T,M̂, L, 8 V; also vagrants to s,c 1 I in June-Aug.

= Breeding phenology.

| J | F | M | A | M | J | J | A | S | O | N | D |

= 2 > (or <, Nov.-Feb.) 3,4 ML,L, flooded G, F; +++ = at Buena Vista L. in 1953.

= Breeding phenology.

```
                 ++
                PP PeeE.....E ee  e
                  y y........Yyy
```

| J | F | M | A | M | J | J | A | S | O | N | D |

= 9 ML,L; ( ) = supplementary records from s.,c.Ore. & w.Nev.; fall peaks of 30,000 reported in Modoc Co., unspecified date.

= Breeding phenology (includes s.Ore.data).

= Occurrences with dates reported at 7,7 L (+M?).

```
-(x X)?  = ?-?-?--(-)? -(- -)  ?
           =    =     ( =)
           =    =     ( =)
         t   ? bc
         bc  ? EE...e
            y NY.....  7777777
          7   77 7   7  77
```

BLACK-NECKED STILT *(Himantopus mexicanus)*

| J | F | M | A | M | J | J | A | S | O | N | D |

= c,s 1 B̂ (increasing recent years, S.F. Bay area) + dikes > M̂,M,ML,L>T; casual to n 1,5 L(Apr.-June,Nov.) & s 1 I (May).

= Breeding phenology.

```
        c   C ? P E....Ee  e
            y  Y.......Y y y
                jJ...Jj
```

GRAPHIC CALENDAR HABITAT DISTRIBUTION

BLACK-NECKED STILT (*Himantopus mexicanus*), continued

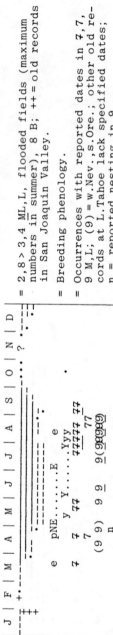

= 2,8 > 3,4 ML,L, flooded fields (maximum numbers in summer); 8 B; ++ = old records in San Joaquin Valley.

= Breeding phenology.

= Occurrences with reported dates in 7,7, 9 M,L; (9) = w.Nev.,s.Ore.; other old records at L.Tahoe lack specified dates; n = reported nesting in 9.

SEMIPALMATED PLOVER (*Charadrius semipalmatus*)

= 1 T>S >> 1 L,V, wet G,F, 1 OK; & known from s 1 OI (Apr.-May,Aug.) & Farallon I., Aug.-Sep.

= 8 B shore > 2,8 L,V, wet G,F.

= Occurrences reported with dates in other districts shown by numerals; (9) = w.Nev.; underline = more than one record.

25.

PIPING PLOVER *(Charadrius melodus)*

| J | F | M | A | M | J | J | A | S | O | N | D |

= individual at Goleta, 1971–early 1974.

= individual(s) at Malibu, 1973–74.

WILSON'S PLOVER *(Charadrius wilsonia)*: At least two occurrences--adult male collected on 29 June 1894 at Pacific Beach & one seen on 11 May 1918 at Imperial Beach, both by Ingersoll. Also, individuals reported without details on San Diego Christmas Count, 29 Dec. 1956 & at Malibu on 12 Dec. 1957.

SNOWY PLOVER *(Charadrius alexandrinus)*

| J | F | M | A | M | J | J | A | S | O | N | D |

= 1 OS>BS>B̂ dikes >> 2 L with barren shores (Los Banos--Woodland; no record, Oct.–Jan.); + 3 L (Elsinore), nesting 1939–40.

= Breeding phenology.

| J | F | M | A | M | J | J | A | S | O | N | D |

beeE........Eee ee
yY.........YFyyJ

= 8 B shores, V, L, + ?F.

| J | F | M | A | M | J | J | A | S | O | N | D |

6 I   I   77   I I   7
(9⁹(9)   y   99+99999⁹⁹⁹⁹
                        E   y

= Breeding phenology (data very incomplete).

= Available records in other districts indicated by numerals, or I = s 1 I; (9) = w. 3 Nev., & Breeding data there (incomplete).

355

GRAPHIC CALENDAR

HABITAT DISTRIBUTION

KILLDEER *(Charadrius vociferus)*

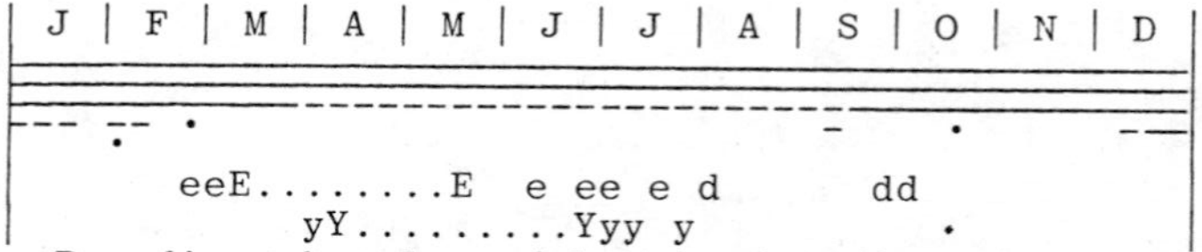

J	F	M	A	M	J	J	A	S	O	N	D

```
            eeE......E  e ee e d    dd
                  yY.......Yyy y         .
```

Breeding phenology (above); d = diversionary
display but no nest or young found.

= 2,5 > 3,4,6,7(few summer),8(few winter)
L shores, G,F (with little or low vegetation &, any bit of water near),V>U
where short G or bare ground included >
1 T,BS,B̂; rarely to OS,OK but irreg.
all year on various I.

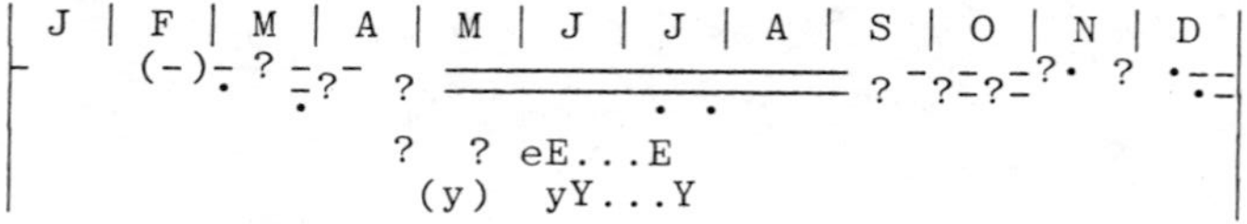

J	F	M	A	M	J	J	A	S	O	N	D

```
(-). ?-?- ?        ? -?-?-?. ?  .-.-
        ? ? eE...E
        (y)  yY...Y
```

= 9 > 7,10 L shores, G,F,V; ( ) = c.Ore.;
also old undated "winter" records at L.
Tahoe.

= Breeding phenology; (y) = w. Nevada.

MOUNTAIN PLOVER *(Charadrius montanus)*

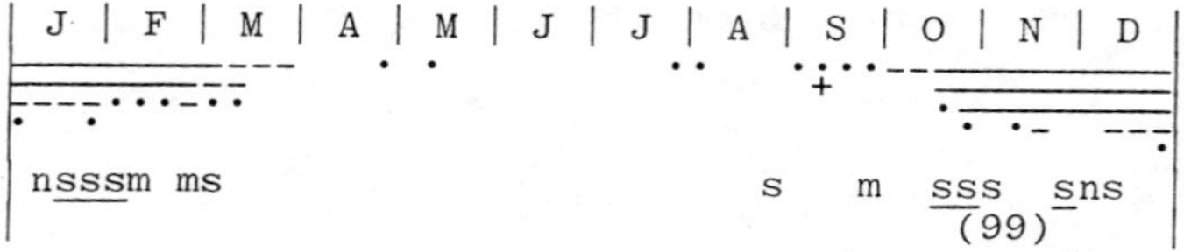

J	F	M	A	M	J	J	A	S	O	N	D

```
nsssm ms       s    m   sss  sns
                         (99)
```

= 2,8 > 4,3 barren mesas or very short G >
bare F; also (irreg.?) s 1 IG (reported
abundant on San Miguel I., 28 Dec.1930).

= Available records in n,c(=m),s 1 G,F,
rarely to S; (9)= w.Nev. (probably occurs in e. Calif. also).

DOTTEREL *(Eudromias morinellus)*: One Calif. record--individual on S.Farallon I., 12-20 Sep.1974 (photo in American Birds 29:23); normal range of the species is in the Old World & it has bred in n. Alaska; casual in w. Alaska, w. Wash.

AMERICAN GOLDEN PLOVER *(Pluvialis dominica)*

J	F	M	A	M	J	J	A	S	O	N	D

= 1, nearby parts of 5,4 > 3 short G>F>
1 S,T > 8 SB,G?,F?,V?,L shores.
= Occurrences reported farther inland in districts shown by numerals & on off-shore islands (= I).

BLACK-BELLIED PLOVER *(Pluvialis squatarola)*

J	F	M	A	M	J	J	A	S	O	N	D

= 1 T+nearby $\hat{B}$,$\hat{M}$, or barren F as roosts
> 8 BS,(+wet F?) > 1,2, coastal parts of
5 > 4 wet G,F > open L shores, open M, 1
I,S>K, 3 L, wet G,F; ** = main migrations.
= (Dates of published records in districts
7 & 9; (9) = w.Nev. In Imperial Valley - Salton Sea area (dist.8) reported by

McCaskie (Calif. Fish & Game 56:89) to reach 500 per day's observation in winter and 1000 in spring and fall but specific dates of such numbers not given.

GRAPHIC CALENDAR   HABITAT DISTRIBUTION

26.

UPLAND SANDPIPER *(Bartramia longicauda)*: At least 8 Calif. records--1 collected at Tule L., Modoc Co., on 8 Aug.1896; 1 seen at Needles Landing on Colorado R., 11 Sep.1952; 1 at Bodega Bay, 15 Feb.1962 (record has been questioned); 1 on S.Farallon I., 22-31 Aug. 1968 (and found dead) & another there on 23 May 1969; 1 near Colton on 10 Sep.1973; 1 on Santa Barbara I., 23 May 1975; 1 in Death Valley, 15 May 1976; and 1 at L.Talawa, Del Norte Co., 13 Sep.1976. Normal range of the species is c. & ne.U.S. west casually to nw.U.S.; migrates to south of U.S. for winter.

WHIMBREL *(Numenius phaeopus)*

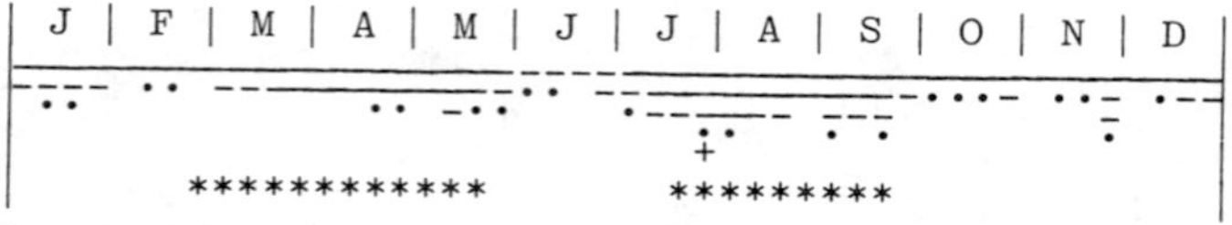
J | F | M | A | M | J | J | A | S | O | N | D
+
****
****

= 1 T,S>K (rare in n 1 in winter) > wet F, G,L,V>I.

= Main migrations (chiefly inland in spring)

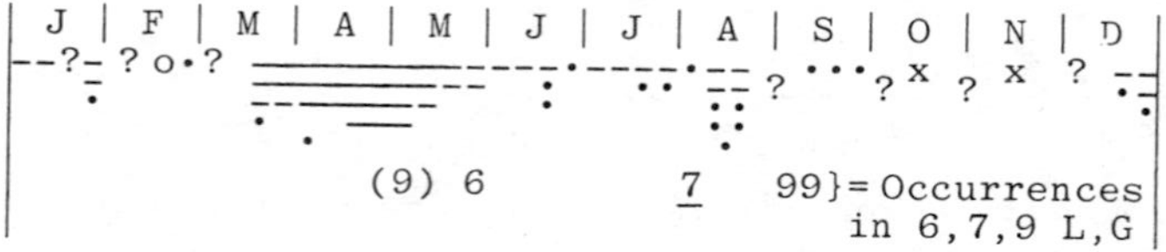
J | F | M | A | M | J | J | A | S | O | N | D
? ? o • ?
? ? x ? x ?
(9) 6
7
99}= Occurrences in 6,7,9 L,G

= 8,2 > 3,4(+5?) wet F,G>L,V, 8 B > 1 I; "thousands" reported in s 2 in Nov. are suspected mostly Long-bills. Up to hundreds per day in winter & > 1000 in fall & spring reported by McCaskie (Calif. Fish & Game 56:90) near Salton Sea but without specific dates for fall.

LONG-BILLED CURLEW *(Numenius americanus)*

```
J | F | M | A | M | J | J | A | S | O | N | D
  · |   ·· |     | --|-- |   |   |   |   |   |   | --·
  · |   ·· |     |   · |·· | * |   |   |   |   |   | ·
· |   |·--·--|-- |· |-- |· |--·--|-- |· |   |
  · | * ****** | ** |   | * |   | * | & ? to....* |
```

= 2,8 > 1 > 4,5, F,G (usually wet) > L,M,
  1 T (+nearby B as roost), 8 B 1 M > 3
  (+6?) wet F,G,L, 1 S,OI.
= Migration dates (observed & probable).

```
J | F | M | A | M | J | J | A | S | O | N | D
x                                              x
? x-xxx xx.. ? ? ? ?.-
        ct
     ee(E) EY       7       7
```

= 9 wet G,M,L.

= Breeding phenology; ( ) = c.Oregon.
= Occurrences reported with dates at 7 L;
  additional "fall" records at L.Tahoe.

MARBLED GODWIT *(Limosa fedoa)*

```
J | F | M | A | M | J | J | A | S | O | N | D
  |   |   |-- |--·--|-- |   |   |   |   |   |
  |   |   |   | ··· |   |   |   |   |   |   |
3  |**| *****|   |**************|   |   |   |
33 |  |      |   |7777777777 37 |   | 3 |
(99  9)      (999999 39) 9 (9      9 9)
```

= 1 T + adjacent B̂, dikes or pools (as
  high tide roosts) > 8 B > 1 S,M >> 1,2,8
  L, wet G,F; also recorded on s 1 OI
  (Sep.-Nov.); & *** = observed migration.
= Occurrences reported at 3,7,7 L, 9 L(+M?);
  (9) = w.Nev.; flocks of up to a few hundred at L.Tahoe & in w. Nev.

HUDSONIAN GODWIT *(Limosa haemastica):* Records of three individuals in Calif.: at Tule L., Modoc Co., 10 Sep.1971; Eureka, 9-10 Aug.1973 (photo'd); and near Daggett, 9 May 1975 (photo'd).

GRAPHIC CALENDAR   HABITAT DISTRIBUTION

BAR-TAILED GODWIT (*Limosa lapponica*): An Old World species breeding also in w.Alaska. Individuals found in Calif. at: Upper Newport Bay, 2 Sep.1962 & (2 on) 9 Sep.1962; Arcata, 11-17 July.1968 (collected, see Condor 72:112); Bolinas Lagoon, 26 Oct.-30 Nov.1973 (Amer. Birds 28:101); Arcata, 17 July-11 Sep.1974; Schooner Bay, Marin Co., 28 Sep.1975; Playa del Rey, L.A.Co., 11 Feb.-2 Mar.1976; Bodega Bay, Sonoma Co., 17 Apr.1976.

SOLITARY SANDPIPER (*Tringa solitaria*)

```
  J | F | M | A | M | J | J | A | S | O | N | D
· ·       ·   · ·   · · ·   ·      —Δ—      I
↑                                   15→     I
?        55    7    7(9)7 9(9)      9
         7     99   7 77777         7
```

= 3,8 > 2,4 > 1,6 small LM or LR or pools in G,W,N>L,V,M; ↑+? = records questioned.

= Published dates of occurrence in 5,7,7, 9, & on s 1 I (="I"); (9) = w.Nevada.

LESSER YELLOWLEGS (*Tringa flavipes*)

```
  J | F | M | A | M | J | J | A | S | O | N | D
· ·     ·oo·  ·o···o         ·:·:·  ·:·
· ·     ·:·   ·:·             ·:·:·  ·:·
       (9)    9      (9)7 7   9
```

= 2,8 > 3,5, n>c, s 1,4,6 pools in G,F, or M>L,V, 1 M>B̂>T,I.

= Occurrences with reported dates in districts 7,9; (9) = w.Nevada.

GREATER YELLOWLEGS (*Tringa melanoleuca*)

```
  J | F | M | A | M | J | J | A | S | O | N | D
            ·o·     ·:·:·:·  ·:·:·  ·:·
                    ·o·              ·:·
(c.Ore.)(9) 9  7 7  777 77   777     77
            7       9999     9999    I7
```

= 2 > (< in midwinter) 8 > 1,3,4,5,6 L, pools in M or G>V, 1 B̂,M>T.

= Occurrences with dates in 7,7,9,L,M,etc. & on S.Farallon I. (="I") in 1968; the Apr. record in 9 = of 500 birds, & species is probably much more regular there than shown.

WILLET *(Catoptrophorus semipalmatus)*

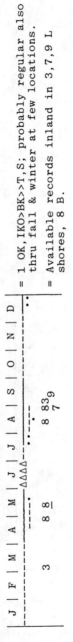

J	F	M	A	M	J	J	A	S	O	N	D

= 1 T,B̂,> 1 M̂>S>K,L,G, 8 B; also recorded in s 1 I(Jan.,June) & c 1 I(July,Aug.).

= Main migration periods.

= 2 > 3,4 > 6 (+5?) L,LM, wet G,F.

= 9 >> n 7 M, wet G, L shores; ( ) = c.Ore.

= Breeding phenology; also territorial or potential nesting activity noted in May-June in 2, n 8 wet G, & 1 M̂ borders.

= Occurrences in 7 L,ML (or wet G?).

WANDERING TATTLER *(Heteroscelus incanus)*

J	F	M	A	M	J	J	A	S	O	N	D

= 1 OK, IKO>BK>>T,S; probably regular also thru fall & winter at few locations.

= Available records inland in 3,7,9 L shores, 8 B.

GRAPHIC CALENDAR

HABITAT DISTRIBUTION

SPOTTED SANDPIPER (*Actitis macularia*)

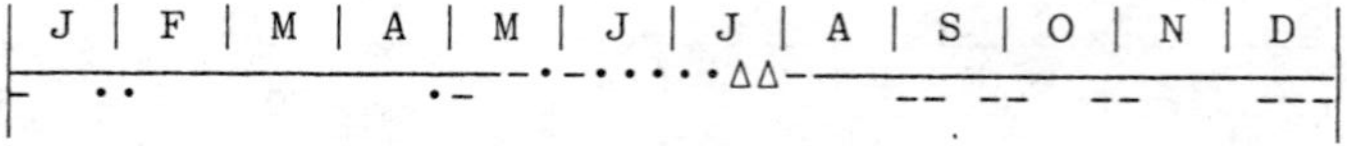

```
J | F | M | A | M | J | J | A | S | O | N | D
                                       --  --   --
 ·     · -·- ·······ΔΔ-            --  --      o--·
  ·-                                              
```
= s>c, n 1 OK,OKI,L,OV, 1,2,3,4,8 > 5,6 L,
  V (non-breeding areas) >>1 T.

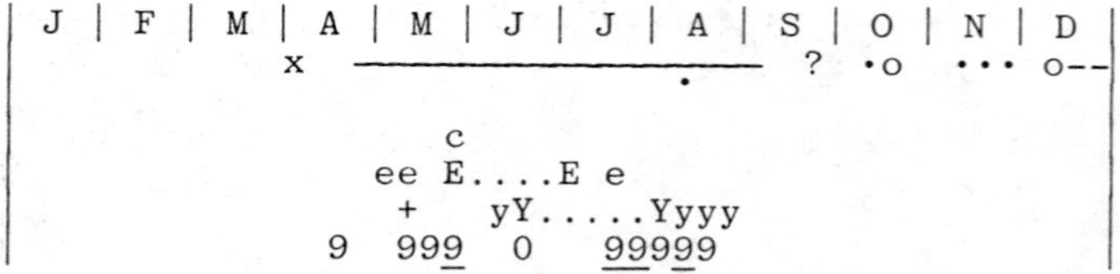

```
J | F | M | A | M | J | J | A | S | O | N | D
 x                       ?  · ·  o      
          ee  c                          
          ee E.....E e                   
        +    yY.....Yyyy                  
        9  999  0   99999                 
```
= 7 > 5 > 7, loc. in 4,6 V,L with wooded
  shores (& nesting also known in c 1
  UB, n 2 VR).
= Breeding phenology (+ = eggs in 3 V in
  1892 & 1900).
= Occurrences reported with dates in 9,
  10(=0) L,V, & probably nests there.

COMMON SNIPE (*Capella gallinago*)

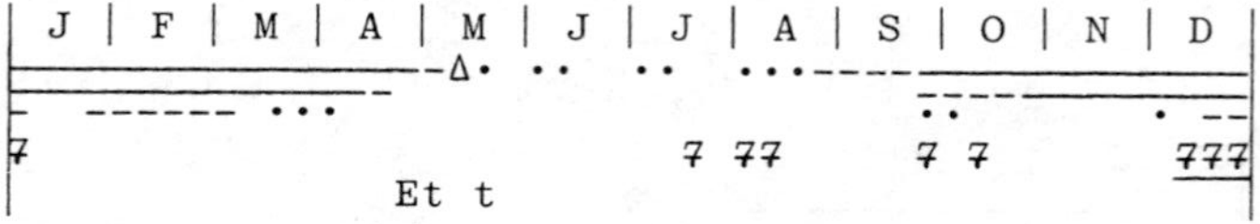

```
J | F | M | A | M | J | J | A | S | O | N | D
 · -·- ·· --Δ· · · · · ·          -- --     
 --------                  · ·  -- --    ???
?                               7  77     
```
= 1,2,4,5 >,3,6,8 wet G, short M>ML,MV >>
  1 upper M, s>c 1 I(M?).
= Occurrences reported in 7 wet G,M.
= Breeding data in e 3 (old), c 2 (recent
  but no eggs or young reported to 1975).

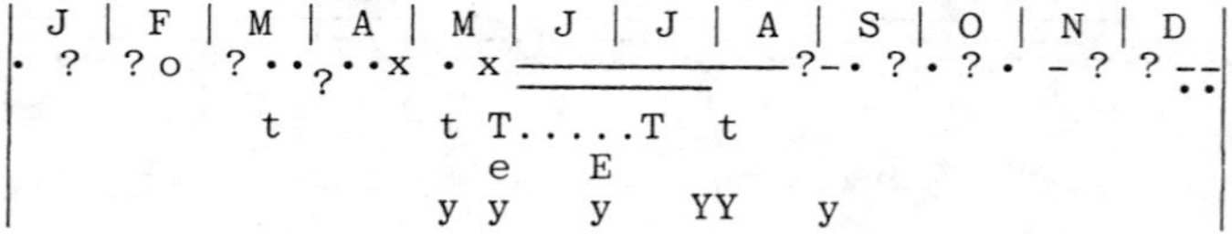

```
        Et t                7  77         
J | F | M | A | M | J | J | A | S | O | N | D
·-?-?o ?··?··x ·x          ?-·?-·?··- ?-·?··
        t     t T.....T t                  
                e   E      -- --           
        y    y  y   YY     y               
```
= 9 > 7 wet G, short M.
= span of reported territorial activity
  (mostly aerial display of males), &
  other Breeding phenology.

27.

362

EUROPEAN JACK SNIPE (*Lymnocryptes minimus*): One Calif. record--female collected 4 mi. NW. of Sutter Buttes, 20 Nov.1938 (see Condor 41:164).

SHORT-BILLED DOWITCHER (*Limnodromus griseus*)

```
J | F | M | A | M | J | J | A | S | O | N | D
--|---|---|---|---|---|---|---|---|---|--+|--
  |   |   | --+-|   |   |   |   | . |   | . |
. |   |   |   |   |   |   |   | . |   | . | --
      ********        ***************
  (9) 2 2 2 2         72222→→2→→
```

= 1 T (+ adjacent B̂, dikes, or M̂ as roosts) >> M̂,S,L,M?, and (spring,fall + stragglers) 8 B>L,V, wet F. Rare in n 1, Dec.-Feb. *** = main migrations.

= Other inland records with dates, & verified by specimen or recent ones by voice or close study of breeding feathering, in districts shown by numerals; (9) = w.Nev. Also noted on c 1 I(banded, Sep.1968), & (this species?) on San Nicolas I., 30 Nov.1963, and San Clemente I., 22-29 May 1894.

LONG-BILLED DOWITCHER (*Limnodromus scolopaceus*)

```
J | F | M | A | M | J | J | A | S | O | N | D
--|---|---|---|---|---|---|---|---|---|---|--
  |   |   | . .o. . ?.|   |   |   |   |   | -- | --
  |   | - | ? |   |   |   | . . | - ? - ? | 7 7 | 7 7
. |   | - | 7 |   |   |   | 7 | 7 7 7 |   | 7 | --
```

= 9 > (Apr.-May; Aug.-Oct.) 2,8 short M, wet F,G>L > 8 B̂, 1,3,4,5,(+6?) L, short M, wet F,G >1 M̂,T,2,3,4(+5,6?),8 V.

= Occurrences reported at 7 L, wet G, or M. Also 4 on Farallon I., Sep.1968.

GRAPHIC CALENDAR   HABITAT DISTRIBUTION

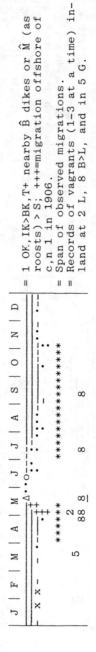

RUDDY TURNSTONE (*Arenaria interpres*)

```
J | F | M | A | M | J | J | A | S | O | N | D
· · · · - - + - - - - - - o - - Δ · · · - - - - - · · · · · · · - - -
                   · ·         · ·           · · · ·
              - · ·                        · · · ·
                                 22  22→   2          (8)      8
         2                      8888 7 888            /=s.Nev.
     8888888888                (9)=w.Nev.
```

= 1 K,S T+B dikes (occas.M) as roosts,
  1 OI; winter numbers larger toward s 1,
  chiefly San Diego Bay; rarely to 1 L,
  wet F.

= Available records inland in districts
  shown by numerals; quite regular on 8 B
  shores in spring.

BLACK TURNSTONE (*Arenaria melanocephala*)

```
J | F | M | A | M | J | J | A | S | O | N | D
                 Δ - - - - o - - - - - -
x x -  - · - · - · ++          · · ·           -
                    +++          · ·        *************
*****                          **********************
      2                               8
  5      88  8                        8
```

= 1 OK, IK>BK, T+ nearby B̂ dikes or M̂ (as
  roosts) > S; +++=migration offshore of
  c, n 1 in 1906.

= Span of observed migrations.

= Records of vagrants (1–3 at a time) inland at 2 L, 8 B>L, and in 5 G.

SURFBIRD *(Aphriza virgata)*

| J | F | M | A | M | J | J | A | S | O | N | D |

? + △ → → →

B   2   B   B   BB

BBBBBB   8

= 1 OK, IKO (& rarely on 1 S).

= Available records in districts 2,8, and (B entries) on c, n 1 BS(+K?); underline = more than one record.

RED KNOT *(Calidris canutus)*

| J | F | M | A | M | J | J | A | S | O | N | D |

2222   222

I(99)   2

= 1 BS>T(+ adjacent B̂ dikes as roosts) >OS, 8 B>L (spring > fall in 8, & no record Nov.-Feb.) >> n(+c,s?) 1 wet G,F.

= Available records in other districts as indicated by numerals; + I = near Farallon I.; (9)= w. Nevada.

SANDERLING *(Calidris alba)*

| J | F | M | A | M | J | J | A | S | O | N | D |

22   222→→2   2   2   3

999   7   7   4

(9)   (9)   (9 9)

= 1 OS>BS>T(+ nearby B̂ dikes or barren flats as roosts),OK>L, 8 B>L shores. Available dated records on L shores.

= (usually open) of other districts as shown by numerals; (9)= w.Nev. Also occurs regularly on 1 OSI>OKI.

SEMIPALMATED SANDPIPER (*Calidris pusilla*)

J	F	M	A	M	J	J	A	S	O	N	D

```
                      m   n   mn→mmm nm     s   9
         3   n                                ?   ?   ?
         8  (8)88888        M  83n→  ?
                             (9)  I→  ?
                                  ?
```

= 33 records in n,c(=m), s 1 T,L,M, on Farallon I.(=I), & in districts 3,9,8 (chiefly B). Additional sight records without supporting details of identification are discounted, esp. reports in winter.

WESTERN SANDPIPER (*Calidris mauri*)

J	F	M	A	M	J	J	A	S	O	N	D

```
                                         ··············?
                                         **********
         *****        **               ***********
                                         **
```

= 1 T + adjacent B̂ and dikes, pools in M̂ or barren flats (as roosts) > BS > 8 B > 1 OS>L, wet F,G,OK,OI.
= Main migrations.

J	F	M	A	M	J	J	A	S	O	N	D

```
                                        ··  - o
         ··                        ·?·  ·
         --                        --   ---
         ··                        ·  7 7 7777
         7   7                     (99)  999999
         799
```

= 2 > 3,4,5 > 6 L,V > wet F,G, sparse M.

= Occurrences reported with dates in districts 7,7,9 (maximum numbers in 9 = thousands in May,Sep.); (9) = w.Nev.
```
9
```

RUFOUS-NECKED SANDPIPER (*Calidris ruficollis*): Primarily e. Siberian, but breeds to w.Alaska. Three Calif. records--1 on Humboldt Bay, 5 May 1969; 1 at Crescent City, 18 June 1974; and 1 collected at s. end of Salton Sea, 17 Aug.1974 (Western Birds 6:111-113).

LEAST SANDPIPER (*Calidris minutilla*)

= 8 B, wet F > 1 M,B̂ edges, 2 > 1,3,4,5,8 > 6 L,V, short M or wet M or wet F or G, 1 T > 1 S>K,OI.

= Observed & inferred migrations.

= Available dated records in 7,7,9 L,M (many thousands in 9 in Sep.).

BAIRD'S SANDPIPER (*Calidris bairdii*)

= c,s>n 1 S (esp. at VO,VB) > 1 T,B̂, 2,8 L (open shores), 8 B; most of the ( ) records are from Christmas Counts and lack published identifying details.

= Available records in other districts as shown by numerals (up to 17 birds to-gether at some inland localities, Aug.-Sep.); also reported on c,s 1 I (May; Aug.-Sep.).

GRAPHIC CALENDAR HABITAT DISTRIBUTION

SHARP-TAILED SANDPIPER *(Calidris acuminata)*

| J | F | M | A | M | J | J | A | S | O | N | D | = 1 T,M or L,M̂?, 2 ML; a total of 26
```
              ....Δ-oo•.••
```
records as of 1976 (all but 2 since 1958), 4 collected, 2 banded; maximum of 4-6 birds near Arcata in late Oct. 1969.

PECTORAL SANDPIPER *(Calidris melanotos)*

| J | F | M | A | M | J | J | A | S | O | N | D | = 1 TG or TM̂,GM, 1,2,4 ML, wet F or G
```
o) • .• • .•••-Δ••   ----...(•o
                    •••
              o     777 79
```
$>$1,3,5,6,8 ML, wet F or G$>$1 T$>$1-5, 8 L,V; also on I in May & Aug.-Sep.

(records in other dists.{7 777 79
The records in ( ) are from s.Calif. Christmas Counts, but lack identifying details.

ROCK SANDPIPER *(Calidris ptilocnemis)*

| J | F | M | A | M | J | J | A | S | O | N | D | = n>c>>s 1 OK, c 1 OKI; individuals often
```
x ---  ---+-+->+
  ^      •
```
remain for months at same locations.
= Occurrences well within n,c 1 B (on BK?).

WHITE-RUMPED SANDPIPER *(Calidris fuscicollis)*: Normal in c. to e. North America. Two verified Calif. records--a bird collected at n.end of the Salton Sea, 6 June 1969; and 1 photographed at s.end of Salton Sea, 16 June 1976. Additional reports of individuals sighted (all lacking verifying details) in "fall" 1962 on Lower Klamath Refuge, early Sep.1965 near Woodland, and on 22 Aug.1966 at Bolinas Lagoon.

CURLEW SANDPIPER *(Calidris ferruginea):*   A primarily e. Asiatic species that has been found in various parts of North America. Recorded four times in Calif. (twice with photographs)--7 Sep.1966 at Rodeo Lagoon, Marin Co.; 16-17 Sep.1972 at Pescadero Creek-mouth, San Mateo Co.; 27-28 Apr.1974 at Salton City, Imperial Co.; & 7-14 Sep.1974 at Bolinas Lagoon, Marin Co. (details of last in American Birds, 29:114).

DUNLIN *(Calidris alpina)*

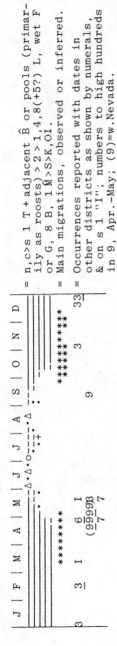

= n,c>s 1 T + adjacent B̂ or pools (primarily as roosts) >2 >1,4,8(+5?) L, wet F or G, 8 B, 1 M̂>S>K,OI.

= Main migrations, observed or inferred.

= Occurrences reported with dates in other districts as shown by numerals, & on s 1 "I"; numbers to high hundreds in 9, Apr.-May; (9)=w.Nevada.

STILT SANDPIPER *(Micropalama himantopus)*

= 8 L (espec. pools), B,V?, wet F?.

= Occurrences reported from 1 (subdivided into s,m=c, n parts) T or M̂,L,OV, & in districts 2,3,4 (numerals). Also known from sc. Ore. & hence expected in 9.

BUFF-BREASTED SANDPIPER (*Tryngites subruficollis*): Normally migrates between the Arctic & Argentina via the Great Plains area of c. U.S. In Calif. individuals found: near Morro Bay, 14 Sep.1923 (collected); at Furnace Creek Ranch, Death Valley, 1-5 July 1935; at Goleta, 10-26 Sep.1964; at Oceanside, 16 Sep.1967; near Arcata, 25 Aug.1970; on Palos Verdes Peninsula, 5-17 Sep.1971; on Catalina I., 30 Aug.1975; & 2 at Salinas River-mouth, 27 Aug.-6 Sep.1976.

RUFF (*Philomachus pugnax*)

= s,c 1 T,M̂,OV,OS>OK, 4 LG (once); some birds have remained at one locality for months (→→ = only such records).

= Occurrences on n 1 G (=1 entries), & in districts 2,9 (latter = Lower Klamath L.).

RED PHALAROPE (*Phalaropus fulicarius*)

= 1 O >> B̂>B,T,L, 8 B>L.

= Occurrences reported with dates in other districts as shown by numerals.

(30.)

NORTHERN PHALAROPE (*Lobipes lobatus*)

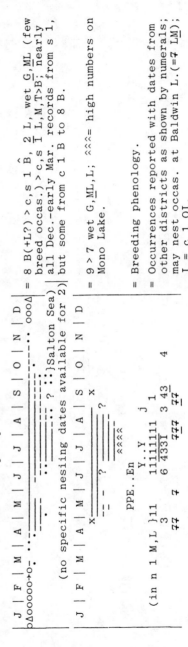

| J | F | M | A | M | J | J | A | S | O | N | D |

= c>s 1 B̂ > 1 O (along & offshore) > B>L, pools in M̂,T; nearly all late Nov.-Mar. records above 1-line level are from s 1 (mostly Orange & San Diego Cos.).
**= Mass migrations, mostly over 1 O.

= 8 B > 9 (no record mid-Oct.- late Apr.) > 2,3,8 L(+V?); <<<=20,000-100,000++ on Mono Lake.

= Occurrences reported with dates on 4,6, 7,7 L.

WILSON'S PHALAROPE (*Steganopus tricolor*)

}Salton Sea)

(no specific nesing dates available for 2)

= 8 B(+L?) > c,s 1 B̂, 2 L, wet G,ML (few breed occas.) > c,s late Dec.-early Mar. records from s 1, L,M,T>B; nearly some from c 1 B to 8 B.

= 9 > 7 wet G,ML,L; <<<= high numbers on Mono Lake.

= Breeding phenology.

= Occurrences reported with dates from other districts as shown by numerals; may nest occas. at Baldwin L.(=7 LM); I = c 1 OI.

371

POMARINE JAEGER (*Stercorarius pomarinus*)

```
 J | F | M | A | M | J | J | A | S | O | N | D
-·--·-·-·--·o·--===o·  Δ··--o------------------oo-- ·
   ·   ·      ·:·                         · · ·
   ·          ·:·                          -  -:-  ·
   ·          ·:·                            -:-  ·
   ·                          9   8           ·
            3
```

| = 1 O (offshore > near shore) >> B; the exceptional Nov. numbers = 3850 in an hr. at 40 mi. S. San Clemente I. in 1964.
| = Records of individuals inland in districts shown by numerals.

PARASITIC JAEGER (*Stercorarius parasiticus*)

```
 J | F | M | A | M | J | J | A | S | O | N | D
-ΔΔΔΔ  ·:·  Δo··ΔΔΔ-Δ·Δ·→ooooΔ·------------------Δo·
                                          ·:·    · · ·
                            7742 224       ·:·
                            8̲8̲8̲8̲8̲8̲8̲8̲8→    ·:·
   2   2                     9   9    9     ·
   4 4→→→→→→]=presumably same)             ·
   bird to reservoir s.of)
   San Jose regularly in 1950's)
```

| = 1 O>OS,OV >> B,L; the exceptional Nov. numbers = 1600 in an hr. at 40 mi. off San Clemente I. in 1964.
| = Occurrences with reported dates inland in districts shown by numerals; reported annually in recent years at 8 B, with up to 8 birds per day.

LONG-TAILED JAEGER (*Stercorarius longicaudus*)

```
 J | F | M | A | M | J | J | A | S | O | N | D
                      ·   ···+++Δ-ΔΔ---    ·
           ·   ·      ·       ^  ^^^
                         ^
```

| = 1 O, mostly well offshore; dates of the few records in n 1 O indicated by ^ .

SKUAS, mostly SOUTH POLAR SKUA (*Catharacta maccormicki*), & few COMMON SKUA (*C. skua*)

```
J | F | M | A | M | J | J | A | S | O | N | D
.   . +.    ..ΔΔo.. ..Δ+. Δo---.--.. .         = 1 O, mostly well offshore; all in c,s
               ^                ^^^               1 except ^ records off Humboldt Co.
                               11 1  1          = Records on 1 B.
```

GLAUCOUS GULL (*Larus hyperboreus*)

```
J | F | M | A | M | J | J | A | S | O | N | D
.---ΔΔΔΔΔΔ  .·.   ..→.    ...→Δ .    .Δ-ΔΔ        = c>n,s 1 (& nearby 4,3) dumps > B,T,O,
                                             3      nearby L; + c,s 1 OI (Feb.,Mar.).
33→→→3→→3→→                              (8)8(8)  = Occurrences at dumps or L farther in-
742 2 8 8 8→→}=same birds remaining                land, districts as shown by numerals;
 8                                                 (8) = s.Nevada.
```

GLAUCOUS-WINGED GULL (*Larus glaucescens*)

```
J | F | M | A | M | J | J | A | S | O | N | D
                          -Δ-.Δ-   ..Δ-Δ---      = c>n,s 1 dumps, nearby B>n,c>s 1 B,U>O
         .:                    |||                 (to far offshore in migrations),S,T,OK,
         --                    |||                 BK,L>near parts of 5,4,3 wet G,F,L.
         --                    ---
         --                    --
J | F | M | A | M | J | J | A | S | O | N | D
o---------→→→·· ooo.→Δ·.        ..Δ→--Δ---        = c 2>n,s 2, inland 3,4 L, wet F or G,V,
77                  +7)        66  7  7            8 B,L,V (+F?).
                                                 = Occurrences on 6,7 L; (+7) = long dead.
```

GRAPHIC CALENDAR

HABITAT DISTRIBUTION

WESTERN GULL   (*Larus occidentalis*)

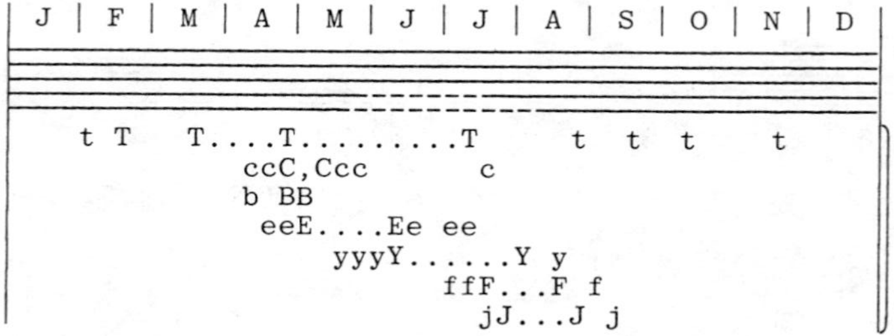

```
 J | F | M | A | M | J | J | A | S | O | N | D

 t  T      T......T..........T         t  t  t  t
              ccC,Ccc            c
           b  BB
              eeE....Ee ee
                  yyyY....:..Y y
                       ffF...F f
                        jJ...J j
```

= 1 O, S, IK or IG, OK, U, UB, outer B, nearby dumps; T̄, c 1 BI >> 1 inner B, T, nearby L, B̂, M̂; summer decline shown = far from large nesting colonies only.

= Breeding phenology.

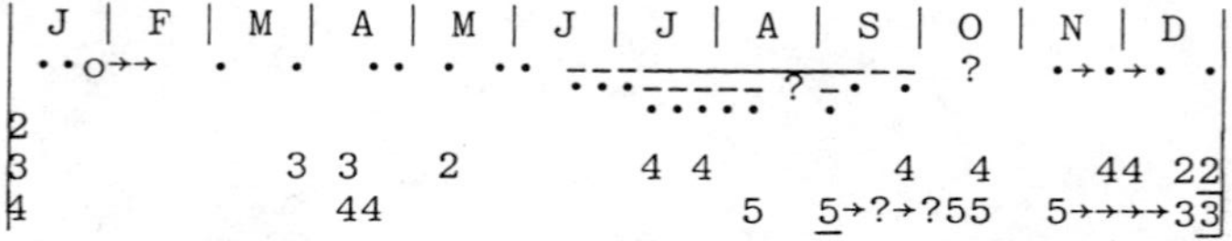

```
 J | F | M | A | M | J | J | A | S | O | N | D
 ..o→→     .   .   .   .   .  :     ?      .→.→.→.        .
                                   ?  -.
2                                    :-.
3    3  3    2        4  4         5→?→?55   44  22
4    44              5    5→?→?255            5→→→→33
```

= 8 B>>V (most Apr.-Sep. records are subspecies *livens* from Gulf of Calif.)

= Occurrences elsewhere inland in districts indicated by numerals; in 5, follow salmon far up major rivers.

HABITAT DISTRIBUTION

...Pidibundus): Five Calif. records (but no specimen collected)--individuals on or near San Francisco Bay, 23-24 Jan.1954, 4 Jan.1956, 26 ...on Tomales Bay, 5-8 Apr.1976. A ...pied at Arcata, 16-23 July 1972 (Amer.Birds 26:901); & ...across Eurasia, with vagrants to e.North America. Similar to but larger than a Bonaparte's ...with red bill and dark underside of primaries.

LAUGHING GULL (*Larus atricilla*)

| J | F | M | A | M | J | J | A | S | O | N | D |

= 8 B (formerly nested on IB), nearby F> V,M,+L?

= Breeding phenology (data incomplete).

= Occurrences reported in s(=1 entries), c(=^) & n(=^) 1 O,B; & flying over 3 U.

FRANKLIN'S GULL (*Larus pipixcan*)

| J | F | M | A | M | J | J | A | S | O | N | D |

= 8 B,F,V,L(+M?) > s>c>n 1 B,B̂, nearby F, G,M,M̂,U>O (but rarely offshore).

= Occurrences with dates reported from other districts as shown by numerals.

HERRING GULL (*Larus argentatus*)

| J | F | M | A | M | J | J | A | S | O | N | D |

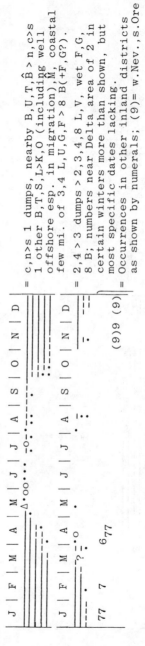

= c,n>s 1 dumps, nearby B,U,T,B̂ > n,c>s 1 other B,T,S,L>K,O (including well offshore esp. in migration),M, coastal few mi. of 3,4 L,U,G,F>8 B(+F,G?).

= 2,4 > 3 dumps >2,3,4,8 L,V, wet F,G, 8 B; numbers near Delta area of 2 in certain winters more than shown, but most specific dates lacking.

= Occurrences in other inland districts as shown by numerals; (9)= w.Nev.,s.Ore.

THAYER'S GULL (*Larus thayeri*)

| J | F | M | A | M | J | J | A | S | O | N | D |

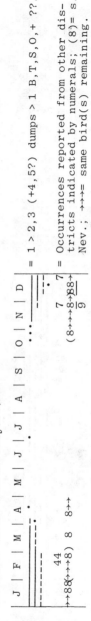

= 1 > 2,3 (+4,5?) dumps >1 B,T,S,O,+ ??.

= Occurrences reported from other districts indicated by numerals; (8)= s. Nev.; +++= same bird(s) remaining.

CALIFORNIA GULL (*Larus californicus*)

J | F | M | A | M | J | J | A | S | O | N | D

= 1 (s>c>n in winter) B,T,U, dumps >L>S,O (chiefly near shore though some well offshore also, esp.in migration),KO > 8 B, 3,8 >2,4 L, wet F or G,U,V; ***=observed migrations.

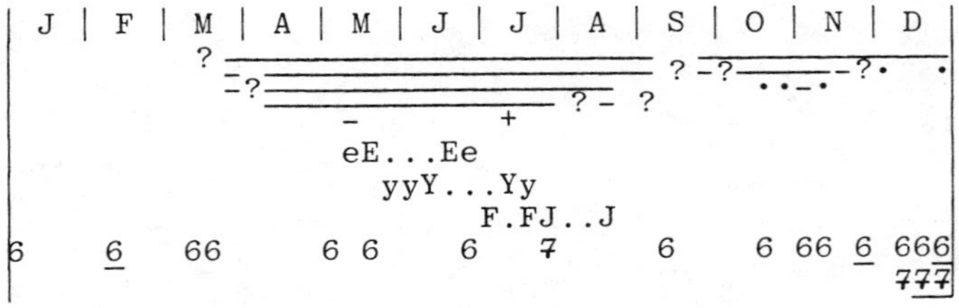

= 9 > n>c>s 7 L,F,G,IL,M,V; also a few "through winter" at L.Tahoe but specific dates mostly lacking.

= Breeding phenology.

= Occurrences reported with dates on or near 6,7 L.

RING-BILLED GULL (*Larus delawarensis*)

J | F | M | A | M | J | J | A | S | O | N | D

= 2,8 > 3,4 F,G,L, 1 $\hat{B}$,T >1,8 B, 5,7>6 L, G,F, 1 S,O (rare offshore & to OI) >1 M, 1-7 M,V,U; also at dumps in any district where larger gulls not too numerous.

******* = observed migrations)

= 9 >> 7 L,IL,ML,F,G.

= Breeding phenology (colonies at Honey Lake and northward).

= n,c 1 B,OV,S,O near shore > OK,O offshore to near I , c s 1B,T, s1 O,B,S, OK,OI, 1,4,5 wet F,G,L,V near B or O.

= Records farther inland, in districts shown by numerals; (8)= s.Nev.; in the Delta area of 2, numbers irreg. to 3-line level in Dec.-Mar.

GRAPHIC CALENDAR (*Larus* ...)

BLACK-HEADED GULL (*Larus* ...)

Apr.1968; photograph...
species abund...
Gull br...

BONAPARTE'S GULL   (Larus philadelphia)

| J | F | M | A | M | J | J | A | S | O | N | D |

= 1 B, B̂, OV>O, T, S, L > 2, 8 > 3, 4 L, V>M, F, G, 1 M, occurs well offshore in migration periods (***); few to none in n 1,2,8 through winter.

= Occurrences reported with dates in other districts as shown by numerals; (9) = w.Nevada.

```
                                          ** ******
         ********                  777            6
      66       6    7
      7 7
      99(9)    9   (9)   9    (99)         (99)   (9)
 6                                                9    9
```

LITTLE GULL   (Larus minutus): Four Calif. records--1 on a pond near Mecca, 16-21 Nov.1968, by many observers & photographed (Audubon Field Notes, 23:109,111); adult at Redondo Beach, 23-25 Dec.1969; adult at s.end of Salton Sea, 3 Dec.1972; imm. near Moss Landing, Monterey Co., 19 Jan.-1 Mar.1975. The species is normal in Eurasia, but stragglers are frequent in e. U.S. It is slightly smaller than a Bonaparte's and similar except for dusky underside of primaries and (in breeding season) a red bill.

BLACK-TAILED GULL   (Larus crassirostris): A species normally found in e. Asia. An adult female was collected at San Diego, 28 Nov.1954, but it is suspected of having arrived there by "other than its own power" (see Calif. Birds, 1:24).

GRAPHIC CALENDAR

HABITAT DISTRIBUTION

HEERMANN'S GULL   (*Larus heermanni*)

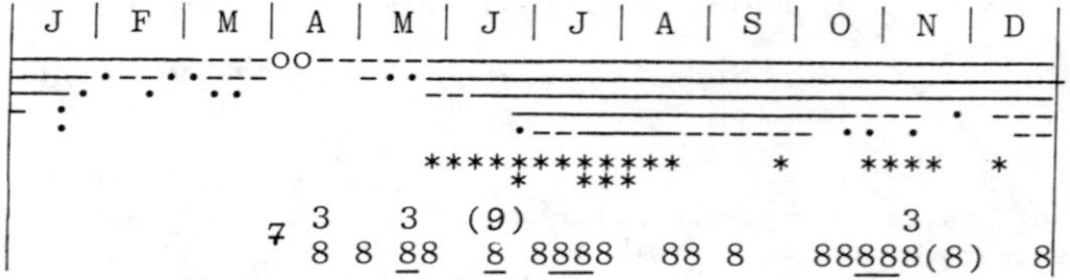

J	F	M	A	M	J	J	A	S	O	N	D

= s,c>n 1 (but few Dec.-May north of Monterey area) O,OK,S>OI>OV, outer B,L>> inner B,L,T.

= Observed migration movements.

= Occurrences inland, by districts as shown by numerals; (8)=w.Ariz.; (9) = w.Nev.; up to 6 individuals at 8 B.

```
                              *  ****    *
                  *************  ****
                                            3
         7  3  3 (9)
            8  8 88   8 8888  88 8   88888(8)  8
```

BLACK-LEGGED KITTIWAKE   (*Rissa tridactyla*)

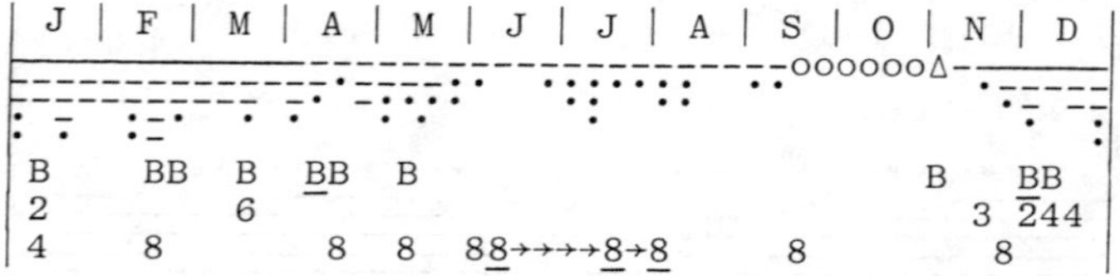

J	F	M	A	M	J	J	A	S	O	N	D

= c,n>s 1 O (offshore > near shore) > OK>S >> L.

= Occurrences reported on 1 B (=B entries) & in districts shown by numerals; →→→= lone birds remaining at 8 B.

```
-ooooooΔ-                  B   BB
  B  BB  B  BB  B                   3  244
  2     6                              8
  4     8        8   8  88→→→8→8  8
```

SABINE'S GULL   (*Xema sabini*)

```
  J | F | M | A | M | J | J | A | S | O | N | D
  .   .  . • • •Δ- - - - • •○  o+++++ = = = - - - - Δ ○ • • •
                            • •=======
                            • •  • • =
                                     •̄
4                     B  B              B  B   B
                      8  8  88    8    3  26  3
                                       888888  8
                                        ─      ─
                                        9      7
```

=	1 O (offshore >> near shore, and rarely to shoreline L).
=	Occurrences with reported dates on c 1 B (=B entries) & in other districts as shown by numerals; 7 birds were together on the Colorado R. in May 1956.

GULL-BILLED TERN   (*Gelochelidon nilotica*)

```
  J | F | M | A | M | J | J | A | S | O | N | D
  Δ++-=xx?+?•- - - - = = = - - - - •○- - -•          ++
      +     +        = = = = - - - - +  • •
                     +++++  - - - - -
  BN        N  N  N                              2?  2
  E         Y     f FF
            1
```

=	8 B,BI,F,V,M(+L?); ++ entries = pre-1960 records.
=	Breeding phenology; irreg., but con-firmed as late as 1972 & 1974.
=	Occurrences in s 1 B and c 2 L,M.
```
```

33

GRAPHIC CALENDAR

HABITAT DISTRIBUTION

FORSTER'S TERN (*Sterna forsteri*)

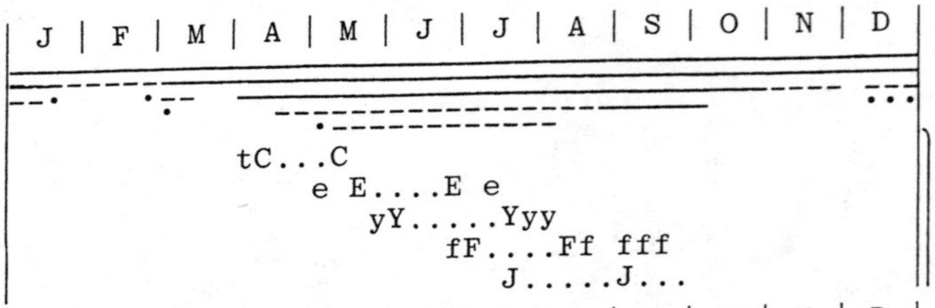

J | F | M | A | M | J | J | A | S | O | N | D

```
                            tC...C
                          e E...E e
                       yY.....Yyy
                    fF....Ff fff
                    J.....J...
                                    oo -oΔo..
        B  ?yB N
      E..E...       f
   _1    1   1  .1 . .1 .1
1?       77  77
```

= c,s (s>c in winter) 1 B,B̂+dikes, islets, OV>O (& to well offshore in migrations), M̂ (nesting upper fringes) > S, T, nearby M; nos. at 4-5 line-level in midsummer only near few large colonies.

= Breeding phenology

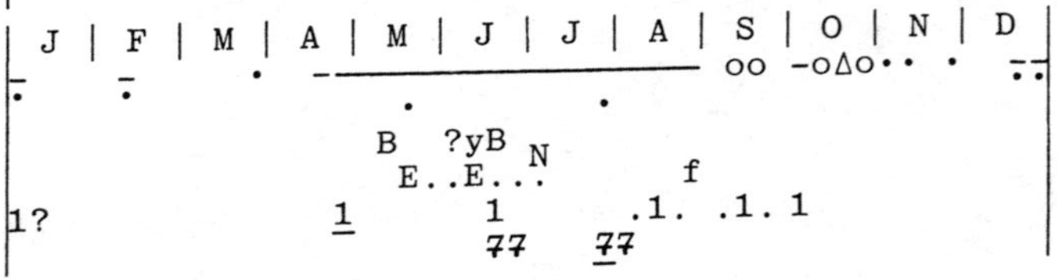

= 9 > 7 (S.Tahoe, mostly) L,ML,V > (Apr.-Oct. & no record in 9,7 other months) 8 B,BVM, L,V > 2,3 L,ML,V.

= Breeding phenology.

= Occurrences reported in n 1 and at 7̄ L.

COMMON TERN (*Sterna hirundo*)

J | F | M | A | M | J | J | A | S | O | N | D

```
                                            ?oΔΔ?
                                    22
                7   7  77777 7 7
                4 3
```

= s>c>n 1 O,OV,B > 1 S, 8 B>L,V > 1 B̂>L; most Dec.-Jan.records are in s 1, and additional winter records near San Diego & at Salton Sea are undated.

= Occurrences reported at 2,3,3,7,7 L(+M?).

ARCTIC TERN (*Sterna paradisaea*)

```
| J | F | M | A | M | J | J | A | S | O | N | D |
| •o--?-        •o•o----====----Δ-              |
|          ^      ^→^ ^^^ ^→^                   |
|                        ==                     |
|                        •• •                   |
```

= 1 O well offshore >> 1 O near shore, S, OV>B; ^ = dates of records on B within the total shown. Additionally, inland records are: 1 seen at Tinemaha Res.,Inyo Co., 13

records are: 1 dead near Bridgeport, 22 May 1973; 1 seen at Tinemaha Res.,Inyo Co., 13 June 1975; and 3 at s.end of Salton Sea, 13 June 1976.

LEAST TERN (*Sterna albifrons*)

```
| J | F | M | A | M | J | J | A | S | O | N | D |
| •) •  •        --== ----======•+=-  •• •  o  (• • o |
|    •  ‡         ++ == ---++++++++‡+‡-  •• •          |
|  ‡      ‡           TeE.....Eee                      |
|                y yY...Y y                            |
|                  f F...F  f                          |
|                  1                                   |
|              3      3  1↦↦13 6        1              |
|            8(8) 8(8) 888‡8↦8 88                       |
```

= s>c 1 BS, OS,B,B̂+islets, dikes; ++ = pre-1950 records; ( ) records = mostly Christmas Counts with inadequate detail.
= Breeding phenology.
= Occurrences reported with dates in districts n 1,3,6,8 (on 8 B>V,L); (8) = s. Nev. or w.Ariz.along Colorado R.

ROYAL TERN (*Thalasseus maximus*)

```
| J | F | M | A | M | J | J | A | S | O | N | D |
| ---+•?•• o +•••••-•-++ •-‡+---•     •••           |
|  •  •           --‡  ‡ •  ---  ‡‡+-‡‡+-++-        |
|  •                       --===-+++++-++          |
| e n f}=Breeding phenology                        |
|        (very incomplete data)                    |
```

= s>>c 1 O,OI, S, nearby or large B, occas. s 1 B̂+dikes (nested 1959, '60 on s.San Diego Bay); ++ = pre-1950 records (rare in c 1 since then); 1 collected on Humboldt Bay, 23 Nov.1918.

383

ELEGANT TERN *(Thalasseus elegans)*

J | F | M | A | M | J | J | A | S | O | N | D

= s,c 1 O,S,OV, nearby or large B,B̂+dikes (nesting at s,San Diego Bay),BS>T; all but about 12 of Dec.-early July records are from s 1, & most of Mar.-June ones from the nesting colony. In addition, up to 44 birds have reached n 1 B,OV (Aug.-Oct.1969, & fewer in Sep. 1972-74).

eT NNE.E

yY..... f f = Breeding phenology

CASPIAN TERN *(Hydroprogne caspia)*

J | F | M | A | M | J | J | A | S | O | N | D

= c>s>n 1 B,B̂+dikes or BI (low,barren) > O, OV,S,L, 8 B, nearby V,L, & former BI 2,4 L,V (Mar.-Nov.); 2-3 line level in Dec.-Jan. mostly from San Diego Bay.

tTTC
eeeE.....E e e
yY.......Y yy y
fffF.......Ff f

= Breeding phenology.

J | F | M | A | M | J | J | A | S | O | N | D

= 9 L,LI (or dikes?) > V,M.

NE.E Y..Yy

= Breeding phenology (reports incomplete).
= Occurrences reported with dates in districts 3,6,7 L & 10 (="0" entry) L.

f
3 3
7 3 3 3
7 0 7 7 76 6 3 33

384

BLACK TERN *(Chlidonias niger)*

J	F	M	A	M	J	J	A	S	O	N	D

= 8 B, 8 > 2 >> 3 L,M, rice fields, nearby moist G,F>V; additional reports of 1000 or more per day in spring, summer, & fall at or near 8 B lack specific dates (see Calif. Fish & Game, 56:95).

Breeding phenology = {see E.Ef   BNN = old records in 2)
                      Y...F F

J	F	M	A	M	J	J	A	S	O	N	D

= 9 > n 7 L,M,F,G; occurrences in 7 mostly at S.Tahoe where it formerly bred.

= Breeding phenology (reports incomplete).

J	F	M	A	M	J	J	A	S	O	N	D

= s >> c 1 B,B,M,OV,O (including well off-shore in spring & fall); +>= same individual remaining, San Diego Bay, 1966.
= Occurrences reported in n 1 (="1" entries), 4,6,7,10 (="0" entry).

BLACK SKIMMER (*Rynchops nigra*)

```
J | F | M | A | M | J | J | A | S | O | N | D
                         o o•••···│··•·    ↑  →•
                            E...E..e       ↑ →•
                            Yf Y F
            11→ 11111→→→→→→→→→→1→→        = 8 B,BIS.
            c        c
→→?  →?  →  1 1 1
```

= 8 B,BIS.

= Breeding phenology.

= Occurrences in s 1 O,S,OV,BS (="1" entries, except two marked "c" from Bodega
  Bay & t.Pinos); →→=presumably same bird
  remaining.

COMMON MURRE (*Uria aalge*)

```
J | F | M | A | M | J | J | A | S | O | N | D
  ----                        -- │
                               •• ?  --
   •  *                •• ***** *****
   *  *       ** ***** *****    ^^^
   ^        ^^^^^^^^^^^^        ^^^
       T...T..TN.....N.....
       e  ee? E....E e
           y Y?Y..Y
           y     f FJ
```

= c,n>>s 1 OKI,OK,O; probably at 5-line
  level all year at S.Farallon I.

= Migration movements reported.

= Occurrences reported on 1 B (chiefly
  outer San Francisco Bay) ^ =2+ records.

= Breeding phenology.

34.

35.

THICK-BILLED MURRE (*Uria lomvia*)

```
 J | F | M | A | M | J | J | A | S | O | N | D
↓→•→→→→→→•.    .   .    .  .••.→→Δ→Δ→?  →?→
```

= c 1 O, harbor; maximum of 3 birds at a
  time; →→=presumably same bird remaining.

PIGEON GUILLEMOT (*Cepphus columba*)

```
 J | F | M | A | M | J | J | A | S | O | N | D
–•–?  o  o–––                    •.••o o•o  –••
                                 •              .
–•                               –––––?––      <
            ttc  CBNE......E    e    ^ ^^^^
                 yyy yY.......Yyy y
                     ff  F..F J
```

= n,c (to San Luis Obispo Co.) > s 1 OK,
  1 OI > n,c>s 1 O (inshore>well offshore);
  ^*^ = vagrants on 1 B (chiefly c.San Francisco Bay).
= Breeding phenology.

MARBLED MURRELET (*Brachyramphus marmoratus*)

```
 J | F | M | A | M | J | J | A | S | O | N | D
–+–•–––––––    •–•  .    ––––––––               +––•
 «    p  P...Pc CPPC.C  Pf  FY F      <        <      <
            ii  i,i iiiiii iii    iiiiiii
(i = dates of birds flying over or to/from H in
 probable or known nesting areas)
```

= n>c>>s 1 O especially near K + N,H near;
  also few records on c 1 B (=^).
= Breeding phenology; probably 5 > s 4 N,H
  for nesting, although the only actual
  nest found was 45 m. up on a tree limb
  at Big Basin, Santa Cruz Co., 7 Aug.1974
  (see Binford, et al., 1975, Wilson Bulletin 87:303-319, for photos and analysis).
```
```

GRAPHIC CALENDAR HABITAT DISTRIBUTION

KITTLITZ'S MURRELET (*Brachyramphus brevirostre*): Normal along e.Siberian to se.Alaska coast. One Calif. record--a weakened juvenile found alive, 16 Aug.1969 at La Jolla & specimen preserved (see Calif.Birds, 3:33-38, including comparison with Marbled Murrelet).

XANTUS' MURRELET (*Endomychura hypoleuca*)

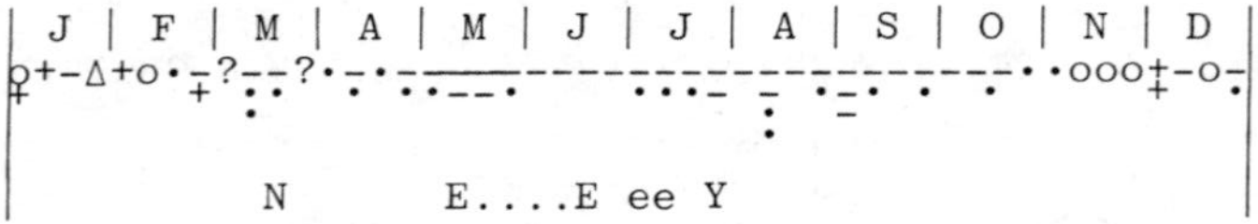

J	F	M	A	M	J	J	A	S	O	N	D

o+-Δ+o•.-?--?·•¦¦---------------------¦¦•.ooo†-o•

p+-Δ+o•.-?--?·•
.:•.•
•:•
.:•

N   E....E ee Y

= s>c>>n 1 O (far offshore to near shore),
  O̲I̲; recorded N.of Monterey Bay, Apr.-Nov.
  & once in San Francisco Bay, early Aug.

= Breeding phenology.

CRAVERI'S MURRELET (*Endomychura craveri*)

J	F	M	A	M	J	J	A	S	O	N	D

··· ++ Δ-
          Δ-
x̲.......+
  †.......+

= s 1 O (with maximum of 30 birds on 9
  Sep.1972).

= c 1 O (near Monterey): x=1972; +.:+ =
  range of dates for specimens before 1911.

ANCIENT MURRELET (*Synthliboramphus antiquus*)

J | F | M | A | M | J | J | A | S | O | N | D = c,n>s 1 O (inshore, espec. near OK, & to 30+ mi. out and near all OI; two Dec. records on S.F.Bay; d = dead, only.

(8) = on L.Mead   (9) = Waifs in s.(=8) & w.(=9) Nevada.

CASSIN'S AUKLET (*Ptychoramphus aleuticus*)

J | F | M | A | M | J | J | A | S | O | N | D = c>n>s 1 OI, nearby O (espec.westward) > O nearer mainland; "+" entries indicate abundance reported on breeding islands at night (chiefly S.Farallon I.).
= Breeding phenology.

PARAKEET AUKLET (*Cyclorrhynchus psittacula*)

J | F | M | A | M | J | J | A | S | O | N | D = n,c 1 O; "+" entries = pre-1940; d = ill or dead in recent years. Also, 2 records of birds dead on s.Calif.beaches--3 N.of La Jolla,28 Jan.1937; & 1 at San Simeon,6 Feb.1955.

GRAPHIC CALENDAR

HABITAT DISTRIBUTION

RHINOCEROS AUKLET *(Cerorhinca monocerata)*

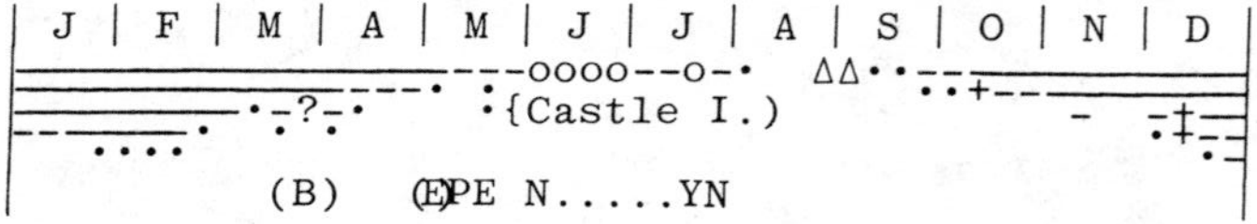

J	F	M	A	M	J	J	A	S	O	N	D

```
                --0000--o--.
            ?-.  :(Castle I.)       ΔΔ--·:·+----
    -?-·
      ·····

(B)    (B)PE N.....YN
```

= c,n>s 1 O (offshore > near shore).
(Breeding data, on Castle I., Del Norte
Co., & S.Farallon I. since 1968; + (B)
from s.Ore. and (E) from Wash. colonies.

HORNED PUFFIN *(Fratercula corniculata)*

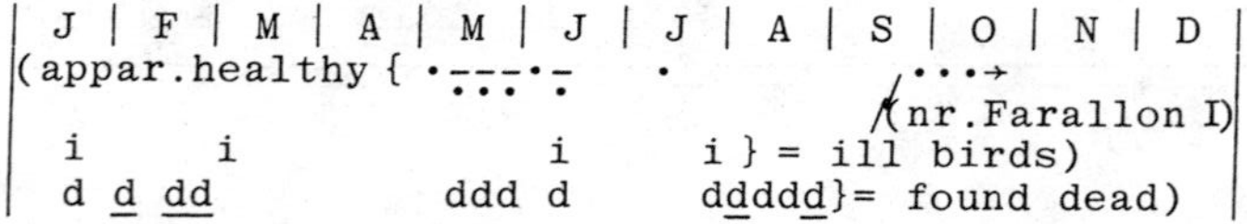

J	F	M	A	M	J	J	A	S	O	N	D

```
(appar.healthy { ·--·--·-           ·--·+···→
                                    (nr.Farallon I)
  i    i            i }= ill birds)
  d  d  dd    ddd d     ddddd}= found dead)
```

= s>c>n 1 O, near OI: most of the healthy
birds in 1975 & 1976 in s.Calif. (maximum of 39 birds on 8 June alone), where
none found prior to 1971.

TUFTED PUFFIN *(Lunda cirrhata)*

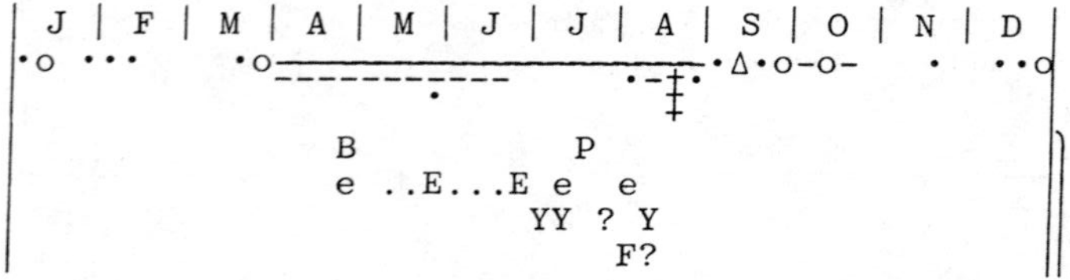

J	F	M	A	M	J	J	A	S	O	N	D

```
·o ····       ·o      ·Δ·o-o-o-   ·    ·    ··o
                      ·-·-----      ·+
        B         P
        e  ..E...E e e
               YY ? Y
                  F?
```

= n,c>s 1 OI>OK, O (offshore > near shore);
"+" entries = 300 reported on Farallon
I. in 1933. Additionally a waif found
on highway e. of Tomales Bay, 19 Aug.
1966.

= Breeding phenology.

36. BELTED KINGFISHER (*Megaceryle alcyon*)

J	F	M	A	M	J	J	A	S	O	N	D

```
        C
     B B_E..E
         yY..Y
```

= 1 OV, OK, O+trees or other elevated perches, s 1,3 > (< in summer) c, n 1,4,5 > 2,6 L,V, +R or other elevated perches; nesting in earth banks near any such.

= Breeding phenology.

Also found in 8 L,V but chiefly in winter; and ranges to 1 OI regularly in winter, and to 7,7,9 at L,V (& nesting locally) but few to none after freeze-ups in winter.

37. AMERICAN DIPPER (*Cinclus mexicanus*)

J	F	M	A	M	J	J	A	S	O	N	D
				?							

```
s        s            ss      ss      s
   B eeB_N..E   e
        y.Y.....Yyyyy
           f    f  F..F
              jj.J
9  4             834
                 7
                        4→→
```

= 7 > 3,5,6, 7̄ > 4 turbulent streams (VK,VR) in mountain & foothill areas W. of the deserts; dispersing birds occas. to L.

= Breeding phenology; s = singing (as noted by author); earlier e,y,f dates are from low altitudes only.

= Occurrences reported in non-nesting locations, in districts shown by numerals.

INDEX

Species are indexed both to the main species accounts (pp. 58-280) and to the graphic calendars and habitat distributions of the Appendix (pp. 300-391). Italicized page numbers indicate illustrations of the entries; color plates are indicated by Pl. number.